Preface

Artificial intelligence (AI) has revolutionized the way forecasting is made in many industries, from finance to retail to healthcare. In today's fast-paced world, accurate forecasting is essential for businesses to make informed decisions and stay ahead of the competition. The article looks at different tools and techniques used in AI for Forecasting and highlights their advantages and limitations. One of the most widely used AI tools for forecasting is machine learning (ML). ML is a subset of AI that allows computer systems to learn from data without being explicitly programmed. Forecasting trains ML algorithms on historical data to identify patterns and relationships that can be used to predict future outcomes. Many ML techniques are available, including regression analysis, decision trees, and neural networks. Each technique has advantages and disadvantages, and the choice of the technique depends on the problem at hand. Regression analysis is a widely used ML technique for forecasting. This involves fitting a line or curve to past data points and using that line or curve to predict future results. Regression analysis can be linear or nonlinear and is especially useful when there is a clear relationship between the independent and dependent variables. Decision trees are another ML technique used for forecasting. A decision tree creates a tree-like model of decisions and their consequences. Each decision node in the tree represents a choice and each outcome node represents a possible outcome. Decision trees help you identify the most important factors, so they are especially useful when multiple factors can affect your results. However, decision trees can be very complex and difficult to interpret, especially if there are many decision nodes.

Chapter 1: Explores the promises and limitations of AI-powered forecasting. It examines how AI can be used to make predictions in various fields and highlights the need for transparency and ethical considerations in AI development.

Chapter 2: Explores the use of a Multilayered Feed-forward Neural network architecture for accurate rainfall forecasting. The approach utilizes historical rainfall data and meteorological variables as input to predict future rainfall patterns. The study evaluates the effectiveness of the proposed method using various performance metrics.

Chapter 3: Explores the use of artificial intelligence techniques for forecasting the stock market index. This includes machine learning algorithms, neural networks, and deep learning models to analyze historical data and make predictions about future market trends.

Chapter 4: Explores the potential of green innovation in the e-vehicle industry as a means to forecast and promote environmental sustainability. The chapter focuses on the development and adoption of electric and hybrid vehicles, along with supporting infrastructure and policies, to reduce emissions and mitigate climate change.

Chapter 5: Explores the differences between traditional and machine learning forecasting techniques, tracing their evolution and the advantages and disadvantages of each approach. The article discusses the potential impact of AI-powered forecasting on business decision-making and the need for data-driven strategies.

Chapter 6: Explores the use of Artificial Intelligence (AI) in workforce forecasting post-COVID-19. This involves analyzing current and past workforce data to predict future trends and make data-driven decisions for workforce planning, talent management, and resource allocation in a post-pandemic world.

Chapter 7: Explores the issue of gender disparity in AI and how it can be perpetuated by unconscious biases. It highlights the importance of creating awareness about this issue and implementing strategies to mitigate bias in AI development and deployment.

Chapter 8: Explores how AI can enhance the customer perspective in forecasting sustainable development of green products. The topic focuses on understanding consumer behavior and preferences to drive eco-friendly product development and marketing strategies.

Chapter 9: Explores the application of AI in weather forecasting and climate behavioral analysis, utilizing ML algorithms to predict atmospheric conditions and identify climate patterns. This technology has the potential to improve disaster preparedness, agricultural planning, and environmental monitoring.

Chapter 10: Explores the ethical and legal implications of using AI to assess climate change, examining the potential benefits and drawbacks of AI-powered climate models, as well as the implications for privacy, accountability, and social justice.

Chapter 11: Explores the potential of AI in workforce forecasting, allowing businesses to predict future workforce demands and make informed decisions. AI can analyze various data sources, including employee performance, market trends, and industry developments, to generate accurate workforce projections.

Chapter 12: Explores the use of AI methods, including machine learning and neural networks, for forecasting stock market indexes. These techniques analyze large amounts of data to identify patterns and make predictions about future market movements, potentially improving investment decisions.

Chapter 13: Explores the comparison between traditional and ML forecasting techniques, analyzing the strengths and weaknesses of both. The study examines the potential benefits of incorporating ML into forecasting processes and identifies scenarios where traditional methods may still be preferable.

Chapter 14: Explores the use of ML models to analyze and forecast demand for paddy and cotton in the Indian state of Andhra Pradesh. Empirical analysis is used to evaluate the effectiveness of the models.

Artificial Intelligence in Forecasting

Tools and Techniques

Editors

Sachi Nandan Mohanty

Professor, School of Computer Science & Engineering
VIT-AP University, Amaravati, India

Preethi Nanjundan

Associate Professor, Department of Data Science
Christ (Deemed to be University), Pune, Lavasa, India

Tejaswini Kar

School of Electronics Engineering
KIIT Deemed to be University, Bhubaneswar, India

CRC Press is an imprint of the
Taylor & Francis Group, an **informa** business
A SCIENCE PUBLISHERS BOOK

First edition published 2024
by CRC Press
2385 NW Executive Center Drive, Suite 320, Boca Raton FL 33431

and by CRC Press
4 Park Square, Milton Park, Abingdon, Oxon, OX14 4RN

CRC Press is an imprint of Taylor & Francis Group, LLC

Library of Congress Cataloging-in-Publication Data (applied for)

ISBN: 978-1-032-50615-9 (hbk)
ISBN: 978-1-032-50618-0 (pbk)
ISBN: 978-1-003-39929-2 (ebk)

DOI: 10.1201/9781003399292

Typeset in Times New Roman
by Prime Publishing Services

Chapter 15: Explores the use of AI for business forecasting and error handling, enabling organizations to make accurate predictions and mitigate risks. AI-powered tools like predictive analytics, ML, and natural language processing can help businesses make informed decisions and achieve their goals.

Chapter 16: Explores the practical benefits of using AI in mental health for more accurate forecasting, including personalized treatment plans, early identification of symptoms, and improved patient outcomes. AI can also help reduce healthcare costs and improve access to care.

Chapter 17: Explores the use of AI in predicting stock market indexes, which involves the analysis of vast amounts of data to identify patterns and make predictions. AI can be used to develop models that take into account a wide range of factors, including economic indicators, market trends, and company performance, to generate accurate forecasts.

Chapter 18: Explores the use of bibliometric analysis to conduct a literature review on the application of AI in forecasting methods. The study aims to provide insights into the intellectual structure and trends of research in this field.

Chapter 19: Explores the development of an enhanced Climate Forecast System through the prediction of effective temperature, which takes into account the combined effects of temperature and humidity. This approach can improve the accuracy of climate forecasts and provide valuable information for decision-making in various sectors.

Chapter 20: Explores the use of ML techniques to forecast future sales demand. By analyzing historical data and external factors, such as seasonality and market trends, these methods can improve accuracy and help businesses make informed decisions regarding production, inventory, and pricing strategies.

Chapter 21: Explores the importance of demand and supply forecasts in supply chain and retail industries. Accurate forecasts can help businesses manage inventory, reduce costs, and improve customer satisfaction. Utilizing data analysis and trend forecasting tools can help companies make informed decisions about their production and distribution processes.

Chapter 22: Explores the use of AI in improving weather forecasting accuracy and analyzing climate behavior patterns. AI techniques such as ML and deep learning can help predict extreme weather events, climate trends, and their impact on the environment, enabling better decision-making and resource management.

Dr. Sachi Nandan Mohanty
Dr. Preethi Nanjundan
Dr. Tejaswini Kar

Contents

Chapter 1

Pursuits of Forecasting

Revisiting the Claims of Artificial Intelligence

Kareti Payaswitha Harshini

1. Introduction

In today's world, forecasting plays a very crucial role in various fields like businesses, science and technology, the economy of a country or a region, and many other areas. In simple terms, forecasting can be defined as predicting or estimating the future based on a keen analysis of the relevant historical data available. The utilization of forecasting has made the procedure of decision-making, which is followed in order to have a better future concerning the circumstances given, an easy task by providing approximate predictions that are likely to take place in the future. The available research done on the areas suggests that forecasting has great significance in the process of evaluation of new ideas (i.e., whether the new ideas are feasible or not, if yes, then to what extent are they feasible), decision-making, and optimal planning to implement the ideas [1].

The forecasts like weather predictions allow us to understand anthropogenic climate change and many other areas where the behavior of weather changes explains or gives rise to new theories of the earth's atmosphere [7]. Also, forecasting in the field of the stock market, which is one of the fastest-growing industries, helps many investors, sellers, and companies with their businesses by providing probable stock prices in the future, based on which the corresponding decisions are made to gain profits or avoid losses.

As mentioned above, forecasting in fields like weather and the stock market are helpful in many ways, but the procedure that we need to follow to do so has many challenges. To say, in the stock market we face challenges because of the dynamic

VIT-AP University, Amaravati, Andhra Pradesh, India. Email: payaswitha@gmail.com nature of the market, which can change at any time due to any reason; the fact that the data is collected from unorganized parts; nonlinear (i.e., the data is not linear or sequential), non-parametric (i.e., not having any certain specific parameters affecting the data) form of the collected data which makes the prediction process complex; the noisy nature of data due to the presence of false data or the missing of essential or right data and chaos created due to the unorganized form of the data [3]. Similarly, in the process of forecasting weather, we face challenges like analysis of huge amounts of data by looking at the satellite imagery, surface data, and precipitation reports to get the most probable prediction possible; the complexity of calculations involved in the numerical models after analyzing the data and usage of dynamic equations in the calculation to make the prediction which is close to the reality [15].

The challenges that are faced during the process of forecasting need to be handled properly to avoid inaccurate predictions or false predictions. This is where AI comes into the picture to tackle these challenges with ease and better accuracy than humans. To predict stock prices, AI uses algorithms that have machine learning (ML)-based approaches like Naive Bayes Bernoulli Algorithm, Support Vector Machine (SVM), Artificial Neural Networks (ANNs), and Linear Regression, which allow us to predict stock prices by overcoming the challenges of dynamic and unorganized nature of data and are capable of handling the noisy data and the created chaos with better strategies. Therefore, resulting in a much more accurate and precise estimation of the probable stock prices. For weather forecasting, AI makes use of ANNs like Multi-Layer-Perceptron (MLP), and SVMs and easily faces the challenges of analyzing vast data collected from satellite imagery, surface data, and other reports, and performs complex calculations with better accuracy and precision. Thus, providing the results of the predicted weather, which is very close to or the same as the actual weather. Therefore, by applying AI in the field of forecasting we can achieve probable predictions which are more accurate and precise when compared to the predictions obtained through other methods. Also, with the help of AI the procedures involved in forecasting are easier to perform and consume lesser time when compared to other methods.

The structure of this chapter is as follows: Section 2 contains the basic concepts of forecasting and its methods of implementation, where Quantitative and Qualitative methods are explained in detail, to the proper understanding of implementing forecasts. Section 3 is based on the traditional methods that were used before the evolution of technology, and on the limitations of using these methods. Section 4 contains a keen description of what is AI. Section 5 gives a clear idea about the classification of AI. Section 6 contains a description regarding the implementation of stock market forecasting using AI-based techniques. Section 7 contains a description of the implementation of weather forecasting using AI-based techniques. Finally, Section 8 provides the conclusion of the chapter.

2. Basic Concepts

In the past, several cultures considered forecasting as something evil and a disrespectful practice, for example, emperor Constantine banned forecasting in his kingdom, as he thought it had negative effects on his people. But, in today's world, forecasting has become an acceptable and respectful activity, which is utilized by various fields of people for several kinds of purposes. Forecasting has reached an impressive level when compared to the past with the help of new evolving technologies in the current world, and also there is an increase in the number of researchers and practitioners who are working on forecasting, which indicates that the importance and value of forecasting have increased a lot in the past years. Yet, it was found that there is a gap between the published research versions of forecasting and the needs expressed in two surveys of academics and practitioners [2].

To deeply understand the limitations and challenges that we would probably come across when forecasting, we need to understand what is forecasting and its classification. This understanding would give us a better idea of why and what kind of challenges would obstruct our goal to forecast and how these could be overcome. In simple terms, forecasting can be defined as the prediction or estimation of the future state of a product or a process based on the analysis of its corresponding historical data. Forecasting is done using mainly 2 types of methods: the Quantitative method and the Qualitative method.

2.1 Quantitative Forecasting

This type of forecasting is used in fields where the product or procedure is already existing for a long enough time with the availability of historical data. It is used for short-range planning and it makes use of statistical methods. Its chances of risk are relatively less when compared to qualitative forecasting, as we have the availability of historical data which can help us in making decisions that are closer to reality in the future. And also, the data considered in this method is objective in nature, so it cannot change from person to person which gives the flexibility of less complicated calculations.

Methods Used in Quantitative Forecasting

(1) Time Series Method

The concept of the Time Series method is suitable for cases where there is the availability of time stamped historical data. This method involves time series analysis, i.e., it involves the process of developing models based on the available data in order to gain an understanding of the data. Here, the desired forecasting is derived from this time series data using statistics.

The components of this method are:

(i) Trend (T)
(ii) Seasonality (S)
(iii) Cyclic feature of the data (seasonality over years) (C)
(iv) Random (R)

These components are used in 2 models: the first one is the additive model, in which the demand is calculated by adding the components T, S, C, and R; and the second one is multiplicative model, in which the demand is calculated by multiplying these components.

The Time Series method is further classified as follows: -

(a) Simple Moving Average (SMA) Method

In this method, all the 'n' demand points are given equal importance, i.e., weightage. It is equated as follows:

$$F_t = \frac{D_{t-1} + D_{t-2} + \cdots + D_{t-n}}{n} \tag{1}$$

where:
F_t = forecast in time period 't'
D_t = demand in time period 't'
n = number of moving averages that need to be determined

'n' value should be high for stable demand because the variations in such demands would not be much over a sufficiently long period whereas it should be low for unstable demand because there would be rapid variations within less amount of time (i.e., can be within few weeks or months). The average will never cross the maximum value of the data points, so there is a high chance that the forecast value which is anticipated using this method may not follow the trend of the data.

(b) Weighted Moving Average Method

In this method, the data points are allotted to different weights, unlike SMAs. These weights are allotted in such a way that the recent data points have more weightage than the past data points, i.e., the values of the weights allotted to the recent data points are more than the values of the weights allotted to the past (not recent) data points.

It is equated as appended below:

$$F_t = \frac{w_1 D_{t-1} + w_2 D_{t-2} + \cdots + w_n D_{t-n}}{w_1 + w_2 + \cdots + w_n} \tag{2}$$

$$w_1 > w_2 > \cdots > w_n$$

(c) Simple Exponential Smoothing Method

This method is suitable for cases in which the demand points don't have any seasonal or trend components in their characteristics. This method calculates the forecast of the next period based on the error which is obtained from the current period, which is mathematically depicted as the following equation:

$$F_{t+1} = F_t + (D_t - F_t)(\alpha) \tag{3}$$

$$0 \leq \alpha \leq 1$$

where:
α = smoothing constant; it's data dependent, for stable demand its value should be low, and for unstable demand its value should be high [i.e., close to 1 or = 1]

D_t = demand of current period
F_t = forecast of current period
The above expression can be rewritten as:

$$F_{t+1} = \alpha D_t + \alpha(1-\alpha)D_{t-1} + \cdots + \alpha(1-\alpha)^{(n+1)} F_{t-n} \tag{4}$$

The coefficients, i.e., the weights of the demands of all the periods, decrease exponentially in this method. So, all the demand points are considered, but with different weightages such that the highest weightage is given to the most recent demand point.

(d) Trend Projection

This method is utilized in cases where the considered data follows a trend. Besides, here the data is captured for the further estimation of the next period's forecast. This method is implemented on the assumption that the trend shown by the considered time series data follows that trend for future periods as well. So, the forecast of the next period is calculated based on an equation (linear or exponential) in which the available data is fitted (Fig. 1).
Example:

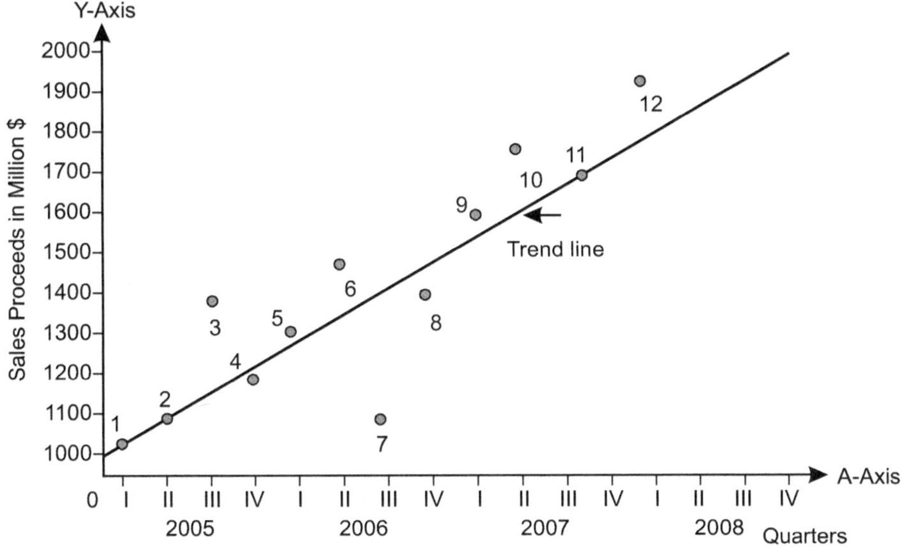

Fig. 1: Example depicting the implementation of Trend Projection.

Equation: Y = a + b(t)
where:

Y = Sales of product
'a' and 'b' are constants
b = annual increase in sales (because it's the slope of the line)
t = time period in years
Here, the above equation is called as the 'Trend Line'.

(e) Decomposition

This method is suitable for the data which has both trend and seasonality components in its characteristics. This method first determines the trend and seasonality components separately from the given historical data, after which it estimates the corresponding future trend and seasonality. And then these future estimates are combined to arrive at the corresponding future forecast either by using the additive model or the multiplicative model.

(2) Causal Method

The causal method makes predictions and forecasts the required information of the given field of interest by considering a range of variables that may influence the future in that field as this method is mainly concerned about the kind of impact these anticipated variables might have on the future of that field it allows us to understand the factors that affect forecasting in an organized way instead of merely considering the available historical data.

The causal method is further classified as follows:

Linear regression

This method is used to derive a linear relationship between the 'Forecast variable' and the 'Predictor variable or variables'

(a) Simple Linear regression

This is a type of Linear regression in which only one "Predictor variable" is considered and a linear relationship between it and the 'Forecast variable' is derived, which allows us to anticipate the required forecasting.

(b) Multilinear regression

This is a type of Linear regression in which multiple "Predictor variables" are considered and a linear relationship is derived between these multiple variables and the "Forecasting variable", which helps calculate the required forecasting.

2.2 Qualitative Forecasting

This type of forecasting is used when a new product or procedure needs to be implemented, where there is no availability of historical data. Since there is no availability of historical data, the data obtained from the surveys and experts' opinions are used for forecasting. This method is used for long-range planning, i.e., for a minimum period of 2 years. It does not use any statistical methods, as the data is gathered from multiple sources. Its chances of risk are relatively more when compared to the quantitative method, as it is solely dependent on the experts' opinions and the surveys conducted, i.e., it is not fixed or constant as opinions change with situations and time. And also, the data considered in this method is subjective in nature, so it can change from person to person, which makes it difficult to analyze the data.

Aspects of information used to derive this forecast:

(a) Expertise

(b) Technical Knowledge

(c) Experience

(d) Intuition

Methods Used in Qualitative Forecasting

(1) Delphi Method

In this method, a set of questions, i.e., a questionnaire, which is based on the product that is about to be introduced, is sent to a panel of experts (of different fields). Then, all the responses obtained from the experts are aggregated into a set of results. Based on these results, another questionnaire is prepared and sent to the experts. This procedure is repeated until a common opinion is obtained from all the experts (Fig. 2).

This method is considered to be advantageous as each expert's opinions are considered and also, we can include a wide range of experts if needed. But it consumes more time and most importantly, its quality cannot be guaranteed, as it depends on the experts chosen.

(2) Executive Opinion

This method involves the collection of opinions from the executives, i.e., the management of the company that needs forecasting. The cases of incorrect forecasts are probable, as there are chances of domination of opinions by one of the executives which would lead to biased opinions.

(3) Salesforce Opinion

This method utilizes the opinions collected from the salespeople, who actually engage and interact with the customers directly and have a real-time experience with the customers, to make predictions and estimations for the required forecasting.

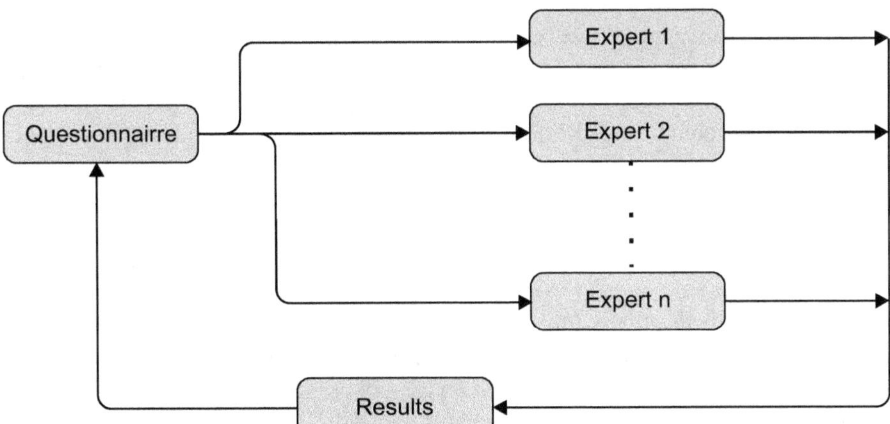

Fig. 2: The procedure followed in Delphi Method.

But the chances of incorrect forecasts are probable, as there might be cases where the salespeople can underestimate the demands to reduce their work or targets.

(4) Consumer/Market Surveys

This method, similar to the Delphi method, involves the circulation of a questionnaire, but among the customers, where the questionnaire is created based on the product of interest. Then all the responses obtained from the customers are aggregated into a single feedback, based on which the required forecasting is derived. This method is considered to be more accurate and reliable but it consumes more time.

Note: Qualitative method-based forecasting gives us the flexibility to deal with a wide range of situations and circumstances, especially in cases like launching a new product, where there is no availability of historical data for that product. Whereas, the Quantitative method-based forecasting does not provide us with this kind of flexibility, as it needs historical data to function properly.

3. Traditional Forecasting Methods: The Methods used Initially

The Traditional Forecasting method was used in the past before the introduction of technology in the field of forecasting and is also being used currently in a few managements based on the requirements. Traditional forecasting methods play an important role in demand planning strategy. The performance of traditional statistical methods is more efficient in the case of businesses with datasheets that are countable and have a finite number of predictive variables. It uses historical observations and estimates future business-based parameters like budget, revenue, or assets. But the implementation of these methods may not always be accurate or successful, as there are chances for them to fail because the past does not necessarily have to represent the future.

There are various types of traditional forecasting methods, a few of them are:

(1) Collective Opinion Method

This method is considered or preferred for the data which does not have any historical data or statistical model, i.e., when introducing new products or procedures. This method forecasts the required information regarding the product of interest, based on the opinions gathered from POS (Point-Of-Sale) owners. But when using this method for forecasting, there is a high probability of a significant variation in accuracy.

(2) Survey Method

The implementation of this method involves conducting surveys with customers, POS owners, and experts in the market, where the information gathered from these surveys is utilized for forecasting. It is considered to be an easy method to implement, as the information is gathered from primary sources. Even though the data considered by this method is based on human opinions where there are chances of human manipulation, an unbiased result can still be achieved when the results are implemented on an overall view considering all the primary sources. But this method is time-consuming and needs customers' cooperation, which is not possible in all cases.

(3) Market Experiment Method

This method is concerned with studying and analyzing the change in demand concerning the change in prices and features of the product. This is done by conducting experiments and surveys among the customers based on the product of interest. These changes in demand are observed under controlled experimenting conditions and the corresponding customers' responses are recorded.

Forecasting was done in the past using the above-mentioned traditional methods. But traditional forecasting methods are not always accurate or successful, as there are high chances for them to fail because of the fact that the future may not necessarily represent the past. There are high chances of error in forecasting if the future is not similar to the past, which may cause a lot of loss and damage for the companies and management who make their decisions based on this forecasting. Most importantly this method cannot predict the turning points in the data, as turning points don't follow the past data. Also, this method relies more on human judgment, which is not constant all the time. And most importantly there might be cases where the overall opinions collected from people are subjected to manipulation by the people who collected them, or the overall results are biased by one of the opinions, therefore leading to inaccuracy and uncertainty in the results.

So, there is a high chance of resulting in incorrect forecasting when utilizing traditional forecasting techniques and it is difficult to overcome all these limitations and challenges just with human labor and strength. This is where AI comes into the picture. AI-based forecasting makes it feasible to overcome the limitations like human manipulation, dynamic data, and collection of data through human labor, which results in high time consumption, possible errors in the data due to human labor, etc. In this chapter we are going to interpret the importance of using AI-based forecasting over the traditional forecasting methods.

4. Revisiting Artificial Intelligence

Being concerned about how AI overcomes the various probable challenges in forecasting done using other methods, first we need to understand what is AI and how it is classified. This knowledge would help us to keenly understand how AI overcomes many challenges with ease and how it could be further utilized to provide alternatives for many other limitations.

In the 1950s, John McCarthy was the first person to coin the term "Artificial Intelligence". Its founder Alan Turing defines AI as: "the science and engineering of making intelligent machines, especially intelligent computer programs" [11]. So, we can say that AI is the capability of a device to judge, understand, produce original thoughts and perform tasks, whereas, these tasks include the capacity for knowledge and the ability to acquire it. Chakraborty says, "In 1956 (during a 2 months' workshop at Dartmouth College), computer scientist John McCarthy perhaps first introduced the term "artificial intelligence" (AI) to articulate the study of intelligence in a computer program that is proficient in thinking non-numerically [12]. Later, McCarthy's high-level computer language (LISP) introduced in the "MIT AI Lab Memo No 1 became a prominent AI programming language.

Humanoid robots have been a monumental achievement in AI since the 1980s." He also says, "For these authors, AI depends on the 4 categories of thinking humanly, thinking rationally, acting humanly, and acting rationally. A system becomes rational if it goes toward the right things. The cognitive procedures of machines set up computer models of the brain like the human brain, as they aim to promote artificial intelligence that can think humanly (following cognitive ways) and accordingly act rationally" [11].

Intelligence is defined by Chakraborty as "Intelligence seems nothing but some rationale-based decisions or choices that stand for reality and has some argumentative grounds behind it. Moreover, intelligence is a kind of toolbox by which an agent can accomplish his/her aim" [11]. So, we can say that intelligence is a part of the contention that helps us to solve problems in the varying optimal way possible. AI challenges human intelligence, so it is mainly dependent on the human brain and cognition. The following characteristics are used to identify the intelligent behavior (these characteristics are present in most of the intelligent systems, and there are chances that certain intelligent systems do not have a few of them) of a system:

(1) Thinking Humanly

(2) Thinking Rationally

(3) Acting Humanly

(4) Acting Rationally

Acting humanly and thinking humanly characteristics indicate the capability of the system to understand and perceive its environment and interact with other humans or systems around it when needed. And thinking rationally and acting rationally characteristics indicate the capability of the system to give justified reasoning for its decisions and behave such that the chances of desired success are maximum.

5. Seeing Through Types of Artificial Intelligence (AI)

AI is a field in which there are a lot of researchers, scientists, and companies working to keenly understand and make significant developments in it to utilize it more efficiently. AI is mainly classified into 2 types they are Type 1 and Type 2, where Type 1 is further classified into 3 types of AI, which are:-

(1) Narrow AI or Weak AI

Narrow AI or Weak AI is concerned with making machines act as if they are intelligent, i.e., the machines don't actually think or represent human minds, but just follow the set of instructions, which are already fed into it, but their way of interaction with the outside world would make it look like they are intelligent. This type of AI allows the machine to perform specific tasks only, therefore its application is confined to just one specific operation or task.

Examples: drive-by-wire cars, speech recognition systems, etc. One of the finest examples of weak AI is self-driving cars since it can cause severe accidents such as how human beings normally did because they don't have the auto correction things in their programming.

(2) Artificial General Intelligence

Artificial General Intelligence is concerned with making machines imitate human sensory movements like being capable of watching, responding, understanding natural language, explaining or delivering external information, etc., similar to a human. This type of AI makes the machine capable of behaving more like a human and also capable of interacting with human beings humanly.

(3) Strong Artificial Intelligence

Strong AI is concerned with making machines think or represent human minds in the future. In this type of AI, the machines will have the cognitive abilities, i.e., to reason, to think and they would also have feelings and emotions and they will perform all the tasks a human can perform. Currently, there are no such machines invented, but research is going on.

Type 2 AI is also further classified into 4 types as follows:

(1) Reactive Machines

Reactive Machines belong to the 'Unsupervised' category of AI, i.e., they don't have any storage memory and cannot learn from their experiences for better decision-making purposes. So, the machines would only operate based on the components and data that are available at that moment. They are programmed only for specific and simple tasks so there is no development with these machines, they just keep repeating the same tasks whenever assigned to them.

(2) Limited Memory

Limited Memory belongs to the 'Supervised' category of AI, i.e., these machines have memory for the storage of data and can learn from their past actions or data. These machines gain this knowledge from experienced data or real-life events. This type of AI allows the machine to build observational data which gives information about many factors related to the task assigned to the machine.

(3) Theory of Mind

Theory of Mind is a type of AI in which the machine is expected to be capable of having thought patterns similar to a human, i.e., generating ideas, creativity, making decisions, etc. Also, the machines are expected to understand that humans have thoughts, feelings, and emotions, and identify, understand and remember emotional responses.

(4) Self-Awareness

Self-Awareness is a type of AI in which the machines would have human-level consciousness, i.e., to think for themselves, have desires, have feelings, and understand their feelings. So, the machines in this type of AI are expected to have self-guided thoughts and have their reactions, and understand their existence, which makes them very similar to humans and allows them to operate effectively among humans with great efficiency and understanding.

Now, let us go through the basic concepts, importance, and implementation of AI-based methods to achieve stock price forecasting and weather forecasting.

6. Stock Market Forecasting Using Current AI Formulae: Relooking at Demand Forecasting

The stock market can be defined as the interaction (i.e., buying or selling) between buyers and sellers of stocks where this interaction is done online. There are a few important factors that affect the sales of a product, the first one is the executive choice or judgment over the product and the other factors are group opinion over the product, relative thinking, sales force composite, and customer market survey. In all these cases there is a need for excessive human labor and time and in the case of the group opinion, there are chances for dominance of a person's opinion on that of others, i.e., human manipulation, which might lead to false results. AI-based quantitative forecasting can prevent the disadvantages of excessive human labor involved during the forecasting and is capable of completely nullifying the chances of human manipulation and therefore providing more accurate and unbiased results.

To understand the application of AI to stock market forecasting, we need to know about demand forecasting as it plays a crucial role in the field of business and the stock market.

6.1 Demand Forecasting

The forecasting which is done based on the information of sales and demand for the products in the market to forecast the probable products' demands and sales in the future is called demand forecasting. This demand forecasting are of 6 types, as appended:

(1) Short-term Demand Forecasting

The forecasting which involves the prediction of the future of a product for a short term (i.e., typically 3–12 months) based on the corresponding available historical data is called as short-term demand forecasting. This type of forecasting is applied often in the field of fashion trends because immediate forecasting techniques play an important role in this field. So, in this case, short-term forecasting makes it possible to give up-to-date information on new increasing trends and the status of the existing trends, based on which the manufacturers and fashion designers would make their plans accordingly.

(2) Long-term Demand Forecasting

Forecasting which involves the prediction of the future of a product for a long term (i.e., greater than a year) based on the relevant historical data available is known as long-term demand forecasting. It considers all the parameters available in a wide range and gives an estimation of the probable nature and status of a product for the long-term future accordingly. This type of forecasting is used in the field of business for better production management. As production management involves the process of managing industrial procedures with proper planning and execution based on the availability of raw materials and many other parameters, long-term forecasting makes the process smooth by giving the estimations of the required raw materials and many other parameters for the long term.

(3) Active Demand Forecasting

The forecasting which often forecasts the outlook for the growth of start-up companies or businesses, where the availability of a significant amount of relevant historical data is not possible is known as active demand forecasting. This type of forecasting makes it easy for the start-up business or company owners to understand their growth and status in the market, which would help them to make decisions accordingly to achieve more growth and save their company or business from failure.

(4) Passive Demand Forecasting

The forecasting that involves the prediction of future sales and demands based on the availability of a significant amount of relevant historical data is called passive demand forecasting. This type of forecasting is not suitable for start-up companies or businesses but is very much advantageous in cases where the companies or businesses which require forecasting have a decent amount of corresponding historical data available so that the forecasting can be implemented with more efficiency.

(5) Internal Demand Forecasting

The forecasting which forecasts the information of the supply chain and demand of the product in the market, based on which the sales-related decisions are made to smoothly carry the business, is called internal demand forecasting. This type of forecasting allows the business owners to get information regarding the requirements in the financial sector and the understanding of the product's supply from an overall outlook, which helps them to get a good understanding of the demand and supply status of the product in the market, based on which they make their decisions.

(6) External Demand Forecasting

The forecasting which forecasts the information regarding the growth, health, outcomes, and competitors of a company is known as external demand forecasting. This type of forecasting makes it feasible for the owners of a company to get a clear idea of the status of their company in the market. This forecast contains a detailed view of the competitor's products and sales along with one's company sales, which helps the owners of the company quickly know the demand shortage or abundance of the products in time.

In the past, when there weren't many technological advancements and the collected data was of low volume and was simple to analyze, traditional statistical methods were used to analyze and make predictions accordingly. And with time, technology has developed a lot and there is a significant increase in the complexity and volume of the collected data, due to which many researchers are working on more advanced technologies to make predictions directly from the collected data [6].

Techniques like moving averages are often used in time series forecasting, to do technical analysis of the collected data and forecast the probable and significant changes in the market. Simple Moving Average (SMA) is one of the moving average techniques, which is often used by long-term traders, as they do not choose to engage much in their trades and hence, are not much worried about the speed of provision of the results. It is calculated by the appended formula:

$$SMA = \frac{A_1 + A_2 + \cdots + A_n}{n} \tag{5}$$

Where, A_n = price of an asset at period n; n = the total number of periods.

Considering over moving average techniques that are familiar with the name of the Moving Average Crossover approach which is often used by stock brokers to predict the trend switches and the direction in which the trend switches, so that they can trade accordingly. In this approach two moving averages, out of which one is a shorter moving average (i.e., moving average calculated over a short period like 5 – 15 days) and the other one is the long-term moving average (i.e., moving average calculated over a long time like 100 – 250 days) are graphed based on separate periods, which tend to cross with time, where this crossing is known as the moving-average crossover. Based on the nature of crossing, the stock brokers are indicated with a 'buy' or 'sell' signal, i.e., when the short-term moving average crosses above the long-term moving average the 'buy' signal is indicated and the 'sell' signal is indicated for the converse case [6].

These techniques can be implemented using the technology of AI, to reduce the time consumption and human labor, and human errors involved in the process. As there are a lot of financial investors around the world who are interested in knowing the predicted prices in the stock market in advance to gain profits and avoid losses, AI has made it feasible to predict the prices in the stock market close to or the same as the actual prices in the future. This is achieved by collecting and analyzing the available past historical data using linear regression and then implementing it using neural network models.

6.2 AI Methods to Forecast Stock Prices

(1) Forecasting Stock Prices using Sentiment Analysis

The collection of tweets made by the public on the social media platform 'Twitter', when analyzed properly as a whole, has information regarding public sentiment. Here, this public sentiment is considered an important parameter affecting the behavior of stock prices. And ML techniques like Support Vector Machine (SVM) and Naive Bayes-Bernoulli classification can be used in the process of sentiment analysis [4].

(a) Sentiment Analysis

Sentiment analysis can be defined as the field of study that is concerned with analyzing people's opinions, attitudes, evaluations, sentiments, appraisals, and emotions towards various entities such as products, organizations, services, individuals, issues, events, topics, and their attributes [5].

A conference paper [4] which is concerned with predicting stock prices based on Sentiment Analysis of tweets made by the public on the social media platform 'Twitter', involves the execution of text processing, training set collection, and training the classifiers. In this procedure, the collected texts (i.e., tweets) are processed and sent to the set collection which is trained such that the corresponding data goes into the respective sets, where these sets are classified by the trained classifiers.

(i) Text Processing

During text processing, the unnecessary URLs, or symbols (which have no meaning) which were considered as 'noise', are removed by using Natural Language Toolkit (NLTK) for Python (in this case). Text processing has 3 subparts, the first one is **tokenization**, where the text is divided by spaces and for each tweet, a corresponding list of such individual words is made. The words in this list are used as features to train the classifiers.

The second subpart is **removing stop-words**, where the above-mentioned list of words are checked with the words in the Python's NLTK library, which has a stop-word dictionary. Any of the words that match with the words in the NLTK library are removed. The third subpart is regarding the Twitter **symbols**, where the words are removed or made sure to be included based on the symbol they are following. The words following the symbol: '@', are always excluded as such words are often the usernames and have no relevance to the required sentiment. And the words following the symbol: '#' are not removed because there is a high chance that these words have useful information about the corresponding tweet's sentiment. Also, URLs are removed entirely as they are inappropriate for sentiment analysis.

(ii) Training Set Collection

A collection of huge numbers of tweets is given as training data for the sentiment analyzer, where two datasets: 'Positive' and 'Negative' are created through Twitter's search API, which was based on the corresponding keywords and emojis:

Positive sentiment query: ":-)", ":)", ":D", "=)", "<3", "like", "love"

Negative sentiment query: "=(", ":(", "dislike", "hate"

The tweets containing one or more of the above-mentioned emoticons would likely be considered as the corresponding sentiment. In this way, the 'Positive' and 'Negative' datasets are done.

(iii) Training the Classifiers

The Naive Bayes Bernoulli and SVM classifiers are used to classify the datasets, as they are the most recent stage of sentiment analysis. Training of classifiers has two subparts. The first subpart is feature extraction, where a unigram feature is created for each unique tokenized word in the list for the classifier. It is implemented such that, if the word 'bad' is present in a negative tweet, the feature for classification would be concerned with the presence of this word in the tweet, and if it contains then the classifier would most probably classify it as negative because the word 'bad' is from a negative tweet.

There would be a requirement for bigrams and trigrams, where the classifier's feature vector is made of two-word and three-word combinations respectively. Because in the case of dealing with words like: "not perfect", the unigram feature extraction would classify it as 'Positive' (because 'not' has neutral sentiment and 'perfect' has positive sentiment), whereas the bigram extraction feature would classify it as a negative tweet. The second subpart is feature-filtering, where the 'n' most significant features are only used for training. In this project, chi-squared test is used to score each unigram, bigram, and trigram in the training set, where the frequency of each feature is found using NLTK.

(b) Applying Machine Learning Techniques

In this project, the classifier is designed such that it either classifies the tweets as 'Positive' or 'Negative', based on their corresponding features and characteristics. Naive Bayes Bernoulli and SVM are the two classifiers used because they are the latest stages of sentiment analysis.

(i) Naive Bayes Bernoulli

Naive Bayes methods are the algorithms that are trained to get a specific output (i.e., supervised algorithms), applied based on Bayes' theorem, where each pair of features are assumed to be independent of the other pairs of features. Bayes' theorem states the appended relationship between y, which is the class variable, and the dependent feature vector containing dependent variables from to :

$$P(y \mid x_1, x_2, \ldots, x_n) = \frac{P(y)P(x_1, \cdots, x_n \mid y)}{P(x_1, \cdots, x_n)} \tag{6}$$

Different Naive Bayes' classifiers differ from each other often due to their corresponding assumptions regarding the distribution of $P(x_i|y)$. In this project, multivariate Bernoulli-based distributions are made for implementing Naive Bayes' Bernoulli classification. Its decision rule is based on $P(x_i|y)$, which in this case is given by:

$$P(x_i|y) = P(i|y)\, x_i + (1 - P(i|y))\, (i - x_i) \tag{7}$$

(ii) Support Vector Machine

SVMs are a collection of methods that implement tasks like classification, regression, and outliers' detection where these methods are implemented on the principle of supervised learning. The SVM Classifier classifies the given linearly separable collection of data in such a way that the maximum gap is present between this separation, which is called the margin. SVM implements the task of classification by constructing a hyper-plane (or hyperplanes, if required) in high dimensional space, which makes a decent separation between the sets thereby forming the margin. The data points which are the nearest and are present on this margin between the hyperplane known as the support vectors. SVM is very much concerned about maximizing the distance between the hyperplane and each training point, for which it considers a parameter vector 'a' such that:

$$Minimize \frac{1}{2}(a * a);\; y(a * x_i + b) \geq 1 \tag{8}$$

where, 'y' represents the class label for positive and negative sets (–1, 1).

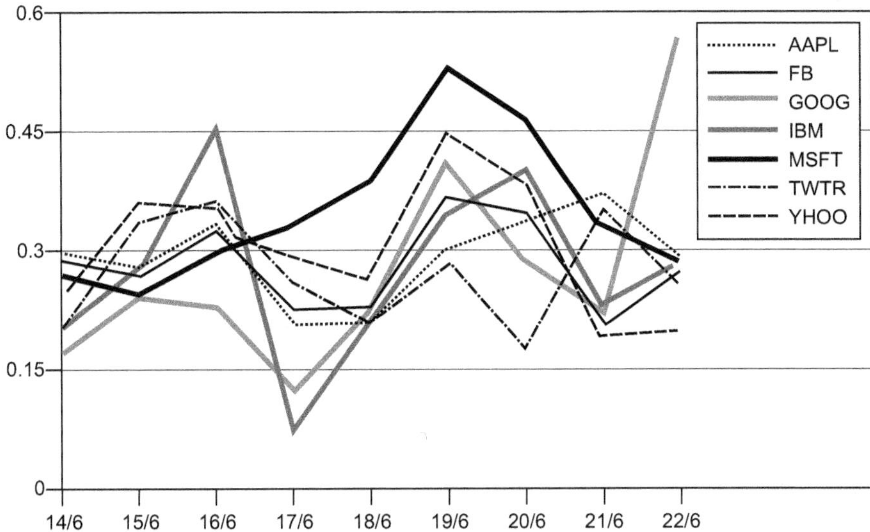

Fig. 3: Sentiment classification of the tweets collected from 7 most popular companies (Apple, Facebook, Google, IBM, Microsoft, Twitter, and Yahoo!) [4].

Fig. 3 depicts the sentiment classification of the collected tweets which are related to 7 of the most popular tech companies (obtained from [4]).

Where the obtained positive sentiment score is normalized with respect to the total number of tweets per day as follows:

$$NormalizedPositiveScore_i = \frac{PositiveTweets_i}{TotalTweets_i} \tag{9}$$

In this way, the sentiment score for the tweets collected for desired companies is obtained. Then, the classifiers were trained and tested with a specific subset of the collected tweets, where the k-fold cross-validation method is implemented to calculate the accuracy of these classifiers.

In this k-fold cross-validation process, the datasets used for training are split into k number of smaller subsets of datasets, out of these subsets, (k–1) subsets are utilized for the purpose of training the classifier, and the remaining one subset is utilized as a test set to calculate accuracy. This training and testing process of these subsets is done k times for every different data partition. In this method, the training process takes place entirely till a specific day, and then for the coming next 'k' days from this specific day, the test process takes place.

The value of 'k' is considered to be 7 in [4], based on which the corresponding average accuracy is obtained. After obtaining the value of accuracy, the values of 'precision' and 'recall' are calculated by using the appended formulae:

$$Precision = \frac{t_p}{t_p + f_n} \tag{10}$$

Table 1: Average effectiveness of Machine Learning techniques for Sentiment classification [4].

Algorithm	**Naive Bayes (%)**	**SVM (%)**
Accuracy	80.60	79.30
Precision	pos: 80	pos: 79.30
	neg: 62	neg: 45.45
	neut: 94.73	neut: 81.25
Recall	pos: 60	pos: 56.66
	neg: 94.73	neg: 83.33
	neut: 74.41	neut: 86.66
F-measure	pos: 68.57	pos: 65.38
	neg: 75	neg: 58.82
	neut: 79	neut: 83.87

$$Recall = \frac{t_p}{t_p + f_n} \tag{11}$$

where, t_p = positive tweets; f_p = false positive tweets and f_n = false negative tweets.

Table 1 [4] contains the obtained values of accuracy, precision, recall, and F-measure (false measure):

Stock Market Prediction

A feature matrix containing the following features is created:

(1) Percentage positive sentiment score

(2) Percentage negative sentiment score

(3) Percentage neutral sentiment score

(4) Close price

(5) HLPCT (High-Low Percentag)e

(6) PCT change (Percentage Change)

(7) Volume

To this feature matrix SVM is applied for predicting the future stock movements. The obtained results when compared to the actual upcoming stock changes, resulted in an accuracy of 87%. The predicted closing price had prediction errors under 10%. Figure 4 [4] depicts the comparison between the predicted and the actual close price for 16 technology stocks that were considered by Yahoo! Finance to be very popular.

Therefore, stock prices can be predicted on the basis of sentiment analysis using AI-based machine learning (ML) techniques as described above.

(2) Forecasting Stock Prices Using Linear Regression and Artificial Neural Networks

(a) Linear Regression

Linear regression is considered as one of the significant methods as it operates on the previous-available data where a linear relationship between the ‘Forecast

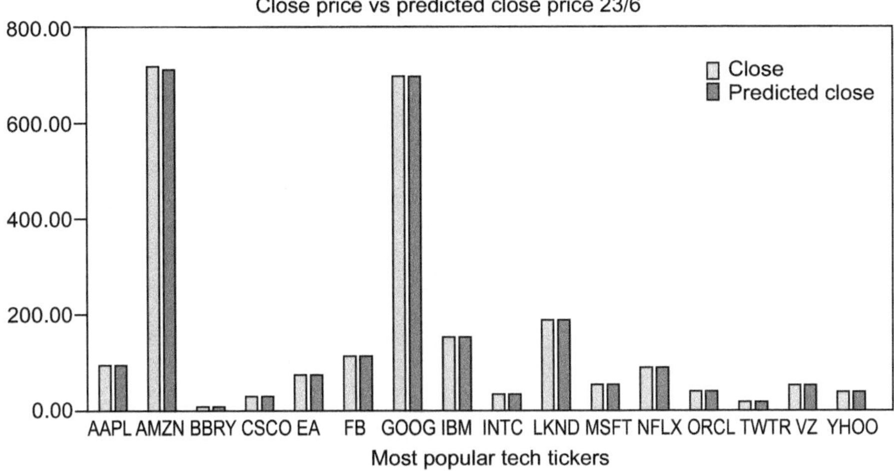

Fig. 4: Graph plot of Actual Close Price vs respect to Predicted Close Price.

variable' and the 'Predictor variable(s)' is derived, which is utilized in the process of quantitative and technical analysis to give us results of the prices in the future.

Linear regression, which deals with only two variables, i.e., one independent variable and one dependent variable, is called as simple linear regression. Linear regression is expressed using the following equation:

$$y_i = \alpha + \beta x_i + \varepsilon_i \tag{12}$$

where, y = dependent variable (along y-axis);
x = independent variable (along x-axis);
α = y-axis intercept/constant;
β = slope;
ε = the residual term/error term

(b) Artificial Neural Networks (ANNs)

The neural networks are considered to predict stock prices as they can operate efficiently even in cases where the relationship between the input and output is unknown. In the book called *Neural Networks:A Comprehensive Foundation* [10], a neural network is defined as follows when considered as an adaptive machine:

A neural network is a massively parallel distributed processor made up of simple processing units, which has a natural propensity for storing experiential knowledge and making it available for use. It resembles the brain in two respects:

1. Knowledge is acquired by the network from its environment through a learning process.
2. Interneuron connection strengths, known as synaptic weights, are used to store the acquired knowledge [10].

As stock prices need to be predicted using linear regression, the Multi-Layer-Perceptron [MLP] model of neural networks is considered. MLP has a minimum of

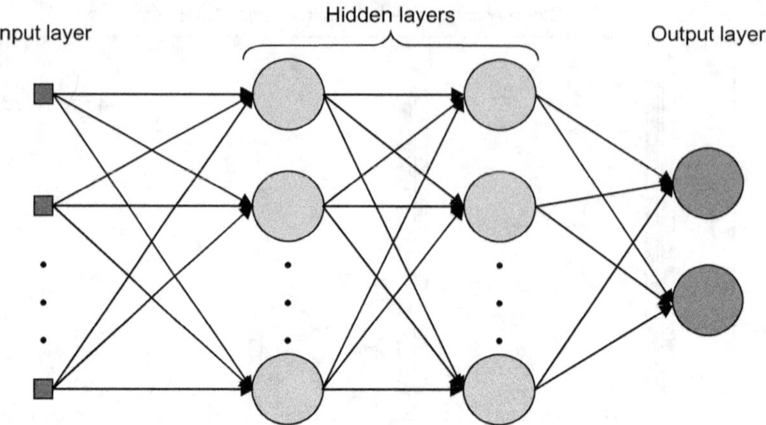

Fig. 5: Multi-Layer-Perceptron Model.

three layers: one input layer, one or more hidden layers, and one output layer, which is depicted in Fig. 5.

The input parameters are obtained from the Input layer and taken by the Hidden layer to sum them according to the weights assigned to each of them. After obtaining this sum, an offset is added to it. Then the finally obtained output values are sent to the Output layer. As a transfer function is used for the calculation of the required output, we could say that the obtained value of the output is the relation between the input level and the number of corresponding input parameters [3].

In this procedure, ANNs which are trained with the error backpropagation algorithm, are utilized, so that the network can learn from its error and make corresponding changes in itself. There are two passes in the error backpropagation algorithm, the first one is the forward pass and the other one is the backward pass. The input vector containing the input parameters is given to the next layer (i.e., the hidden layer(s)) in the forward pass, and its corresponding effect is propagated sequentially layer after layer in the network, where the weights assigned to each of these parameters are fixed. And finally, the outputs are obtained at the final layer, i.e., the output layer. And in the case of a backward pass, the calculation of an error signal is done by obtaining the difference between the desired response and the actual response. Based on this error signal, the weights are adjusted (in such a way that the error is less). Then, the error signal is sent from the final layer, i.e., the output layer of the network where its effect is propagated in the backward direction from the output layer to the input layer sequentially [10].

A paper [3], which is concerned with the implementation of ANNs to predict stock prices, involves the research and work done on the neural network model which was trained for 60 days with the available data, based on which the model was expected to predict the stock prices for the next day (i.e., the 61st day), the obtained results of various stocks are as follows:

[The doted graph represents the predicted price for 60 days and the thick-line graph represents the actual price for those 60 days. In the graphs depicted (Figs. 6–8), the number of days is represented along the x-axis and the corresponding price is represented along the y-axis.]

(a) Apple share price:

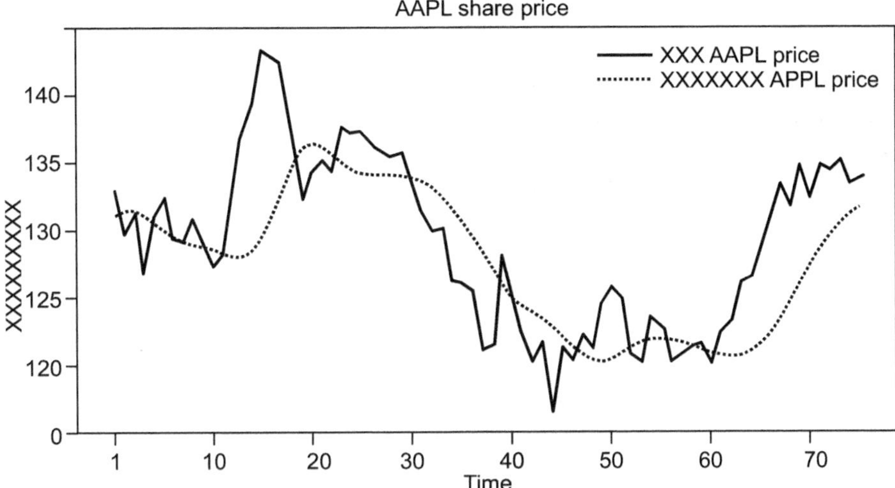

Fig. 6: Graph containing the plots of both predicted and corresponding actual price of Apple share.

Based on the 60 days of training and prediction, the model had predicted the price for one day in the future as:

Today's price =133.50 USD

Tomorrow's predicted price= 131.63713 USD

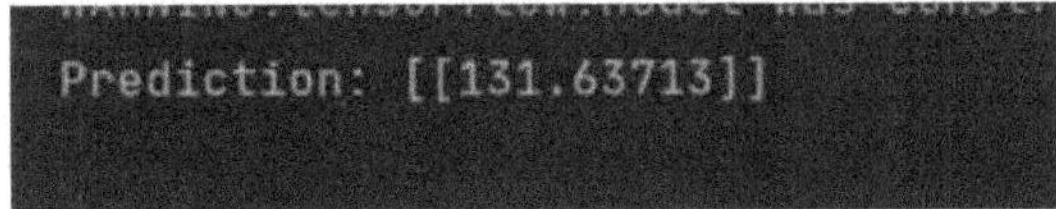

(b) Google Share price:

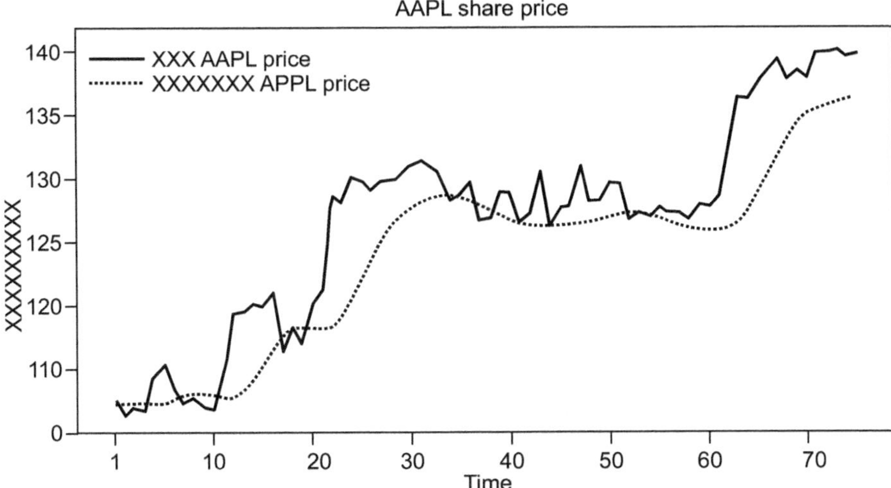

Fig. 7: Graph containing the plots of both predicted and corresponding actual price of Google share.

Similarly, for google share, the model had predicted the price for one day in the future after being trained for 60 days as follows:

Today's Price = 2,278.35 USD

Tomorrow's predicted price = 2228.3293 USD

```
Prediction: [[2228.3293]]
```

(c) TATA Motors price:

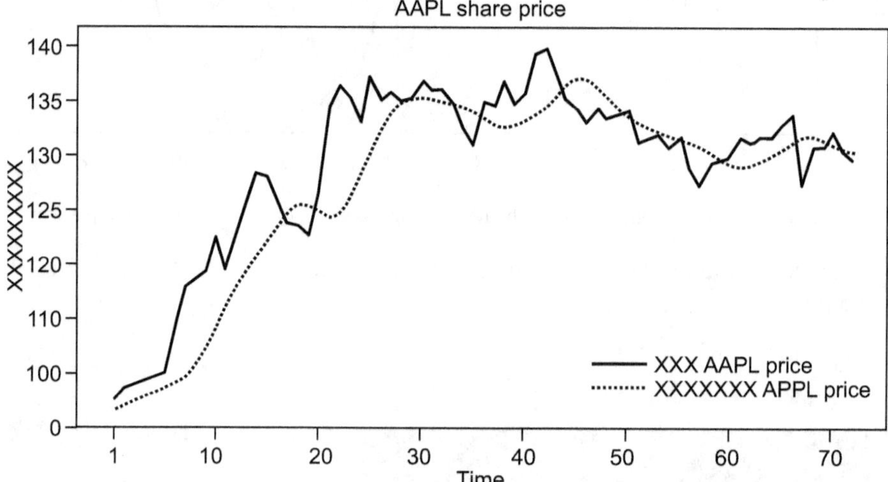

Fig. 8: Graph containing the plots of both predicted and corresponding actual price of TATA motors price.

Similarly, the predicted price for the TATA Motors by the model after its 60 days training and prediction is obtained as follows:

Today's Price = 298.05 Rs

Tomorrow's Predicted price = 300.35170 Rs

```
Prediction: [[300.35107]]
```

All these obtained results suggest that applying AI-based ML methods to predict stock prices has significant and positive results, from which it could be understood that AI-based forecasting results in more accurate and effective forecast of stock prices [3].

7. Weather Forecasting Using AI-based Techniques

Weather forecasting is a process that involves the prediction of the probable changes in weather, rainfalls, storms, high temperatures, etc., which can be utilized for research purposes of various studies, helpful for the preparation of disaster management in advance to face the possible problems predicted. Weather forecasting also involves the prediction of individual predictions like rainfall prediction, storms prediction, etc., out of which prediction of rainfall is dealt with more often when compared to the other predictions. Rainfall prediction involves the process of predicting the precipitation probability in a region and estimating the intensity of the rainfall. Supervised learning methods like regression are used to predict the results in a discrete amount by deriving a relationship between the independent and dependent variables, where this relationship is further used by the predictive analysis [9].

There are many such methods involved in the forecasting of weather, which requires a lot of time and energy if implemented by humans, so AI is used in the field of weather forecast, as knowing about future weather conditions in advance would be advantageous in many ways. As many decisions or plans for great tasks like setting the Olympic games dates (i.e., to plan the games such that there aren't any interruptions or effects of weather on the games) or for general tasks like setting family trips' dates (to avoid bad weather conditions during the trip time) and many other important tasks like planning military-based activities (i.e., to take action over the enemies or protecting the country with better planning as the weather plays an important role in this procedure), sending precautions to public according to the possible severity of weather (e.g., fishermen are warned not to go for fishing or asked to finish their fishing within a period when the weather forecast results show severity), etc., are done based on weather forecasting to have smoother and safer future.

In this way, many people from several fields would be benefited from weather forecasting. But the prediction of weather involves very complex calculations containing dynamic equations and huge storage and analysis of data, which can be performed by AI with great efficiency and speed. AI allows us to get the prediction of weather that is nearly accurate and close to the actual weather in the future, which is useful for many people like farmers, fishermen, Military officers, Olympic players, and research scientists who are trying to discover theories based on weather and many other people around the world.

In AI-based weather forecasting, AI-based techniques like ensemble neural networks are used for ensemble weather forecasting, SVM is used for regression (which is called Support Vector Regression (SVR)) in the process of rainfall prediction and the MLP model is used for weather forecasting.

(1) Forecasting Weather Using An Ensemble of Neural Networks

(a) Ensemble Forecasting

In general, forecasting of weather is implemented using two approaches: the first approach is the empirical approach, which is preferred in cases where the prediction of local-scale weather with a huge amount of data is required, and the second approach is the dynamical approach, which is suitable for predicting the weather phenomenon on a large scale. Both these approaches involve the storage and analysis of a huge amount of data and complex calculations, which are difficult and would consume a lot of time if implemented through traditional techniques. ANNs give us the flexibility to solve various kinds of nonlinear problems which may be too complex to solved using traditional methods or strategies.

Now, we are going to look at the application of ensemble neural networks for forecasting weather. An ensemble neural network involves the training of a collection of neural networks for the same task, where the number of neural networks is finite. The ensemble forecasting process involves the forecasting of a range of forecasts, which are considered from slightly different initial conditions, and also each of the models has a slightly different formulation for their modeling, etc. This difference between the considered forecasts can be utilized to predict the future weather states over their corresponding range. The main advantage of using an ensemble-based neural network model is, it reduces the variance or imbalance of our network which is utilized for forecasting. This type of forecasting allows us to deal with the uncertainty in the forecasts and it reduces the chances of error.

An article [8] which is on forecasting of weather with an ensemble of neural networks, involves the utilization and analysis of various models (like MLP, Elman Recurrent Neural Network (ERNN) model, etc.) for weather prediction. The considered ensemble neural network's weights are calculated and assigned such that they are directly proportional to the closeness of the respective outputs with the corresponding target values.

Basic ensemble method (BEM) output is obtained by using the appended formula below:

$$f_{BEM} = \frac{1}{n} \sum_{i=1}^{n} f_i\,(x) \tag{13}$$

where, $f_i\,(x)$ = the output of the ith network and n = number of networks.

Whereas the generalized ensemble method (GEM) involves the calculation of the weights for each output such that it reduces the Mean Square Error (MSE), and it is given by:

$$f_{GEM} = \sum_{i=1}^{n} \alpha_i f_i\,(x) \tag{14}$$

The error of a network f_i is obtained by finding the difference between f(x) and $f_i\,(x)$, i.e.,

$$\varepsilon_i\,(x) = f(x) - f_i\,(x) \tag{15}$$

The correlation matrix is given by:

$$C_{ij} = E[\varepsilon_i\,(x)\,\varepsilon_j\,(x)] \tag{16}$$

The MSE is obtained using the formula given below:

$$MSE[f_{GEM}] = \sum_{n=1}^{i} \sum_{j=1}^{n} \alpha_i\,\alpha_j\,C_{ij} \tag{17}$$

where, the value of α_i is assumed such that the MSE reduces regarding the target function f, and sum to 1. The optimal value of α_i satisfying the required condition can be given by:

$$\alpha_i = \frac{\sum_{j=1}^{n} C_{ij}^{-1}}{\sum_{k=1}^{n}\sum_{j=1}^{n} C_{kj}^{-1}} \tag{18}$$

Initially, the weights of the backpropagation network are assumed randomly and then trained such that the resulting MSE is minimum. As the initial conditions of individual forecasts in the ensemble forecast are different, they result in different weights leading to different results. So, these individual neural networks are integrated to get more generalized results.

Based on the assumption that the individual neural networks of the corresponding individual forecasts have identical configurations and different initial conditions, the MSE between F(x) and the conditional expectation, E[D|X = x] is considered as shown below, in order to derive the corresponding bias and variance components which are equated as follows:

$$E_D\,[(F(x) - E[D|X = x])^2 = B_D\,(F(x)) + V_D\,(F(x)) \tag{19}$$

where, B_D (F(x)) is the bias squared, given by:

$$B_D\,(F(x)) = (E_D\,[F(x)] - E[D|X = x])^2 \tag{20}$$

and the V_D(F(x)) is the variance, given by:

$$V_D\,(F(x)) = E_D\,[(F(x) - E_D\,[F(x)])^2] \tag{21}$$

At the output of these ensemble neural networks, a simple ensemble average is used as the combiner. From the analysis done, it was found that the bias of the ensemble-averaged function F_i(x), which is based on the multiple classifiers, is exactly the same as F(x), which is based on a single neural network. And the variance of the F_i, i.e., ensemble average function, is less than the F(x).

The weather data collected from the Regina Airport by the Meteorological Department, Environment Canada in 2001, is split into 4 seasons as follows:

Winter (from December to February)

Spring (from March to May)

Summer (from June to August)

Fall (from September to November)

Hourly seasonal data was utilized for training purposes. The training data sets for the Winter, Spring, Summer, and Fall seasons were taken from December 1, 2000 to February 25, 2001; March 1 to May 5, 2001; January 1 to August 6, 2001; and September 1 to November 9, 2001, respectively. The hourly data sets of the typical days: November 10, August 7, May 6, and February 26; were chosen as these days are few of the severe weather days during the seasons, as they show significant variation in their parameters like temperature, wind speed, etc., within a time period of 24 hours. All the considered data sets were converted into values between –1 and 1, as neural networks perform more efficiently when dealing with normalized data.

The architectures of MLP, Hopfield Model (HFM), and ERNN models were decided based on a trial-and-error approach. The training of these networks was done for a constant number of epochs, where the observation of the error gradient was done for these epochs. A one-step secant learning algorithm was considered for the training of MLP and ERNN, where the activation functions for these networks were chosen to be hyperbolic-tangent-sigmoid and log-sigmoid for the hidden layer units and a pure-line for the output layer units. For the Radial Basis Function Network (RBFN) model, the hidden layers were chosen to be 2, and the neurons selected were 180. The Gaussian activation function is considered for this network's hidden layer units and the pure-line function for the output layer units.

Figures 9–11 [8] depict the comparison of the considered neural network-based models, i.e., MLP, ERNN, RBFN model, and HFM with their ensembles, i.e., weighted average (WA) and Winner Take All (WTA) ensemble methods, for forecasting temperature, wind speed, and relative humidity in the summer season:

From the efforts and analysis done, it was found that the time taken by the MLP and ERNN models to be trained was in the range of 5–30 minutes, whereas for the RBFN and HFM models to be trained, it just took a few seconds. From the obtained experimental results, it was observed that the humidity prediction done for the winter and spring seasons had the lowest Mean Absolute Percentage Error (MAPE) values. Whereas, in the case of summer and fall seasons, the prediction done for temperature had the lowest MAPE. In an overall view, humidity prediction results were the least accurate in all the four seasons, and among all the models considered, the RBFN model's forecast was the most accurate.

So, in this way, the neural network-based ensemble models were built and applied for forecasting the weather of southern Saskatchewan on an hourly basis. The results of this article [8] suggest that ensemble neural networks can be utilized for weather forecasting as they can be trained effectively. And it was observed that significant prediction accuracy was observed for most of the considered models, from which we can say that implementation of AI-based techniques for weather forecasting are far better and more efficient than the utilization of the traditional methods.

Similarly, a journal paper [9] which is basically on the annual and non-monsoon rainfall prediction modeling, involves the utilization of MLP and SVR, where this annual and non-monsoon rainfall prediction is done in the state Odisha (India). Regarding the implementation and results, the input data had 4 weather parameters, which were analyzed and verified carefully, after which Min-Max normalization was applied on them to normalize them in the range of 0 to 1.

This normalized data of each of the 8 non-monsoon months was created as 8 different inputs for MLP and SVM, for which separate monthly multiple regression was carried. This normalized data was then utilized in the Weka3.8.3, which is a data mining tool, in such a way that 80% of the normalized data was used for training and the remaining 20% was used for testing. For the SVM model, 'SMOreg' (Sequential Minimal Optimization regressor) option was selected, where 'PUK' (Pearson VII

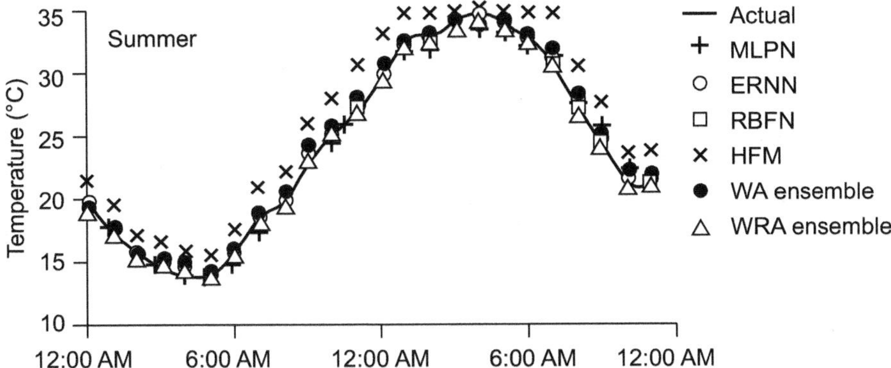

Fig. 9: Comparison of the neural networks with their ensembles for temperature forecasting.

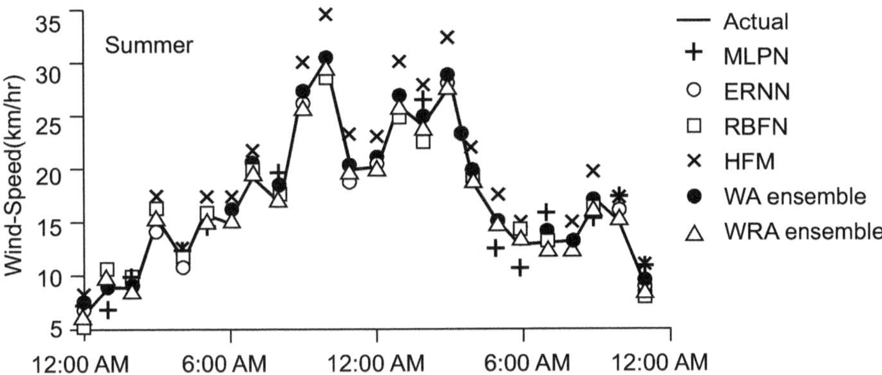

Fig. 10: Comparison of the neural networks with their ensembles for wind speed forecasting.

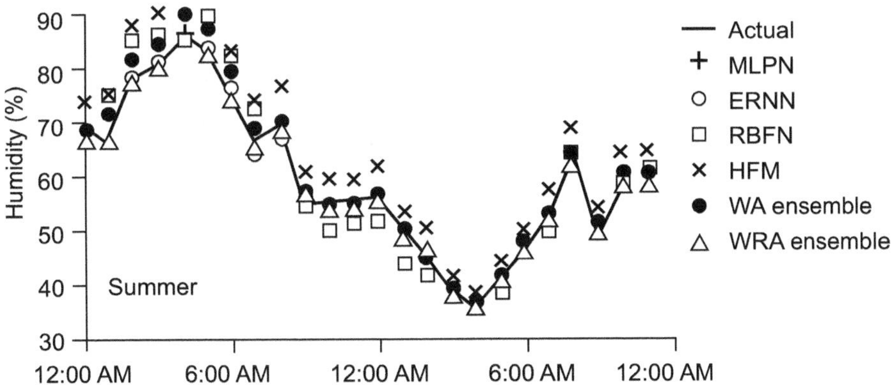

Fig. 11: Comparison of the neural networks with their ensembles for humidity forecasting.

function-based Universal Kernel) was the corresponding kernel that was selected. The corresponding results of the SVM are in the form of the Root Mean Squared Error (RMSE), Correlation Coefficient, and Mean Square Error (MSE), based on which the observation for the SVM model is done. Similarly, for the MLP model, the parameters like RMSE, Correlation Coefficient, and Mean Absolute Error (MAE) was observed and analyzed. Tables 2–5 [9] appended below depict the results obtained from the Weka data mining tool:

In this way, the prediction of rainfall in Odisha (India) was predicted using the AI-based techniques, i.e., MLP and SVM, where there are estimations and expectations on the further scope of these techniques in the future, with other advancements and development like using a time series prediction can be utilised for avoiding ambiguity or considering more parameters based on the availability and the possibility of collecting them.

Table 2: Annual Comparative study of Regression result using MLP and SVR.

Model	Correlation Coefficient	MAE	RMSE
SVM	0.9989	0.085	0.1024
MLP	0.9913	0.169	0.1874

Table 3: Non-monsoon rainfall result for different SVM kernels.

S. No.	Non-Monsoon Months	SVM	R	MAE	RMSE
1	JANUARY	RBF Kernel	0.69	0.05	0.09
2	FEBRUARY	PUK Kernel	0.68	0.09	0.13
3	MARCH	POLY Kernel	0.62	0.12	0.15
4	APRIL	POLY Kernel	0.59	0.20	0.22
5	MAY	RBF Kernel	0.76	0.10	0.13
6	OCTOBER	POLY Kernel	0.80	0.13	0.17
7	NOVEMBER	POLY Kernel	0.79	0.07	0.11
8	DECEMBER	RBF Kernel	0.71	0.07	0.13

Table 4: Non-monsoon rainfall regression result for different MLP models.

S. No.	Non-Monsoon Months	R	MAE	RMSE
1	JANUARY	0.78	0.0971	0.1355
2	FEBRUARY	0.79	0.0981	0.1368
3	MARCH	0.82	0.1077	0.1377
4	APRIL	0.40	0.2047	0.2550
5	MAY	0.75	0.0882	0.1061
6	OCTOBER	0.83	0.1811	0.2084
7	NOVEMBER	0.70	0.1213	0.1454
8	DECEMBER	0.83	0.0716	0.0909

Table 5: Non-monsoon comparative regression study between SVM and MLP.

S. No.	Non-Monsoon Months	Accurate Model for Rain Fall Prediction
1	JANUARY	SVM RBF Kernel
2	FEBRUARY	SVM PUK Kernel
3	MARCH	MLP (2,2)
4	APRIL	SVM POLY Kernel
5	MAY	MLP (2,2)
6	OCTOBER	SVM POLY Kernel
7	NOVEMBER	SVM POLY Kernel
8	DECEMBER	MLP (3,3)

8. Conclusion

The notion of forecasting looks statistical because it underlies some questions which are based on the probability of stability over prediction besides the ones with direct calculation and reasoning. The proviso for forecasting in different fields is increasing with time, as it helps in general tasks like planning our work in day-to-day works (i.e., scientific or non-scientific) and more complicated tasks like formulating the executive plans for organizations, as the anticipation of production level, material requirement, workplace, investors and many other factors of the organization is done based on the corresponding forecasts done. The most vital part of forecasting in organizations and management rests on the manufacture and demand policies, where the corresponding forecasting is expected to be close to or the same as the actual value.

As accuracy and perfectness of forecasting are grounded on probabilities and as its important aspects include observation, analysis, and calculations based on the corresponding vast amount of data, AI-based techniques can be utilized to promote more accuracy and efficiency along with the consumption of less time and human labor. As AI-based forecasting has twofold dimensions based on the methodology, where one is qualitative and the other one is a quantitative method, it helps us to make use of quantity and quality methods together by forecasting over different time range horizons like the short-range, medium-range, and long-range. Applying AI-based techniques gives us the flexibility to deal with a vast amount of data with more accuracy and efficiency, with the guarantee of not being manipulated by humans, as there are high chances of human manipulations in traditional techniques due to the significant involvement of humans.

As picked up in this chapter, many works suggest that AI-based techniques like Multi-Layer-Perceptron, Support Vector Machine, Support Vector Regression, Ensemble Neural Networks, etc., provide results that are way faster, unbiased, accurate, and efficient when compared to the results of traditional techniques. Hence, it could be understood that AI-based techniques deploy a significant and advantageous role in the field of forecasting owing to the flexibility and facilities they provide.

Acknowledgments

My thanks go to the editor of the volume, Professor Sachi Nandan Mohanty, for giving me the chance to write a chapter on this area of AI forecasting and making helpful suggestions, and to Professor Sanjit Chakraborty for guiding me on the right path.

References

[1] Byrne, Cristina L., Amanda S. Shipman, and Michael D. Mumford. The effects of forecasting on creative problem-solving: an experimental study. *Creativity Research Journal,* 22(2): 119–138, 2010.

[2] Scott Armstrong, J. Research needs in forecasting. *International Journal of Forecasting,* 4(3): 449–465, 2018.

[3] Gadgil Ambarish Shashank, Harsh Shailesh Dange, Aditya Fakirmohan Desity, Spurti Shinde, and Prasanna Hemant Asole. Stock market prediction through artificial intelligence, machine learning and neural networks. *International Conference on Artificial Intelligence and Machine Vision (AIMV), IEEE,* 2021.

[4] Kordois John, Symeon Symeonidis, and Avi Arampatzis. Stock price forecasting via sentiment analysis on twitter. *Proceedings of the 20th Pan-Hellenic Conference on Informatics,* Article No. 36: 1–6, 2016.

[5] Liu Bing. *Sentiment Analysis and Opinion Mining.* Morgan & Claypool Publishers, 2012.

[6] Kardile Rucha, Trishna Ugale, and Sachi Nandan Mohanty. Stock price predictions using crossover SMA. *9th International Conference on Reliability, Infocom and Optimization (Trends and Future Directions)*, 2021.

[7] Scher Sebastian. *Artificial Intelligence in Weather and Climate Prediction.* Doctoral Thesis in Atmospheric Science and Oceanography at Stockholm University, Sweden, 2020.

[8] Imran Maqsood, Muhammad Riaz Khan, and Ajith Abraham. An ensemble of neural networks for weather forecasting. *Neural Comput. & Applic*, 13: 112–122, 2004.

[9] Zhang Xiaobo, Sachi Nandan Mohanty, Ajaya Kumar Parida, Subhendu Kumar Pani, Bin Ding, and Xiaochun Cheng. *Annual and Non-Monsoon Rainfall Prediction Modelling Using SVR-MLP: An Empirical Study from Odisha.* Vol. 8, IEEE, 2020.

[10] Simon Haykin. *Neural Networks: A Comprehensive Foundation,* (2nd Edn.), Prentice Hall International, Inc., New Jersey, 1999.

[11] Chakraborty Sanjit. Can humanoid robots be moral? *Ethics in Science, Environment and Politics,* 18: 49–60, 2018.

[12] McCarthy, J., Minsky, M.L., Rochester, N. and Shannon, C.E. A proposal for the Dartmouth summer research project on Artificial Intelligence, August 31, 1955. *AIMag*, 27(4): 12, 2006.

Chapter 2

A Multilayered Feed-Forward Neural Network Architecture for Rainfall Forecasting

Manikandan Rajagopal[1,*] and *Ramkumar S*[2]

1. Introduction

The amount of rain received in a particular demographic region in a given time interval is called the rainfall. Rainfall is a natural and complex process and has significance in different domains including agriculture, transport, disaster management, and natural calamities resilience [1]. Abnormal rainfall affects every facet of humans and all other living beings of the world and also has a great impact in wellbeing and financial disruptions of a country. Accurate rainfall predictions at regular time intervals are always important to issue warnings about likelihood of any disaster about to happen. This also provides people a time for strategic planning in their work and precautions at time of adversity [2]. It is worth noting that rainfall forecasting does not only have an impact in day-to-day life, but more importantly for tropical countries like India where the chief occupation being agriculture and also for various other industries. It largely helps in disaster management and recovery process as well. The rainfall being a variable over time, geography and atmospheric conditions makes the forecasting considerably difficult [3]. Rainfall forecasting keeps a person informed about the likelihood of rainfall the forthcoming day, week, or month which enable long-time planning and on the other way; hourly prediction helps for short-

[1] Associate Professor, Lean Operations and Systems, School of Business and Management, CHRIST (Deemed to be University), Bangalore, ORCID: 0000-0001-7915-0180

[2] Associate Professor, Department of Computer Science, School of Sciences, CHRIST (Deemed to be University), Bangalore. Email: ramkumar.s@christuniversity.in, ORCID: 0000-0002-6224-6167

* Corresponding author: manikandan.rajagopal@christuniversity.in

term planning such as enforcing traffic measures. Literature has seen various studies in this domain using predictive machine learning (ML) algorithms such as neural networks (NNs), Genetic algorithms, and Fuzzy-based systems [4].

The spatial and temporal distribution of rainfall depends on many sets of variables such as atmospheric conditions, wind parameters, climatic conditions, etc. Also, the local level features such as intense rain in-between two intervals of less intense rain significantly affect the time series. These local level features can never be fixed in a given slot that makes the prediction still more complex [5]. Over time, forecasting on atmosphere and related entities have been done through physical methods where the current state of related parameters are collected and future values are calculated numerically using derivatives of liquid dynamics. However, rainfall is characterized by hydrological parameters such as wetlands, surface level water, soil level, etc. There are three major classifications in hydrological models, namely, the physical, metric-driven, and conceptual models. Physical models are more based on science theory and conceptual model helps to visualize the real-time scenario based on hydrological parameters. Although these methods are good in performance, these could not be effective when used at different time intervals with different climatic conditions that varies during modeling [6]. Hence, the best possible solution is to choose a model that is specific to a region under study which has no hydrostatic parameters. It has been seen that the closed models are best fit for complex and nonlinear modeling of a hydrological problem. Hence, metric-driven or the computational models are considered best suited for predicting atmospheric entities in time series bound events.

NNs are structures used in computational theory. These work like the human brain, i.e., like the neural processing in human brain. It interrupts the given input and generates output once it interprets the data [7]. A typical NN is depicted in Fig. 1. Adaption to an environment is achieved by adjusting the internal connections of neurons through algorithms. NN models are to be trained with possible inputs

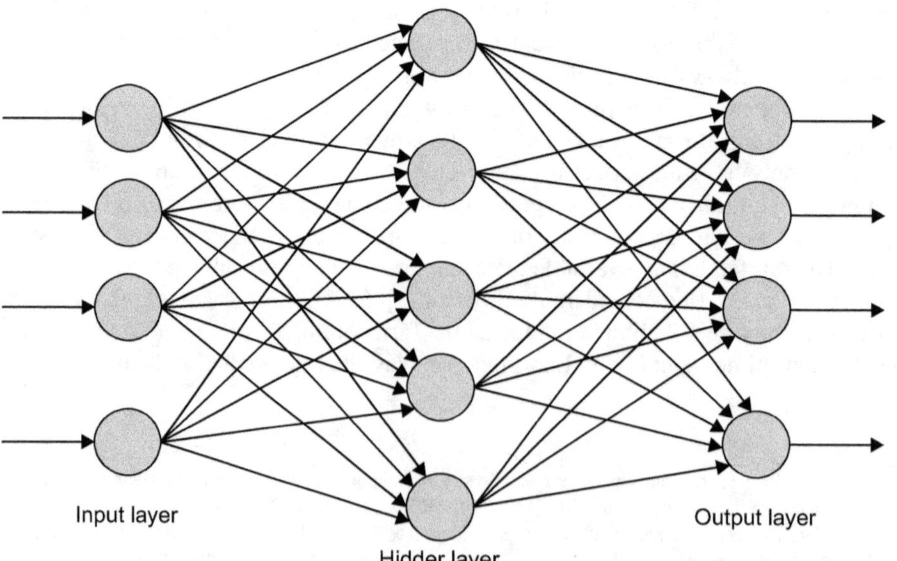

Fig. 1: Simple Neural Network.

to autonomously behave based on environmental changes. The very fact that these changes its behavior based on environment makes it the best fit developing model for prediction where large time-series data is present.

In this chapter, an NN-based predictive model is proposed for rainfall forecasting. The prediction process involves data preparation, developing, implementation, and evaluation of a Multilayered feed forward neural network (FNN) for rainfall prediction. The chapter is aimed to provide a decision-support system to the meteorological departments for prediction of rainfall and recommend course of action thereof. The research hopes to technically contribute to the research domain of artificial intelligence (AI) for forecasting that can be used for providing efficient rainfall forecasting to targeted public.

2. Model Selection

NNs are computational elements containing neurons that operate in the local environment and which communicate with other neurons [8]. These networks receive inputs and process them and thereafter produce output signals with every neuron interconnected with at least one other neuron with a weight coefficient which signifies the importance of a connection [9]. With FNNs, the connections among the units do not tend to be circular while in others, it forms a direct cycle. This advantage of FNN makes it more suitable for developing predictive models [10]. Other models like the multilayered perceptron (MLP) or the SLP are considered as siblings of FNN where in each layer has more than one neuron connected in parallel. SLF consists of one layer which can receive input and produce output while MLP contains more layers, namely, the input, output, and in addition the hidden layers as well. Each of the neurons in the said layers gives at least a single output that can connect with all of the inputs [11]. The proposed network, thus, can give accurate correlation amidst linear parameters and nonlinear parameters which makes it more suitable for forecasting. With different models, after being studied under different techniques, it is found that the MLP techniques are more promising for rainfall prediction as its behavior is more like a human in addressing the linear problems, a typical character that rainfall prediction exhibits. It is observed that MLPs are more suited for problems associated with prediction. The advantage of having the capability to do regression when prediction in real time data are supplied as input and the flexibility of such techniques to learn mapping or correlation between input and output are notable. Taking the advantages of ability of handle non effectiveness in prediction, this research selects the deep neural network in the shape of MLP for predicting rainfall.

3. Research Backround

Prediction of future trends and value from a recorded time series data plays a vital role in improving the service quality. For example, precise predictions shall greatly impact the quality in medicine, engineering, business analytics, meteorology and more important, the financial positions [12]. The prediction process should involve suitable statistical methods and computational models for identifying the corelation among the dependencies. There should be a consistent thread between the past and the future for the model to scale up to longer value [13]. There are large number of researches conducted in time series data predictive modeling and forecasting. There still exists a notable gap in research and challenges due to the complexity of the problem. In rainfall forecasting using physical models, complexity grows due to atmospheric parameters on which rainfall depend. Data availability in the required format and scales also affect the process of prediction [14]. Hence, advanced and efficiency model are in demand to meet the challenges of complexity for capturing the high end nonlinear process in climate.

The area or site to be forecasted is also significant and to be considered as a key feature for prediction. Single-step forecasting for a time series data are challenging while using multi-step forecasting has implementation overheads due to errors which get accumulated, and uncertainty [15]. In general, the multistep process is implemented through recursive and direct approaches. In case of the recursive model, a single model for the time-series data are developed and consecutive prediction happens using the previous predictions. In case of direct methods, a distinct model is established for time series data for every site, and prediction happens through tailor-made models. The choice of method involves a deliberation in biases and variances between the two [16]. Traditionally, multistep forecasting is achieved through recursive models. Fully observed models were initially used for rainfall forecasting which depend on two kinds of variables. Although they are simple, the memory shortages happen to be a great constraint [17].

In the beginning, time series predictive modeling was done through ARIMA (autoregressive moving average) and SARIMA (seasonal autoregressive moving average) [18]. However, these consider the problem as linear and have minimal capacity to handle nonlinear parameters.

Recent advancement in computational intelligence has paved the way to use AI techniques, particularly pattern matching, for rainfall forecasting [19]. NNs perform nonlinear correlation between the input and output. In [20], the researchers have predicted the local regions to receive rainfall using MLP using a hybrid back propagation (BP) and PASO (personal assistant search optimization). It has been seen that ANN has superior performance when compared to that of conventional modeling. Multi Layered FNNs have been recently studied for developing accurate predictive models. In [21], a first ever application was developed using the concept for long-term prediction problems with the help of hybrid variant of data mining.

A multilayered ANN normally has an input, output, and hidden layers. The usefulness of hidden layers lies in performing intermediate calculations prior mapping the input and the corresponding output [22]. Before the advent of deep learning techniques, problems with more hidden layers were rare. The additional hidden layers serve the purpose of handling large and complex time series data. These ANNs with more hidden layers form a deep network and comes in the category of deep learning. Deep learning architectures are found to have capabilities to extract high order abstracts from input with the help of multi-processing layers with both linear and nonlinear functions [23]. Having gained an adequate knowledge through the literature, it is found that a number of approaches were used for the forecasting problem. Among these, deep learning methods are found to be more effective and promising in machine learning (ML) algorithms for predictions involving time-series data. Hence, this research focuses on building an MLP – FNN for predicting the rainfall.

4. Literature Review

Accurate models for rainfall forecasting are important in various spheres of life. Conventional statistical methods do not have the ability to do prediction for long-term forecasting. The advent of ML in the forecasting domain has been a hot topic of research in recent times. This section provides the gist of prominent researches that are done in rainfall forecasting using NNs, their methodology, and results.

Zhang et al. [32] proposed the Dynamic and Regional DRCF (Digital Regulation Cooperation Forum) approach combined with MLP network for short-term rainfall prediction [24]. The proposed method used Principal Component Analysis for reducing the dimension of 12 physical parameters that act as an input for the MLP. Next, a Genetic algorithm is proposed for determining the MLP's structure. The nearby sites are taken based on the site to be forecasted. To resolve the issue of clustering caused due to the extended perception's range, the proposed DRCF model is then enhanced with various dynamic approaches. The result shows that the model is better with low root mean squared error (RSME).

Yerin Kim et al. (2022) [13] presented a Genetic adversarial network for rainfall prediction for very short-term forecasting for a range between 15 minutes to 5 hours. The proposed method was trained and then tested using the famous KMA method and the CAPPI (constant altitude plan position indicator) data [25]. The proposed model is tested for its efficiency through metrics such as Deduction probability, False Alarming ratio, and critical success factor, respectively. The method proved efficient for hourly rainfall prediction. The power of proposed methodology was then extended to other forecasting systems as well.

Venkatesh et al. [26] proposed a system for rainfall prediction using GAN (generative adversial network) and LSTM (long short-term memory). The proposed model used a generator and a CNN as a part of discriminator and a network algorithm is used like a generator. It is observed that the proposed technique LSTM was considerably favorable for time series prediction [26]. The experiments showed that the system provided 99% accuracy in predicting rainfall in short range of time.

Khan et al. [28] introduced a scheme for bringing out the effectiveness of the Deep Learning (DL) technique for augmentation in short-range rainfall prediction that exploits the advantage of GCM (general circulation model) for simulating the variables in meteorological avenues [27]. The proposed model is then applied in 15 various locations with various climatic conditions and varying atmospheric parameters. The proposed method is compared with the traditional deep learning techniques such as SVM and SVR and it is found that the proposed method efficiently captures the corelation among the atmospheric variables and this has proven an advantage in bringing out a unified approach for bringing out hidden features.

Chang Hooziang (2022) proposed deep learning technique for improvisation in performance of physical models [28]. Here, the relationship in transformation of the output and the actual data are learnt through a GAN which are used for making corrections in the model. Experiments are conducted on the data collected for about 9 months and from radars for demonstrating the fact that making corrections in forecasting data would significantly improve the performance specifically in case if the rainfall is more. The proposed technique gives a practical way to combine models and data-driven DL methods.

Abebe et al. [29] made an attempt for investigating the use of AI methods for predicting the monthly rainfall. The possible avenues of application of ANNs and ANFIS (adaptive neuro-fuzzy inference system) were studied for a period of 10 years [29]. The experiments show that the ANFIS models are more effective for long-term prediction and ANNs are effective for short-term predictions.

Apart from the significant methods discussed above, there are numerous data driven predictive models used for the prediction of rainfall using nonlinear methods through different optimization methods [30]. Data-driven methods are rather simple to be used and requires less computational time and effort that of GCMs. The models are used for effectively handling the nonlinear complexity. ANN are modern methods for addressing the signals and has been used as a calculation measure for solving problems that concerning hydrological resources. Other models such as fuzzy algorithms and Gas were having logical problems which make it not suitable for long term predictions [31]. The ANFIS techniques are found to have good ability for integrating the fuzzy logics as stated in [32], however, they are not suited for short-term predictions.

It is observed from the literature that considerable research is being carried out in the predictive analysis particularly in climatic and weather forecasting. NNs are found more attractive particularly in solving nonlinear problems which is the exact problem the short-term rainfall prediction faces.

5. Proposed Methodology

5.1 Data Preparation

The performance of any predictive model significantly depends on how well the data is prepared for analysis. Large variations in the results arise out of poor preparation of data. In order to develop a robust predictive model, adequate training, testing and validation have to be performed for the proposed MLP network. Data is first selected and cleaning is done in order to represent the nonlinear variables in a way which the proposed network can process them efficiently. The online open access data repository of Indian Meteorological department data service portal [33] (https://dsp.imdpune.gov.in/) is taken for performing the study. The data has complete record of different climate related variables for 8 major cities in India which includes Bangalore, Delhi, Bombay, Nagpur, Hyderabad, Kanpur, Pune, and Jaipur. Data segregation is done across multiple years further drilled down to monthly and weekly data. The dataset was produced with the help from www.worldweatheronline.com API. The data set contains data for 10 years from 2011 till 2020. This data was found more sufficient for model training and to predict the rainfall for short term like weekly, monthly, and hourly. The screen of the data source is shown in Fig. 2.

The data features are shown in Table 1.

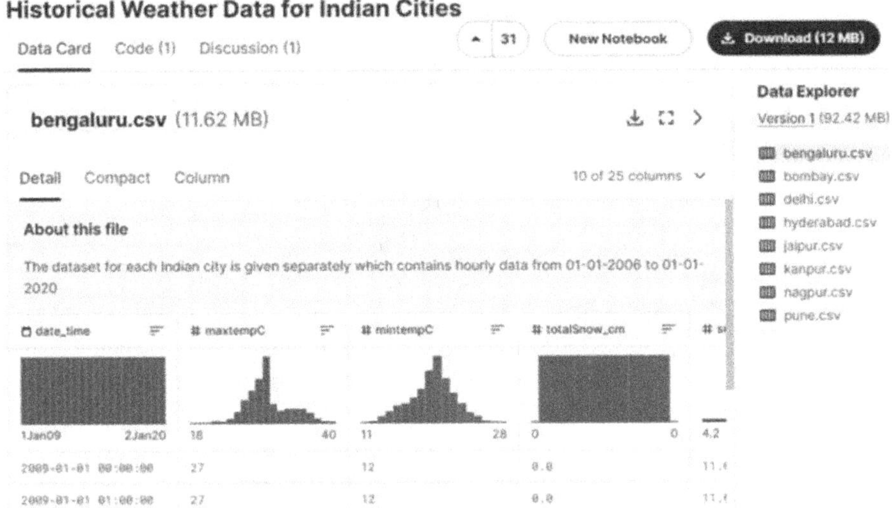

Fig. 2: Snapshot of the data source.

Table 1: Data Features.

Symbol	Feature	Unit of Measure	Symbol	Feature	Unit of Measure
DT	Date and Time	Time stamp	CC	Cloud cover	Numeric
MX_T	Max Temp	^{0}Celcius	H	Humidity	g/kg
MN_T	Min Temp	^{0}Celcius	P	Pressure	g/kg
TS	Total snow	Numeric	T	Temp	^{0}Celcius
UVI	UV Index	Numeric	V	Visibility	Numeric
MR	Moon rise	Time stamp	WD	Wind degree	^{0}Angle
MS	Moon set	Time stamp	WS	Wind speed	KMPH
SR	Sun rise	Time stamp	SH	Sun hour	Time stamp
SS	Sun set	Time stamp	UVI	UV index	Numeric
DP	Dew point	Numeric	IL	Illumination	^{0}Celcius
WC	Wind chill	Numeric	HI	Heat index	^{0}Celcius
WD	Wind dust	Numeric	T	Feels like Temp	^{0}Celcius

The variables present in the data set directly correspond to the features of the climate. It was found that there are variables which represent the mere indication of the occurrence of a given weather phenomenon. The data was converted into a table representation for making it convenient to be used with other tools. Data cleaning was done to deduct noise, and missing value imputation through Random Forest is done for filling out the missing data. Before proceeding with the cleaning process, an inspection was carried out to find out how better the data can be prepared for best use by the proposed network. The excel sheet was prepared and each variable and its data are manually verified. Variables such as MR and MS were removed from the dataset as it does not contribute to the prediction of rainfall. Of the 8 cities, Bangalore is taken as the site forecasting the rainfall. The choice of city was made on the basis of selecting the least missing value of 0.21% (Bangalore). The transformation was done using Min_Max technique as depicted in Equation (1) as this type of transformation are largely used in meteorology when compared to Z-rank transformation for having the advantage of high accuracy.

$$d_{norm} = \frac{d_i - d_{min}}{d_{max} - d_{min}} \tag{1}$$

The minimum maximum transformation of a given value is termed as T_{norm} provides the value that needs to be transformed T_i, and the maximum value T_{max} and the minimal value T_{min} for the value ranging to T_i. In this study, T_i correspond to the value of data belong to one among the selected variable, and T_{min} and T_{max} represent the minimum and the maximum value of the selected variables. Normalization is then done on the data and the complete data are split into various sets. These are the training and testing data. The third set called the validation was also prepared to help validating the performance of the proposed model to the daily effects to be validated.

In general, much importance is given to training set with more data for avoiding the overhead of over fitting in any predictive model [34]. The ratio of training and testing is kept in accordance with [35]. This also depends on the nature of the problem. From the literature, it is understood from various meteorological forecast predictions in tropical countries have the percentage concept of allotting training and testing data like 70–30 ratio [36].

Climatic prediction studies which have used an ANN also adopt a scenario where the coverage of data over time is used for assigning training and testing data [37]. In [38], the authors have used 158 years of climatic data for analysis where 99 years of data were taken for training and remaining for testing purpose. In [39], the 14 years data is segregated into twofold 2 years data as training and rest for testing resulting in a 80:20 ratio. Further, in [40], the authors have partitioned the entire data in accordance to the time coverage as well. In 36 years of data, 11 formed the training set and the remaining as a testing set. The segregation of training and testing data for this research is shown in Section 6.

5.2 MLPNN Implementation

Figures 3 and 4 present the proposed architecture with the number of hidden neurons changed. It shall be seen from the figure that the proposed MLP network has 9 input neurons with variables, Date and Time, Min Temp, UV Index, Humidity, Pressure, Temp, Wind speed, Illumination, and Heat index .

A couple of NNs are implemented where the first network having 40 hidden neurons uses the scaled conjugate gradient (SCG) algorithm with HTA activation [41]. The second network has 60 hidden neurons that uses SCG and the activation function being sigmoid. Selection of activation function was made based on the range it falls. The signals from the proposed network were then multiplied with the weights and are then mapped to 24 outputs indicating the forecasting for

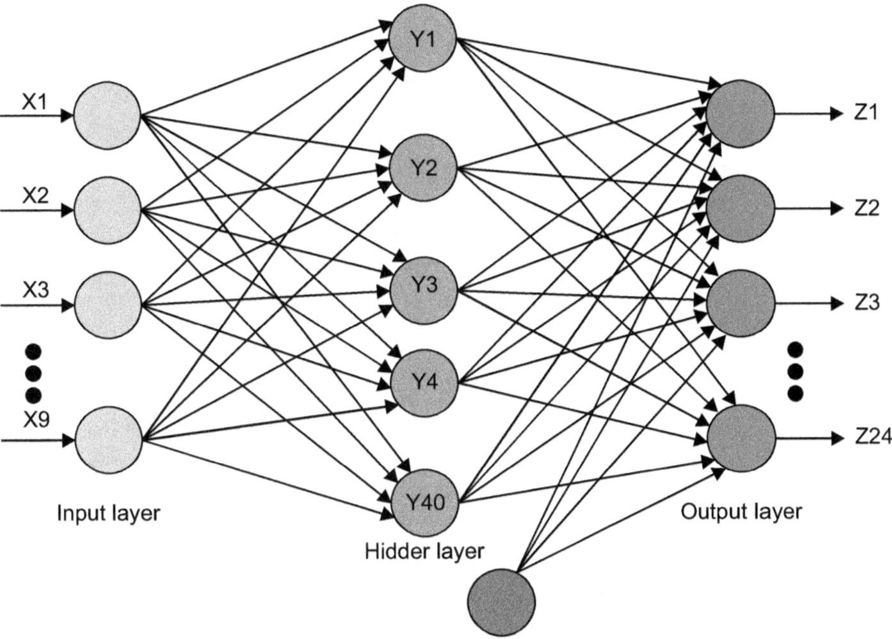

Fig. 3: Proposed architecture with 40 hidden layers.

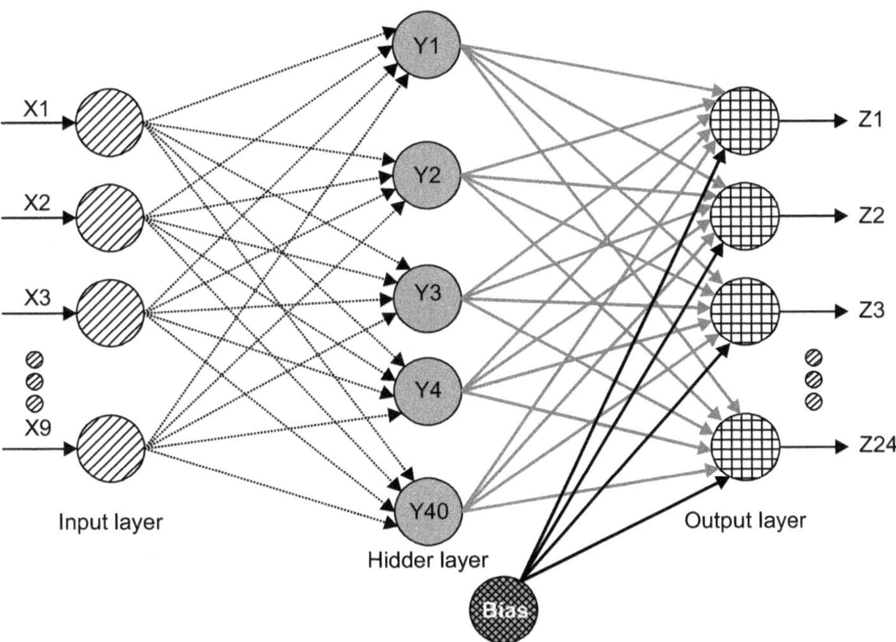

Fig. 4: Proposed architecture with 60 hidden neurons.

24 hours of the day. While the test and train data results are generated, the proposed model is implemented via software with the aid of ENCOG, an ML framework. The framework supports graphics unit to process the time series which is vital in this research as the weather data are large.

Further, this research used ENCOG framework as it has inbuilt functions to converge the data that are out of the rainfall forecasting system into the present system in an efficient way. It helps to generate data from both the training and testing data, compatible to be used with other tools. The model is validated using the data for the year 2019 and for the month of June. The input was fed into the NN for producing a day's rainfall forecast. The data of June 1 to 10 is provided as input for the prediction of rainfall for June 11. The results are then compared to the original values for rainfall. The accuracy was verified by giving the same input to forecast the rainfall on June 20. The process was then iterated for prediction of rainfall from June 11th till June 31st. A measure of error is vital while testing a predictive model. The value of error on each day is segregated and also measured for it to be compared with the actual results. This actually means that the June 11th forecast error is averaged, and for day 2 and so on till the time interval of 20 days is reached. The most significant criteria for evaluating the accuracy of a predictive model are the error measurement. MAE, RSME, and MAPE (mean absolute percentage error) are the general metrics used to measure the accuracy of time series predictive analysis [42]. As the data contains many 0 values, the MAPE was not found suitable for measuring accuracy.

Both RMSE and MAE are used in the validation as it can be handled together for understanding the error in forecast results. Equation (2) shows the way the larger deviations are projected from prediction to actual making the distribution of errors more visible. .

$$MSE = \sqrt{\frac{1}{n}\sum_{i=1}^{n}(d_i - p_i)^2} \tag{2}$$

where, *f* stands for the value which is forecasted and stands for the actual result. RMSE calculates the error with respect to the mean of squares pertaining to the deviation and here, the original rainfall and the predicted value. Minimal values of error indicate that the model is consistent. The magnitude of error is given as MAE and calculated using the Equation (3)

$$MAE = \frac{1}{n}\sum_{i=1}^{n}|p_i - d_i| \tag{3}$$

where, *a* is the actual value and *P* is the predicted rainfall.

The output obtained from each of the network model is then subjected to denormalization and are compared with the original values of rainfall through a linear graph. For every model, the values that are denormalized in each day's prediction are plotted against their corresponding original rainfall on the same day. These graphs are then visually inspected. The plots were made taking into the actual and observed values on a day basis for the entire duration of 21 days, i.e., from June 11 till June 30 and are shown in Section 6.

6. Results

6.1 Preparation of Data

The time period considered for data preparation is 10 years, i.e., from 2010 till 2020. The site chosen is Bangalore. There were a total of 24 variables as presented in Table 2 on which variable such as snow, dew which reported 0 values were removed. The non-contributing variables such as moon rise and set were also removed. The derived variables such as sun hours, moon hours were also not considered for the experiment. Table 2 shows the data without imputation and Table 3 shows the data after imputation, respectively, for the first five days of the month of June 2019, respectively.

Table 2: Before Imputation.

T	TS	UV	WC	WD	H	P	HI	DP
24.2	31.2	11	6	6.45	70	1000.14	12.6	987.5
23.1	32.1	14	5.6		81	1012.23	13	976.2
26.3			4.8		69	1087.21	14.2	923.5
					72			956.1

Table 3: After Imputation.

T	TS	UV	WC	WD	H	P	HI	DP
24.2	31.2	11	6	6.45	70	1000.14	12.6	987.5
23.1	32.1	14	5.6	7	81	1012.23	13	976.2
26.3	32	12.5	4.8	7	69	1087.21	14.2	923.5
25	32	12.5	5	7	72	1100	14	956.1
25	32	12.5	5	7	81	1100	14	955
25	32	12.5	5	7	81	1100	14	955

6.2 Data Normalization Results

As discussed in Section 3, data are normalized using min–max normalization. Tables 4 and 5 depict the data before and after the normalization process. The data after normalization are then fed into the proposed network.

6.3 Data Partition Results

Partitioning of training and testing data are done on the basis of the years rather than percentages. Years ranging from 2010 to 2019 were taken for training data and the year 2020 was taken as the test data. Validation data was taken for the month of June 2020. In order to maintain consistency with the data, each of the partition was put in a separate container for easier prediction and representation. Data slicing on year and month wise was done manually using the excel sheet. The data taken for experiment is segregated as training and testing data. The training data started from January 1 of 2010 till December 31, of 2019 and the testing data was taken from January 1,2020 to December 31, 2020 as shown in Table 6. The validation data was taken from the authentic source of Indian Meteorology department of June 2020. Data taken for validation also went through the cleaning process.

Table 4: Non-normalized data.

T	TS	UV	WC	WD	H	P	HI	DP
24.2	31.2	11	6	6.45	70	1000.14	12.6	987.5
23.1	32.1	14	5.6	7.2	81	1012.23	13	976.2
26.3	24.1	18	4.8	4.12	69	1087.21	14.2	923.5
27.1	28.2	23	7.1	5.21	72	1084.37	12.7	956.1
22.7	29.1	29	6.9	6.1	78	1046.3	13.3	952.1
21.8	26.3	13	7.4	7.85	80	1071.2	11.1	945.3

Table 5: Normalized data.

T	TS	UV	WC	WD	H	P	HI	DP
0.561	0.531	0.7814	0.6	0.018	0.9431	0.4345	0.1400	0.0121
0.432	0.310	0.7654	0.6784	0.012	0.7874	0.2765	0.1789	0.0131
0.423	0.443	0.6790	0.6654	0.011	0.9467	0.5465	0.2065	0.0129
0.345	0.312	0.7765	0.7102	0.016	0.6781	0.2341	0.1075	0.0185
0.208	0.241	0.7210	0.8689	0.017	0.6583	0.3461	0.1621	0.0124

Table 6: Data Partitioning.

Data Set	Period	Years	Total Records
Training	1.1.2010 to 31.12.2019	9	4567
Testing	1.1.2020 to 31.10.2020	1	366
Validation	1.6.2020 to 30.6.2020	0.1	30

6.4 MLPNN Implementation Results

As discussed earlier, the validation data for the model are taken from June 1, 2019 till June 30, 2019 and are provided as input to the appropriate MLP model for obtaining the daily forecast. The measure of error is computed using MAE and RMSE. Figure 5 demonstrates the error values of the validation set of the proposed model for the SCG activation function. It is seen that the lowest value of error is achieved on June 4th prediction with MAE as 0.00723 and RMSE as 0.01054. On the other hand, the largest value of MAE was observed for June 14th with 0.01642 and of RSME, the largest error value was reported for June 13th with 0.02321. It is observed that the average MAE and RSME for the proposed model was 0.01276 and 0.01562, respectively. It is also observed that more than 60% of the error measurements were less than the average for both the MAE and RSME which proves the consistency.

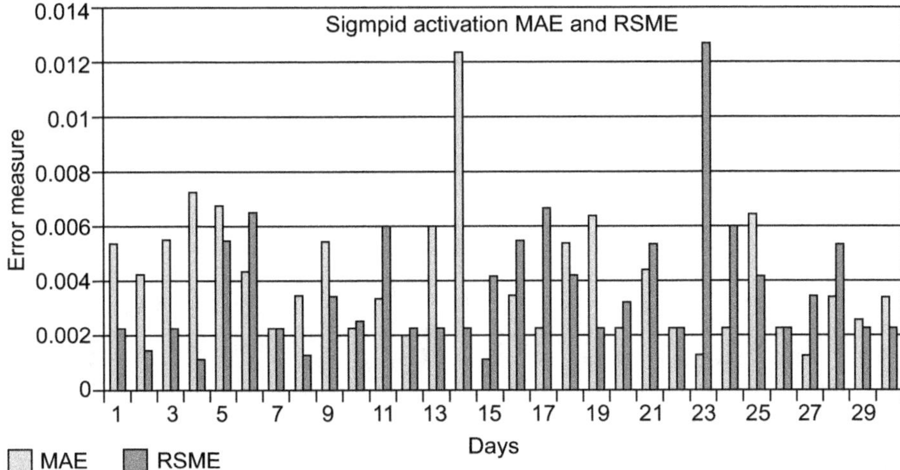

Fig. 5: Error Measurement—SCG Activation.

It is seen from Fig. 6 that for the activation function Sigmoid, the minimum error measurement for MAE is obtained for June 14 with value 0.01237, while June 23has the minimum RMSE of 0.0128. The error measurements were found to be greatest for the 1st June for both MAE and RMSE with values measuring 0.01438 and 0.03165, respectively. The average MAE and RMSE was observed to be 0.01324 and 0.01345, respectively. It is observed that more than half of the results are below the average error value similar to SCG activation function. Hence, the count of total correct prediction is more than that those of the incorrect ones.

It is observed from the average error measurements that SCG activation performs better results for forecasting than the sigmoid. From Fig. 7 which shows the values for all 24 hours for a particular day June 13, 2019, SCG was found to be more consistent with the range of error measurements falling between 0.01 – 0.03 whereas sigmoid ranges between 0.01 – 0.40 were wider.

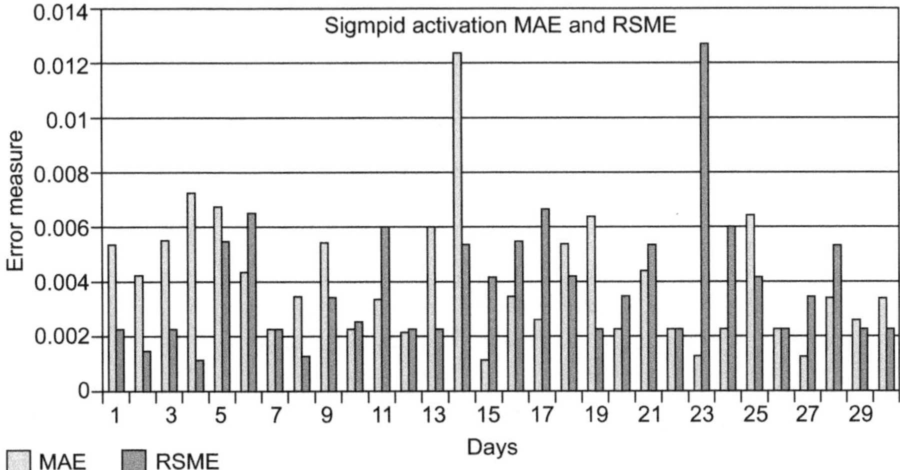

Fig. 6: Error Measurement—Sigmoid activation.

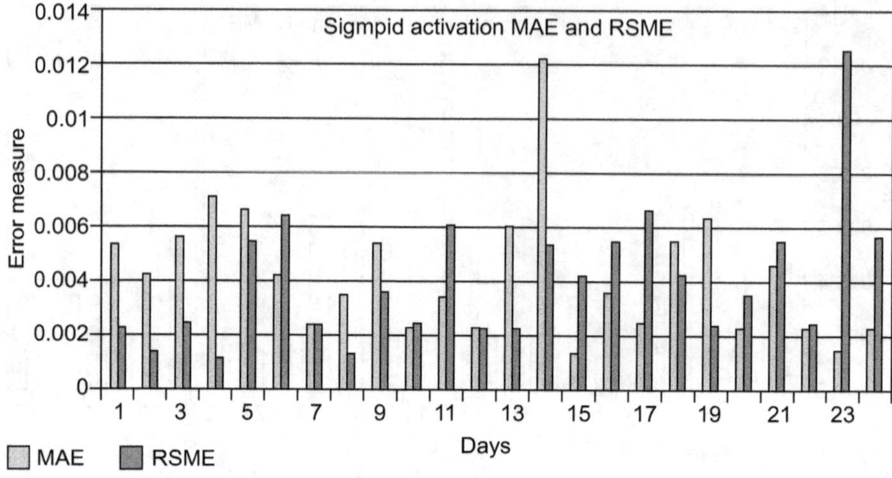

Fig. 7: Comparison of average MAE and RSME of activation function.

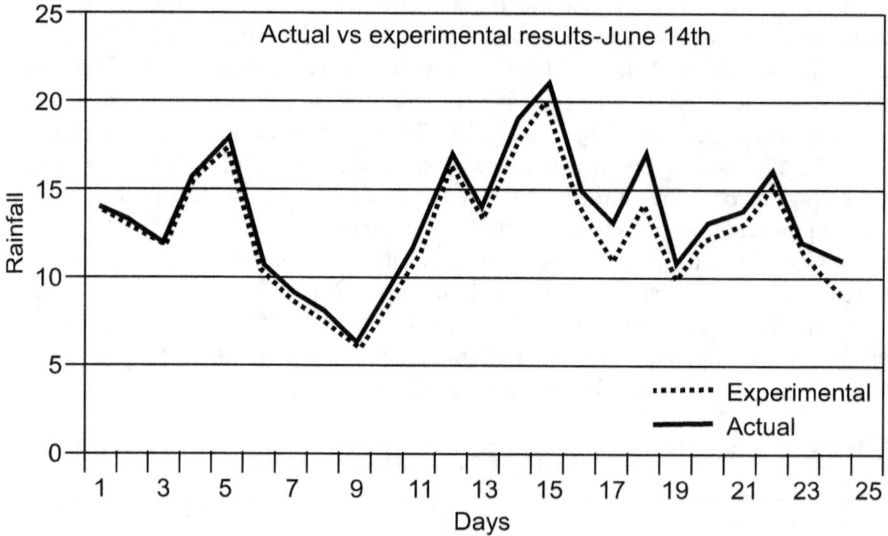

Fig. 8: Actual and Experimental result – June 14, 2019.

Figures 8 and 9 provide the accuracy plot for June 14 and June 30, respectively, in terms of the SCG activation function. The results predictions are discussed for SCG only as it has higher accuracy. The results from Sigmoid function are ignored. It is seen that the plot for June 14 followed a regular pattern as that of the predicted values and June 30 has a greater deviation from the actual results. This indicates that performance of the proposed model has a definite pattern with that of the actual rainfall received.

It is also essential to see the prediction results for the validation set on day 1 of the month of June 2019. This helps in providing the pattern of outputs across the different hours of the day. The skewness of the output shall also be observed for

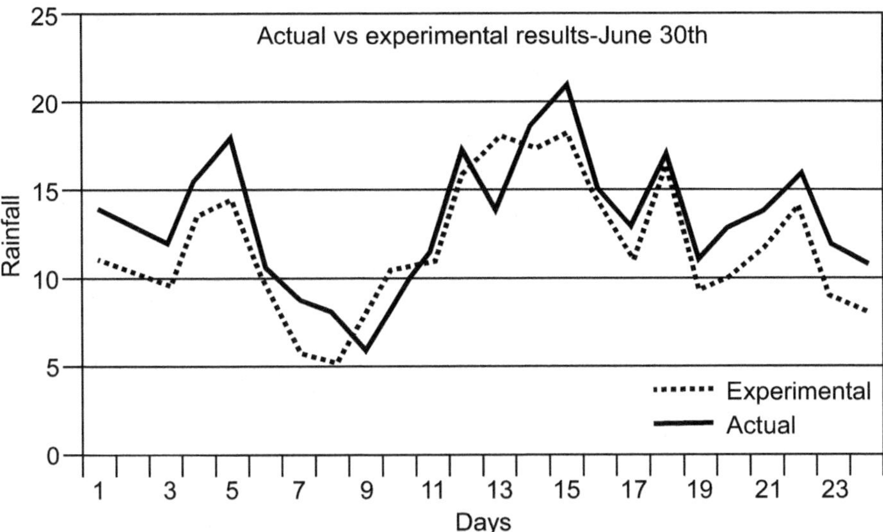

Fig. 9: Actual and Experimental Result – June 30.

indicating as which model to be re-trained. It was observed that for SCG activation, the 11th, 16th, and 18th of June had less accuracy and hence required retraining. It was seen that the model's performance was also good in terms of visual prediction.

7. Limitations

The study was conducted for forecasting the level of water with SCG activation and it is found that the results were not accurate as the validation set was extended. The accuracy kept decreasing to the end of the validation range when compared to the actual results. Interestingly, the error measurements were holding good for the training set. It was concluded that since the error measurements were good in training, the same need not be the case with that of validation set. The plots may have variable results as that of actual measurements. Further, based on the given problem and the scope, the model may have to be retrained for making the model more accurate. The other limitation was that although the parameters are kept constant, the day beyond the validation range was also considered, but this time , the results were not found to follow the pattern of actual measurements of rainfall received.

8. Conclusion and Future Directions

The research was carried out with intention of using the power of AI in forecasting for a time series data. The climate and related factors play a major role in the daily activities of humans. Forecasting rainfall on a given day plays a significant role in short-term planning and management. A MLP FNN model is proposed in this chapter for the prediction of a day's rainfall. The data considered are 10-year-old day-wise rainfall forecasting for the city of Bangalore, India. ENCOG was used for technical implementation of the model. The proposed model had 40 and 60 hidden layers for SCG and Sigmoid activation function. The average MAE and RMSE for SCG activation was found to be 0.01276 and 0.01562, respectively, which is comparatively better than the results from Sigmoid function which yielded a mean MAE of 0.01438 and 0.03165, respectively. But, both the activation functions performed their predictions to an acceptable level. Overall, the findings from the research shows that the proposed model has a significant potential for rainfall forecasting for a given day provided appropriate measures for preparing the data, choice of network architecture, and activation functions and validations are performed.

The future scope of research is to consider a much shorter range of data for improving the accuracy of the ANN models. The need of exploring the various methods for the choice of parameters needs to be performed as this will help to find the reliable neural network for forecasting. A universal model for selecting the methodology for data segregation among testing and training shall be arrived.

Acknowledgments

My thanks go to the editor of the volume, Professor Sachi Nandan Mohanty, for giving me the chance to write a chapter on this area of AI forecasting and making helpful suggestions, and to Professor Sanjit Chakraborty for guiding me on the right path.

References

[1] Thorndahl, S., Einfalt, T., Willems, P., Nielsen, J.E., Veldhuis, M.C.T., Nielsen, K.A., Rasmussen, M.R., and Molnar, P. Weather radar rainfall data in urban hydrology. *Hydrol. Earth Syst. Sci.*, 21(3): 1359–1380, 2017.

[2] Seliga, T.A., Aydin, K., and Direskeneli, H. Disdrometer measurements during an intense rainfall event in central Illinois: Implications for differential reflectivity radar observations. *J. Climate Appl. Meteorol.*, 25(6): 835–846, June, 1986.

[3] Kusiak, X., Wei, A., Verma, P., and Roz, E. Modeling and Prediction of Rainfall Using Radar Reflectivity Data: A Data-Mining Approach. *IEEE Transactions on Geoscience and Remote Sensing*, 51(4): 2337–2342, April 2013. doi: 10.1109/TGRS.2012.2210429.

[4] Barnes, P., Kjeldsen, T.R., and McCullen, N. Video-based convolutional neural networks forecasting for rainfall forecasting. *IEEE Geoscience and Remote Sensing Letters*, 19: 1–5, 2022, Art no. 1504605. doi: 10.1109/LGRS.2022.3167456.

[5] Nair, A., Singh, G., and Mohanty, U.C. Prediction of monthly summer monsoon rainfall using global climate models through artificial neural network technique. *Pure Appl. Geophys.*, 175: 403–419. 2018. https://doi.org/10.1007/s00024-017-1652-5.

[6] Singh, P., and Borah, B. Indian summer monsoon rainfall prediction using artificial neural network. *Stoch. Environ. Res. Risk Assess*, 27: 1585–1599, 2013. https://doi.org/10.1007/s00477-013-0695-0.

[7] Chandrasekaran, S.S, Elayaraja, S., and Renugadevi, S. Damages to transport facilities by rainfall induced landslides during November 2009 in Nilgiris, India. *In: Proceedings of Landslides Science and Practice,* Vol. 6: *Risk Assessment, Management and Mitigation, Springer,* Nilgiris, India, pp. 171–176, 2013.

[8] Hou, S.S., Li, A., Han, B., and Zou, P.G. An early warning system for regional rain-induced landslide hazard. *International Journal of Geological Sciences*, 4(3): 584–587, 2013.

[9] Dunne, S., and Ghosh, B. Weather adaptive traffic prediction using neurowavelet models. *IEEE Transactions on Intelligent Transportation Systems*, 14(1): 370–379, March 2013. doi: 10.1109/TITS.2012.2225049.

[10] Ahuna, M.N., Afullo, T.J., and Alonge, A.A. Rain attenuation prediction using artificial neural network for dynamic rain fade mitigation. *SAIEE Africa Research Journal*, 110(1): 11–18, March 2019. doi: 10.23919/SAIEE.2019.864314.

[11] Kang, You, M., Lee, D., and Uyeda, H. Rainfall estimation by S-band polarimetric radar in Korea. Part I: Preprocessing and preliminary results. *Meteorol. Appl.*, 21(4): 975–983, Oct. 2014.

[12] Ortiz-Garcia, E.G., Salcedo-Sanz, S., and Casanova-Mateo, C. Accurate precipitation prediction with support vector classifiers: A study including novel predictive variables and observational data. *Atmospheric Research,* 139: 128–136, 2014.

[13] Kim, Y., and Hong, S. Very Short-term rainfall prediction using ground radar observations and conditional generative adversarial networks. *IEEE Transactions on Geoscience and Remote Sensing,* 60: 1–8, 2022, Art no. 4104308. doi: 10.1109/TGRS.2021.3108812.

[14] Devi, S.R., Arulmozhivarman, P., Venkatesh, C., et al. Performance comparison of artificial neural network models for daily rainfall prediction. *Int. J. Autom. Comput.* 13: 417–427, 2016. https://doi.org/10.1007/s11633-016-0986-2.

[15] Khan, M.I., and Maity, R. Hybrid deep learning approach for multi-step-ahead daily rainfall prediction using GCM simulations. *IEEE Access,* 8: 52774–52784, 2020. doi: 10.1109/ACCESS.2020.2980977.

[16] Marzano, F.S., Rivolta, G., Coppola, E., Tomassetti, B., and Verdecchia, M. Rainfall nowcasting from multisatellite passive-sensor images using a recurrent neural network. *IEEE Transactions on Geoscience and Remote Sensing,* 45(11): 3800–3812, Nov. 2007. doi: 10.1109/TGRS.2007.903685.

[17] Sourav Dey Roy, Anindita Mohanta, Dipak Hrishi Das, and Mrinal Kanti Bhowmik. Cloud pattern classification for rainfall prediction using convolutional neural network. *In: Proceedings of the 8th International Conference on Networking, Systems and Security (NSysS '21). Association for Computing Machinery.* New York, NY, USA, pp. 83–90. 2021. https://doi.org/10.1145/3491371.3491379.

[18] Fu JieTey, Tin-Yu Wu, and Jiann-Liang Chen. Machine learning-based short-term rainfall prediction from sky data. *ACM Trans. Knowl. Discov. Data,* 16(6), Article 102, December 2022, 18 pages. https://doi.org/10.1145/3502731.

[19] Ali Haidar, and Brijesh Verma. 2017. A hybrid genetic algorithm for climate input features and neural network parameters selection. *In: Proceedings of the Genetic and Evolutionary Computation Conference Companion (GECCO '17),* Association for Computing Machinery. New York, NY, USA, pp. 281–282. https://doi.org/10.1145/3067695.3076038.

[20] Rivero, R., Pucheta, J.A., S. Baumgartner, J.S., Laboret, S.O., Sauchelli, V.H., and Patiño, H.D. Short-series prediction with BEMA approach: application to short rainfall series. *IEEE Latin America Transactions,* 14(8): 3892–3899, Aug. 2016. doi: 10.1109/TLA.2016.7786377.

[21] Hasu, V., and Koivo, H. automatic rain and wind measurement fault identification in mesoscale weather station networks. *IEEE Instrumentation and Measurement Technology Conference.* Victoria, BC, Canada, 2008, pp. 489–494. doi: 10.1109/IMTC.2008.4547085.

[22] Crown, M.D. Validation of the NOAA space weather prediction center's solar flare forecasting look-up table and forecaster-issued probabilities. *Space Weather,* 10(6): 1–4, 2012.

[23] Moran, K.R., Fairchild, G., Generous, N., Hickmann, K., Osthus, D., Priedhorsky, R., et al. Epidemic forecasting is messier than weather forecasting: The role of human behavior and internet data streams in epidemic forecast. *The J. Infectious Diseases,* 214, suppl._4, pp. S404–S408, 2016.

[24] Zhang, P., Jia, Y., Gao, J., Song, W., and Leung, H. Short-term rainfall forecasting using multi-layer perceptron. *IEEE Transactions on Big Data,* 6(1): 93–106, 1 March 2020. doi: 10.1109/TBDATA.2018.2871151.

[25] Kim, Y., and Hong, S. Very short-term rainfall prediction using ground radar observations and conditional generative adversarial networks. *IEEE Transactions on Geoscience and Remote Sensing,* 60: 1–8, 2022, Art. no. 4104308. doi: 10.1109/TGRS.2021.3108812.

[26] Venkatesh, R., Balasubramanian, C., and Kaliappan, M. Rainfall prediction using generative adversarial networks with convolution neural network. *Soft Computing, 25,* pp. 4725–4738, 2021. https://doi.org/10.1007/s00500-020-05480-9.

[27] Jeong, C.H., and Yi, M.Y. Correcting rainfall forecasts of a numerical weather prediction model using generative adversarial networks. *J. Supercomput.,* 79: 1289–1317, 2023. https://doi.org/10.1007/s11227-022-04686-y.

[28] Khan, M.I., and Maity, R. Hybrid deep learning approach for multi-step-ahead daily rainfall prediction using GCM simulations. *IEEE Access*, 8: 52774–52784, 2021. doi: 10.1109/ACCESS.2020.2980977.

[29] Abebe, W.T., and Endalie, D. Artificial intelligence models for prediction of monthly rainfall without climatic data for meteorological stations in Ethiopia. *J. Big Data*, 10(2), 2023. https://doi.org/10.1186/s40537-022-00683-3.

[30] Ouma, Y.O., Cheruyot, R., and Wachera, A.N. Rainfall and runoff time-series trend analysis using LSTM recurrent neural network and wavelet neural network with satellite-based meteorological data: Case study of Nzoia hydrologic basin. *Complex Intell. Syst.*, 8: 213–236, 2022. https://doi.org/10.1007/s40747-021-00365-2.

[31] Ghamariadyan, M., and Imteaz, M.A. Prediction of seasonal rainfall with one-year lead time using climate indices: a wavelet neural network scheme. *Water Resour. Manage.*, 35: 5347–5365, 2021. https://doi.org/10.1007/s11269-021-03007-x.

[32] Zhang, X., Mohanty, S.N., Parida, A.K., Pani, S.K., Dong, B., and Cheng, X. Annual and non-monsoon rainfall prediction modelling using SVR-MLP: an empirical study from Odisha. *IEEE Access*, 8: 30223–30233, 2020. doi: 10.1109/ACCESS.2020.2972435.

[33] Appel, K.W., Gilliam, R.C., Davis, N., Zubrow, A., and Howard, S.C. Overview of the atmospheric model evaluation tool (MET) v1.1 for evaluating meteorological and air quality models. *Environ. Modelling Softw.*, 26(4): 434–443, 2011.

[34] www.worldweatheronline.com.

[35] Cozzi, L. Weather models as virtual sensors to data driven rainfall predictions in urban watersheds. *EGU Gen. Assem. Conf. Abstracts*, 15: 111–115, 2013.

[36] https://dsp.imdpune.gov.in.

[37] Glorot, X., and Bengio, Y. Understanding the difficulty of training deep feedforward neural networks. *13th Int. Conf. Artif. Intell. Statist.*, 9: 249–256, 2010.

[38] Collobert, R., and Weston, J. A unified architecture for natural language processing: deep neural networks with multitask learning. *25th Int. Conf. Mach. Learn.*, pp. 160–167, 2008.

[39] Ergezinger, S., and Thomsen, E. An accelerated learning algorithm for multilayer perceptrons: Optimization layer by layer. *IEEE Trans. Neural Netw.*, 6(1): 31–42, Jan. 1995.

[40] Hundecha, Y., Bardossy, A., and Werner, H.W. Development of a fuzzy logic-based rainfall-runoff model. *Hydrol. Sci. J.*, 46(3): 63–76, 2001.

[41] Talei, A., and Chua, L.H.C. Influence of lag time on event-based rainfall-runoff modeling using the data-driven approach. *J. Hydrol.*, pp. 223–233, 2012.

[42] Kisi, O., and Shiri, J. Precipitation forecasting using wavelet-genetic programming and wavelet-neuro-fuzzy conjunction models. *Water Resour. Manag.*, 25(13): 3135–3152, 2011.

CHAPTER 3

Forecasting the Stock Market Index Using Artificial Intelligence Techniques

Sunita Kumar,[1,*] *Shiv Onkar Deepak Kumar*[2] and *Sridharan A*[3]

1. Introduction

If the stock market would have a predictable to maximum accuracy, then every stockbroker and investor would have been billionaire. But it is not the ground truth. In a one-to-one interaction with stock analysts, who mention that the stock market is unpredictable and that is why their role is important, else everything would have been black and white.

The Efficient Market hypothesis (EMH) indicates that the nature of the stock market is intrinsically complex and the belief that the stock market depends hugely on past data is not always correct. Considering this, the forecast of the stock market falls into the category of shooting arrows in the dark [6].

1.1 Motivation

To predict the unpredictable is the main motivation of this chapter. The stock market has been unpredictable from very beginning and will continue to be so in the future. There will be continuous efforts to defy this belief and work towards forecasting the stock market.

[1] Associate Professor, School of Business and Management, Christ University, Karnataka, India 560076, ORCID iD: 0000-0002-0628-1873

[2] Head of Data Science, Data Science, Straive, Chennai, Email: shivonkar@yahoo.com, ORCID iD: 0000-0002-6399-5119

[3] Professor, School of Business and Management, Christ (Deemed to be University), Bangalore. Email: sridharan.a@christuniversity.in, ORCID No: 0000-0002-0753-3585

* Corresponding author: sunita.kumar@christuniversity.in

1.2 What is a Stock Market Index?

A stock market index is a business sentiment measure of that country because it consists of a subset of stocks. These stocks are leaders of their respective sectors. If these stocks are doing well, then those sectors must be doing well and if all sectors are doing well then the country's economy must be doing well.

In this chapter, the stock market index of National stock exchange (NSE) of India, will be leveraged to explain various concepts, examples, and forecasting techniques.

The stock market index has different weightage of different sectors and stocks. For NSE, Nifty 50 is a well-known index. It consists of the following sectors and top stocks [8] (See Tables 1 and 2).

There are other well-known exchanges such as BSE (India), Nasdaq (US), The New York Stock Exchange (NYSE), The Tokyo Stock Exchange (TSE), and The London Stock Exchange (LSE), and many more. Each of them has indexes and stock composition (different %).

Table 1: Nifty 50 Sector Representation.

Sector	Weight (%)
Financial Services	37.16
Information Technology	14.38
Oil Gas & Consumable Fuels	12.99
Fast Moving Consumer Goods	8.58
Automobile and Auto Components	5.51
Metals & Mining	3.98
Healthcare	3.87
Consumer Durables	3.06
Construction	3.02
Telecommunication	2.5
Power	1.91
Construction Materials	1.78
Services	0.76
Chemicals	0.51

1.3 Movement of the Stock Market Index

To understand the movement, the analysis of all factors impacting the movement need to be analyzed. That will help to go deeper and understand the trends and patterns.

As per various research papers and analysis, many factors impact the stock market index. Those are historical prices [9], consumption-based pricing [6], intermediary pricing [6], interest rates, exchange rates, growth rates of industrial production, consumer purchasing capacity, company specific information such as income statements and dividend yields [10], news [6, 9, 10].

Table 2: Nifty 50 Top constituents by weightage.

Company Name	Weight (%)
Reliance Industries Ltd.	11.36
HDFC Bank Ltd.	8.53
ICICI Bank Ltd.	8
Infosys Ltd.	7.21
Housing Development Finance Corporation	5.89
Tata Consultancy Services Ltd.	4.19
ITC Ltd.	3.61
Kotak Mahindra Bank Ltd.	3.45
Larsen & Toubro Ltd.	3.02
Hindustan Unilever Ltd.	2.89

1.4 Forecasting Approach

The technical approach leverages data presented in Fig. 1. It tries to build a model such that the price is dependent on these values. Besides these, it tries to accommodate crowd psychology (hype, panic, excessive optimism, and pessimism) [11].

The Fundamental approach, also, leverages data presented in Fig. 1. Besides these data, it leverages macro data extensively. The macro data mentioned includes GDP, change in exports' and in imports' trend. It helps to forecast economic conditions, which is a strong influencer on stocks, and hence the stock market index.

Micro economic factors	Macro-economic factors
i. Monetary policies ii. Growth rates of industrial production iii. Unemployment rate iv. Company income v. Company dividend vi. Consumption-based pricing	i. Interest rates ii. Exchange rates iii. Consumer purchasing capacity iv. Global economic data v. Similar Country's index
Historical factors	**Miscellaneous**
i. Historical prices ii. Volume trend iii. Technical (PE, PB) iv. Support and Resistances v. Future and Options	i. News ii. War like situation iii. Slow down iv. Crowd psychology v. Overall sentiment

Fig. 1: Factors impacting stock market index's price.

1.5 Chapter Map

Section 1, 'Introduction', touches upon various points in brief. Section 2, 'Methodology', explains the objective for this chapter and approach to achieve the result. Section 3, 'Literature Review', highlights various work done by previous researchers. Section 4, 'Forecasting techniques', explains various strategies already used by different researchers and organizations. Section 5, 'Important Factors in Forecasting Stock Market Index', is all about feature engineering and prepares the base for data analysis. Section 6, 'Data Analysis', is about the detailed steps taken for AI. Section 7, 'Conclusion', discusses the future steps for improving the work.

2. Methodology

Objective: Use latest AI Techniques and Multidimensional Approach to Forecasting the Stock Market Index.

The chapter has considered the following practical steps. First, understand the forecasting need from investor and trader point of view. Please note that investor and trader are different and, do not getinto details as that is covered in the trading book, and is outside the purview of this chapter. Second, to get authentic data from the NSE website [8], which is freely available. Third, do EDA and visualize the data from a multidimensional angle. Fourth, do feature engineering. Fifth, try to find out the right set of data and timeframe instead of using the whole data. Sixth, use DL techniques to find the best model with minimum mean absolute percentage error (MAPE). Iterate above and tune the models to get the best forecast (Fig. 2).

It is secondary data-based to understand the DL's applicability in stock market index forecasting.

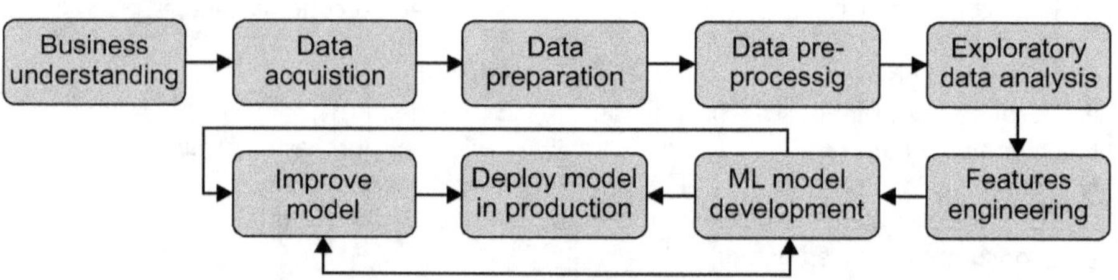

Fig. 2: High-level steps for forecasting.

3. Literature Review

Can you forecast stock market index? This question is asked consistently by the investor and trader in many seminars. There is constant study and research to answer this question. Here are few instances.

The stock market index is time series data, at least numerically. Many times, it shows the behaviors exhibited by standard time series data as shown in Fig. 3. The data (Fig. 4) has cycle, periodicity, trend, and seasonality. Considering these, many researchers have tried to forecast using Standard time series techniques MA, AR, and ARIMA [15, 16, 17].

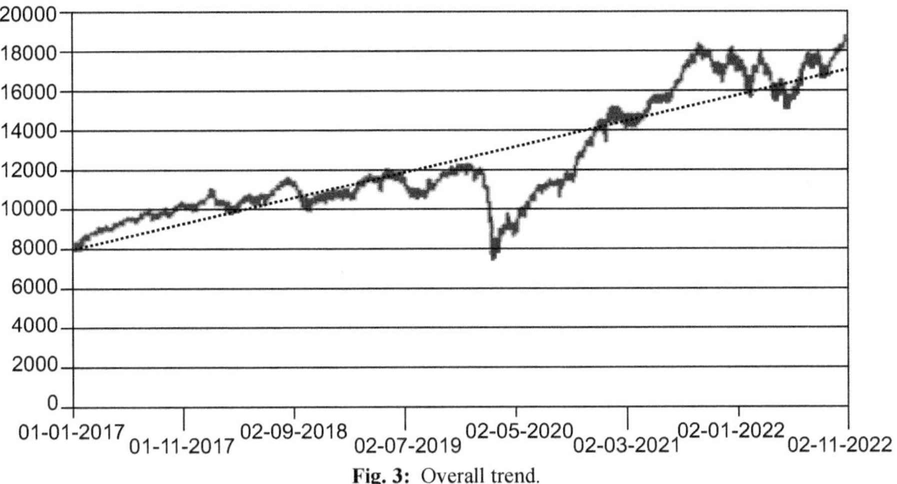

Fig. 3: Overall trend.

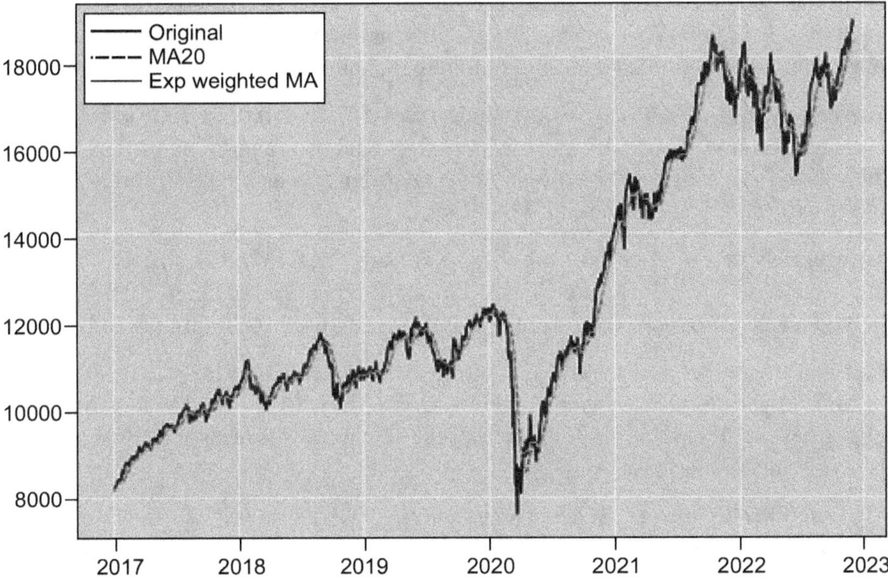

Fig. 4: Trend with MA20 and exponentially weighted moving average.

By nature, the stock market index is nonlinear data. By default, the nonlinear data is complex and there is no straightforward way to forecast it. There have been multiple techniques for nonlinear regression. It has its pros and cons. For the stock market index, support vector machines (SVMs) have been used and promised to a certain extent [18, 19, 20].

Deep learning (DL) has been latest entry in forecasting the stock market index. There are various techniques in DL. All have been tried out by various researchers. The Long short-term memory (LSTM) [23, 24] and Neural Network (NN) [21, 22] have primarily given promising results. The LSTM takes care of sequential data

Table 3: Interpretation of MAPE values [12].

MAPE Value	Interpretation
< 10	Highly accurate forecasting
10 – 20	Good forecasting
20 – 50	Reasonable forecasting
> 50	Inaccurate forecasting

(time series aspect), and the NN takes care of the nonlinear aspect. The NN builds feature engineering automatically and is useful in forecasting.

The current paper uses the paper [25] as the latest base paper although it uses NN instead of LSTM. The NN with its latest improvements is the main theme of the current paper.

4. Forecasting Techniques

There has consistently been an attempt to build a perfect technique to forecast the stock market index. When one technique fails, another takes birth. This sequence has been continuing and this paper is a step in that chain.

Moving average (MA) comes in windows of 10, 20, 50, and 200 days. You can calculate for any days and these days are the industry standard and often heard by analysts in their analysis. The moving average often acts as a support for indexes and a long call can be taken from the MA bases.

Autoregressive integrated moving average (ARIMA) consists of "AR: Autoregressive", "I: Integrated", and "MA: Moving average". It combines three techniques and produces the forecast. It is generally used for future price forecast.

Vector auto-regression (VAR) is a multivariable regression technique. It depends on the analyst on how many parameters to use from Fig. 1 in this technique. The appropriate number of variables increases the accuracy of forecasting. VAR is mainly used in Fundamental analysis.

Artificial intelligence (AI): The Deep Learning (DL) is subset of Machine learning (ML) and ML is a subset of AI. All these three terms are used for one another in simple talk.

Machine learning (ML) has many ML techniques which have been used for forecasting the stock market index. Few of them are Random forest (RF), XgBoost (xgb), and SVM. These have increased the forecast accuracy, although it still not perfect.

Deep learning (DL) has many DL techniques that have been used for forecasting the stock market index. Few of them are LSTM and NN. The inherent capabilities of feature engineering make these techniques better than ML. The DL has an inherent capability to process large amount of data and extract trends and patterns. Are these perfect? The whole chapter will explore the DL technique (NN).

Error metric used is for time series data; the industry has put faith in Mean Absolute Percentage Error (MAPE) besides MSE and RMSE. Since it is percentage, it is applicable for any range of prices.

5. Important Factors in Forecasting Stock Market Index

The first question comes—"What are all the important factors necessary to forecast the Stock market index?" Before answering it, can we define "all important factors". If there would have been answer of "all important factors", then this chapter would not have been written because forecasting would have been already one.

There are two categories of "important factors"—Business defined, and AI defined. The former is covered here and latter is covered in the "Data analysis" section.

Business defined important factors

- Trends: If you ask a seasoned stock analyst for one advise then they will say, "Go with trend". In simple terms, a trend is all about the market going up or down with a certain degree of conviction. This chapter is also about identification of the trend (Figs. 4a,b).
- Moving Average: It is one of the most important term that the analysts use in their analysis. The industry standard Moving averages are MA10, MA20, MA50, MA200. Here number 10 means average window duration for calculation of average. Similarly, for other number prefixes.

During analysis for last 3 years, it was observed that trend is not always in any one direction. The trend seems to reverse and tries to settle around the 'mean'. It is also called "mean reversion" phenomenon. The settling can happen anywhere MA10, MA20, MA50, MA200.

- Random value: As per various studies [14], the stock market shows a perfect trend and the future does not depend on past prices. This phenomenon and price is called martingales in the probability theory. The forecast is the next higher number, probably the small number but in a higher direction. It is also known as a sub-martingale and also as a random walk with a minor drift upside. It has been proved many times. If you see the price, during some defined duration, the price is little higher. Few exceptions are there for stock specific cases, but index price exception may be rare (See Figs. 3, 4, and 5).
- Price at right valuation: The investor and trader often think or want to buy at price which is less for the value shown (or inherent) in stocks. Because, as per history these are situations that will get recognized in the future and price will go up. There are various parameters and few of them are price to book (PB) and price to earnings (PE) (Fig. 6).
- Candle trend: The definition of 'Candle' is available in many books and internet, hence it is not discussed here. Please understand the following concepts in detail:
    - o Candle body trend

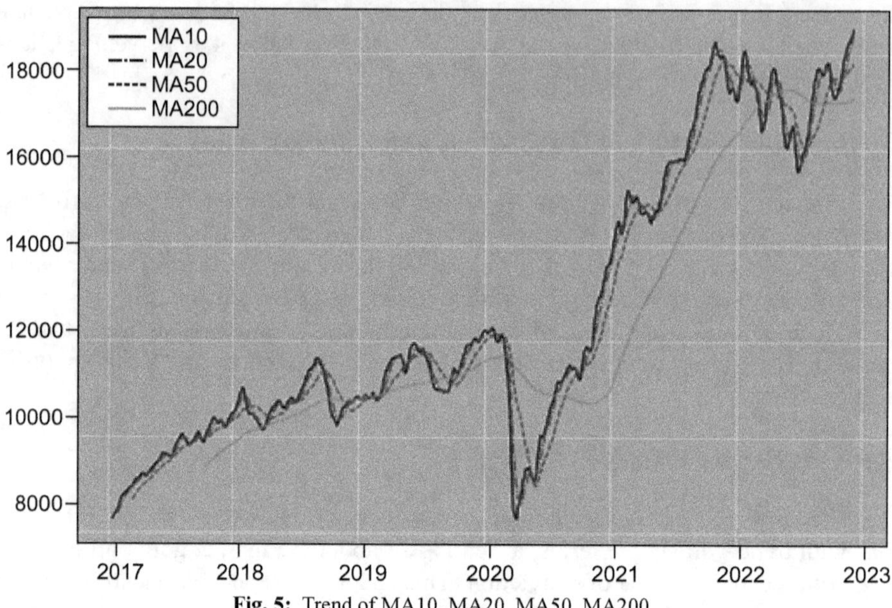

Fig. 5: Trend of MA10, MA20, MA50, MA200.

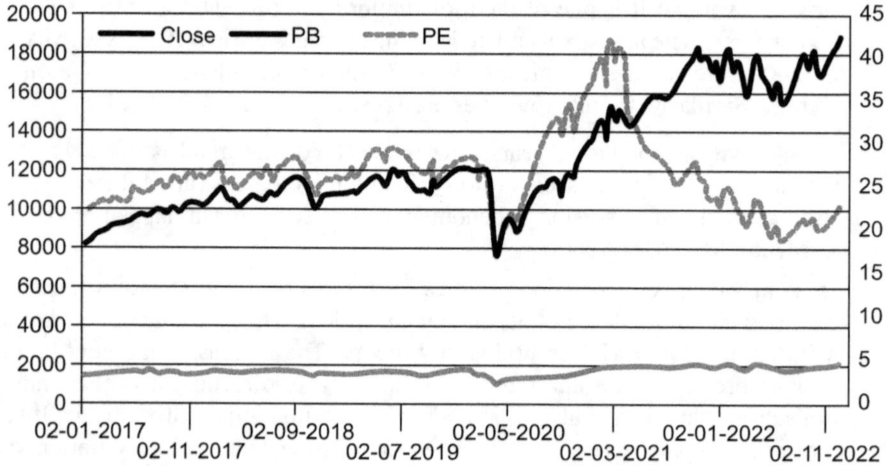

Fig. 6: Nifty 50 close price w.r. to PB and PE (on secondary y axis).

- o Candle body type
- o Candle color
- o Candle tail trend

The Candle concept is well accepted across the industry. The above points add value in forecasting. The following candle graph of NSE index NIFTY has all the above points which need to be interpreted together for forecasting. For lack of space in the graph, only the last 3 months (Oct.– Dec. 2022) (Fig. 7) have been used.

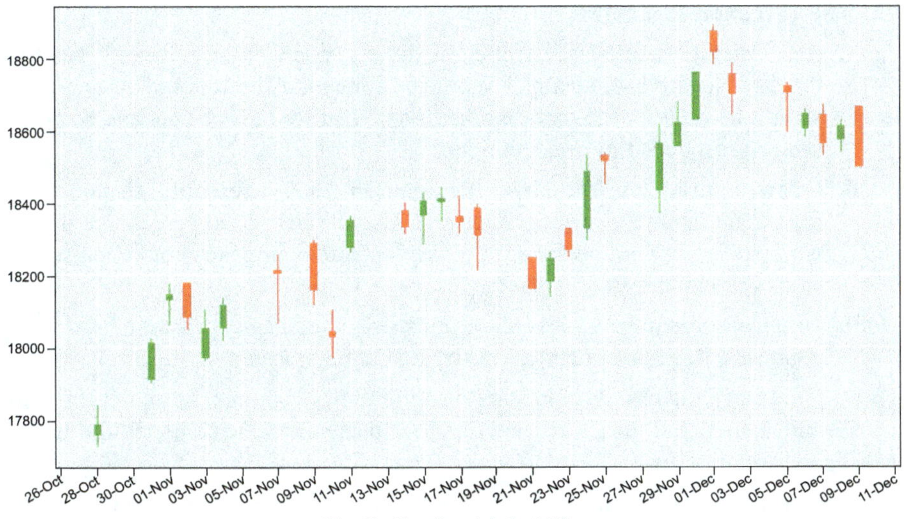

Fig. 7: Candle stick for Nifty.

6. Data Analysis

There are multiple steps for data analysis.

6.1 Data Acquisitions

The first major challenge is to get data as per Fig. 8. Please note, these are just adopted by the author and not necessarily the best way to get data. The other possible ways must be explored. Let us start one by one as follows:

A. For the last 3 – 5 years, the stock market index data from NSE website [8].

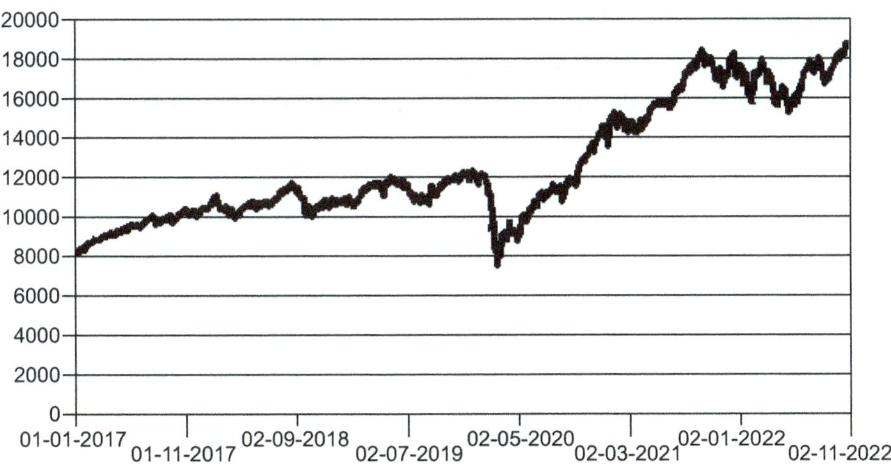

Fig. 8: First view of NIFTY 50 Close price.

B. Micro-economic factors

i. **Monetary policies**: The decoding is done in this way – Positive: 1, Neutral: 0.5, and Negative: 0. Ideally, it should be done for all sectors and the aggregate should be used for the stock market index. But for lack of time and resource, sector-level decoding was not done.

ii. **Growth rates of industrial production**: It is available for quarterly (sometime monthly in few countries). Use the published percentage data as is. Ideally, it should be done for all sectors and aggregate should be used for stock market index.

iii. **Unemployment rate**: Use the published percentage data as it is. In few countries, it is not available. Try to find data that is employment related.

iv. **Company income**: Use income of minimum 3 companies from each sector and average out for index level use. Try to use the same company for the last 3 years.

v. **Company dividend**: Use income of minimum 3 companies from each sector and average out for index level use. Try to use same company for the last 3 years.

vi. **Consumption-based pricing**: Use the published percentage data as it is. In a few countries, it is not available. Try to find data that conveys consumer spending or purchasing.

C. Historical factors

i. Historical prices: Data for the last 3 years.

ii. Volume trend: Volume is available in index data.

iii. Technical (PE, PB): These are available in index data.

iv. Support and Resistances: It needs to be calculated.

v. Future and Options: For the last 3 years, from NSE website [8].

D. Macro-economic factors

i. Interest rates: Use the published percentage data as it is. In India, use Cash reserve ratio (CRR), and in other countries use similar data consistently for 3 years.

ii. Exchange rates: Use local currency vs US $ exchange rate.

iii. Consumer purchasing capacity: Use the published percentage data as it is.

iv. Global economic data: Although it is subjective, use any one indicator for all the 3 years.

v. Similar country's index: Singapore index opens few hours before NIFTY and gives an indication of NIFTY moves.

E. Miscellaneous

i. News: Use sentiment of business prospects within country.

ii. War like situation: If major war is going on (like Russia and Ukraine), then take value as 1, else 0.

iii. Slow down: Is slow down going on?

iv. Crowd psychology: If majority are selling, then 0; and if majority are buying, then 1.
v. Overall sentiment: If there is strong rating or +ve news then 1, for any normal days 0.5 and – 1 for any – ve news.

6.2 Data Cleaning

The next step is to combine all the above data into one file and conduct the following steps.

The data taken from NSE websites are mostly cleaned. The data derived need validation from past when a similar situation had occurred.

If any missing data is there, then try another source for getting those data. If still missing, then take average of previous and next available data.

6.3 Feature Engineering

From raw data, calculate PE, PB, MA10, MA20, MA50, MA200, R1, R2, R3, S1, S2, S3, RSI, PIVOTS, LB, UB...

The above is an indicative variable list and many more will be added based on the business analyst.

6.4 Exploratory Data Analysis

The following analysis of all numeric variables are performed:

a. The basic decomposition of data
b. The trend analysis
c. Corelation of all features at a time. Draw heat map and see if there are many strong corelations. If yes, then collinearity removal can be planned during modeling.
d. Outlier and anomaly detections. Keep option open to substitute these with expected variables because outliers impact forecasting (Fig. 9).

6.5 Model Building

The result is an ensemble of three models as follows:

i. The DL technique Feedforward Neural Network (FFNN) is constructed with features listed in sub section "Data Acquisition". It isa regression model and forecasts the value of stock market index just before the market opens because the model needs other features as input variables and all of them are available before the market opens. In practical scenarios, few may not be available, then move with default values.
ii. Time-series forecast using DL technique LSTM: It is built with univariate data of the stock market index.
iii. Time-series forecast using 'fbprohet': It is built with univariate data of the stock market index.

Finally, the result was a weighted ensemble of 50% FFNN, 25% LSTM, and 25% 'fbprohet'.

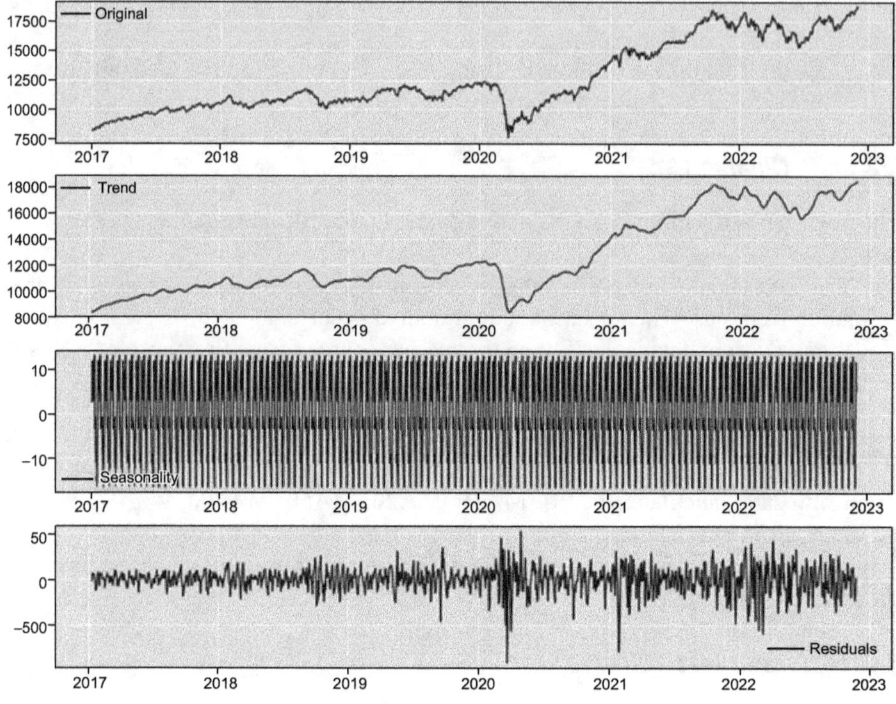

Fig. 9: Detail inside the time-series data

Once the model is stable and consistent for different samples of data then the model tuning was performed. The parameters were tuned as per industry standards.

6.6 Error Metric

The MAPE, MSE, and RMSE are used to calculate error. The MAPE is primarily used to compare the accuracy for given model and across models. The residual plots (Q-Q plot, Normality of Residual and Homoscedasticity) are used to understand why the error was high. The analysis is fed to data preparation and model building. It is repeated till consistency is achieved.

7. Conclusion

If there would have been a straightforward way to forecast the stock market index, then this chapter would not have been written. The stock market index is impacted by many factors, and hence the forecasting approach must be multidimensional. The main challenges are to identify factors impacting the stock market index. It requires very good finance knowledge and analytical skill so that subjective terms can be converted to numerical terms which become the input for various AI techniques.

As the in-depth functional knowledge increases, few more factors will be identified and fed into the model. It will enhance the accuracy of forecast. Similarly, the more technology advances, the accuracy of forecast will enhance too.

References

[1] https://core.ac.uk/download/pdf/39667613.pdf.

[2] https://www.sciencedirect.com/science/article/pii/S2666827022000378.

[3] https://www.investopedia.com/articles/07/mean_reversion_martingale.asp.

[4] https://www.kdnuggets.com/2020/01/stock-market-forecasting-time-series-analysis.html.

[5] https://www.frontiersin.org/articles/10.3389/fenvs.2022.917047/full.

[6] https://en.wikipedia.org/wiki/Efficient-market_hypothesis.

[7] https://en.wikipedia.org/wiki/Stock_market_index.

[8] https://archives.nseindia.com/content/indices/ind_nifty50.pdf.

[9] Kolarik, and Rudorfer, G. *Time Series Forecasting Using Neural Networks*. Department of Applied Computer Science, Vienna University of Economics and Business Administration, 1090: 2–6, 1997.

[10] Fama, E. *Foundations of Finance*. New York: Basic Books, 1976.

[11] Balvers, R., Cosimano, T., and McDonald, B. Predicting stock returns in an efficient market. *Journal of Finance*, 55: 1109–1128, 1990.

[12] https://stephenallwright.com/interpret-mape/.

[13] https://en.wikipedia.org/wiki/Efficient-market_hypothesis.

[14] https://en.wikipedia.org/wiki/Ma.rtingale_(probability_theory).

[15] Lusk, E.J., and Neeves, J.S. A comparative ARIMA analysis of the 111 series of the Makridakis competition. *International Journal of Forecasting*, 3: 329–332, 1984.

[16] Box, G., and Jenkins, G. *Time Series Analysis: Forecasting and Control.*, Holden Day, San Francisco. 1970.

[17] Hamilton, J., and Lin, G. Stock market volatility and the business cycle. *Journal of Applied Econometrics*, 11: 573–593, 1996.

[18] Mukherjee, S., Osuna, E., and Girosi, F. Nonlinear prediction of chaotic time series using support vector machines. *Proceedings of the IEEE Workshop on Neural Networks for Signal Processing, Amelia*, pp. 511–520, 1997.

[19] Tay, F., and Cao, L. Application of support vector machines in financial time series forecasting. *Omega*, 29: 309–317, 2001.

[20] Mills, T. Non-linear time series models in economics. *Journal of Economic Survey*, 5: 215–241, 1990.

[21] Kim, H., and Shin, K. A hybrid approach based on neural networks and genetic algorithms for detecting temporal patterns in stock markets. *Applied Soft Computing*, 7: 569–576, 2007.

[22] Chen, K., Zhou, Y., and Dai, F. An LSTM-based method for stock returns prediction: A case study of China stock market. *In: 2015 IEEE International Conference on Big Data (Big Data),* pp. 2823–2824, 2015.

[23] Nelson, D.M., Pereira, A.C., and de Oliveira, R.A. Stock market's price movement prediction with LSTM neural networks. *In: 2017 International Joint Conference on Neural Networks (IJCNN)*, pp. 1419–1426, 2017. IEEE.

[24] Bhandari, H.N., Rimal, B., Pokhrel, N.R., Rimal, R., Dahal, K.R., and Khatri, R.K.C. Predicting stock market index using LSTM. *Machine Learning with Applications*, 9: 100320, 2022. https://doi.org/10.1016/j.mlwa.2022.100320.

Chapter 4

Forecasting of Environmental Sustainability through Green Innovation of E-Vehicle Industry

Jobin Jacob,[1,*] *Pancinovia Neelu*[1] and *Arun C Antony*[2]

1. Introduction

E-mobility sustainability forecasting is getting more detailed with study, taking into account social cost in addition to technological, economic, or environmental factors. One solution for reducing greenhouse gas emissions is to implement green innovation in the transportation sector. The citizenry's view and acceptance of electric cars must be improved, more research into the social cost of these innovations is required. Consequently, the transportation industry might decarbonize more quickly. Another approach to do it is to advocate for more all-encompassing green innovations that can enhance sustainable development. Using *Our Common Future*, published in 1987 by the World Commission on Environment and Development [1], the commission emphasized the importance of sustainability while integrating social and economic development. Additionally, it recommended that governments take environmental factors into account while making decisions. The significance of sustainability was then increased and institutionalized, which meant that nations began passing laws that promoted sustainability. Consumer awareness of sustainability is rising largely

[1] School of Business and Management, CHRIST (Deemed to be University), Bengaluru, Karnataka, India, Email: pancinovia.neelu@christuniversity.in

[2] Christ College, Vadgaon Sheri, Off Pune Nagar Road, Pune, Maharashtra, India. Email: fr.arun@christcollegepune.org

* Corresponding author: jobin.jacob@christuniversity.in

from an economic and environmental standpoint. This also has an impact on the transportation industry and poses significant environmental, social, and economic difficulties. However, given that it generates close to 5% of the GDP and employs almost 11 million people, transportation is crucial from an economic standpoint.

Decarbonization and green mobility are emerging as potential options with electric vehicles due to social demand. Researchers have taken a number of steps and initiatives to improve the sustainability of the electric vehicle industry. Analyzing the patterns in the growth of the electric vehicle industry and its societal costs enhances pertinent industry research. The main arguments for the necessity for sustainable development are "vehicle exhaust emissions", "climate change", and "integration". This study examines consumer citizenship behavior toward green innovation which leads to green marketing systems from a broad perspective in response to the important difficulties in sustainable development. The driving forces underlying industry development highlight the various motivations and tactics that direct the sector toward sustainability. The responsibilities and efforts required from several consumers, as a result of regulatory consequences, driving demand-side variables, and supply-side demands for technological advancement. As a result, the electric vehicle sector and its green innovation for sustainable development and unique ideas to other important research disciplines may benefit from consumer synergy and social cost optimization. The term electric vehicle (EV) refers to vehicles with an electric drivetrain and plug-in charging, as well as fuel cell electric vehicles (FCEV). In the EV industry's sustainable development, long-term social costs and environmental benefits are sought, but the widespread use of EVs is constrained by considerable uncertainty. Examples include geopolitical differences in policy preferences, greenhouse gas emissions, end-of-life waste, and technological bottlenecks in the supply chain.

Researchers have identified a variety of factors that have an impact on the commercialization and development of EVs, including patents, prototypes, R&D (research and development), fuel efficiency, subsidy policies, social costs, and investor interest. Only a small number of relevant studies have thoroughly examined the growth of the business and the societal costs, focusing instead on particular aspects of the EV industry and its consumer citizenship behavior on green marketing. Over the past 10 years, the subject has seen an increase in publications, and as technology innovation continues, researchers have opened up a variety of new research areas. The trends of industrial development and societal cost were thoroughly reviewed using the data collected on publishing outputs, publication patterns, topic categories, and global productivity. As a result, the approach might be effective in the face of complexity, diversity, and unpredictability in the growth of the growing EV industry from a standpoint of green innovation. Future EV sector success in terms of green innovation and sustainable development will heavily depend on whether important industries (such as governments, the general public, and suppliers) keep their commitments, get involved, make an effort, and support good consumer citizenship. Therefore, additional research is required to thoroughly examine the key players in the industry and to assist its sustainable development. The term 'stakeholder' is used to refer to any group or person who influences or has the potential to affect the accomplishment of an organization's goals. This idea was developed to address the issues of sustainable value creation, capitalist ethics, and management attitude. Stakeholder theory, though it has its roots in traditional shareholder theory, aims to create a more balanced distribution of the rewards from operating results between shareholders and non-shareholders. Researchers use consumer citizenship behavior to assess a wide range of issues related to business, such as mergers and acquisitions, corporate governance, and corporate social responsibility. With the consumer engagement system, the study takes a systematic approach to assist the industry's sustainable growth, taking into account development patterns and stakeholders' needs. The system examines the driving factors (such as environmental concerns, energy security, corporate strategy, and industrial strategy) behind the global industry development as well as the important players and their key responsibilities in this developing sector's sustainability.

2. Variable Relevance

2.1 Green Innovation

In the past 10 years, the phrase "green innovation" has gained a lot of forecasting popularity. It was done with the purpose of raising public awareness of important environmental concerns. The protection of the environment, support for renewable goods, and increased use of energy are characteristics of green innovation. Simply because it is a hue associated with natural leaves, 'green' was included. Green innovation specifically refers to energy generation methods like solar energy and wind energy. Green has evolved to signify 'renewable' as well. The sun is the primary source of renewable energy since it can emit energy indefinitely without endangering the environment. Green innovation's true level of 'greenness' is frequently questioned. The truth is that it is essentially innocuous to the environment. By encouraging the usage of 'greener' alternatives like EVs, it helps to minimize air pollution. As a result, the need for energy will gradually replace that for nuclear power facilities. Another query is, "Can the entire pieces of a car, notably the battery, be totally recycled?" The answer is no; regrettably, there is no technology that does not in some way hurt the environment. Several factors influence how interested corporations are in green innovation. First and foremost, the government is emphasizing environmental concerns and encouraging the use of green technologies. Consumer desire and preference for green innovation-based goods is the second element. As a result, it is anticipated that customer preference will shift toward green innovation's goods in the near future. The EV is one such product with a fairly straightforward build. An electric motor that rotates the wheels is powered by a battery, and a controller controls how much power is sent to the motor. The EV design may also be used in things like electric trains and elevators, expanding its use beyond simply autos.

2.2 Advancement and Social Cost

It is a well-known truth that, in contrast to local rivals, foreign businesses incur additional expenses, such as taxes, transportation costs, marketing expenses, customs, as well as cultural and societal distinctions that often benefit domestic businesses. As a result, multinational corporations (MNCs) must invest in green innovation as well in order to optimize efficiency in global marketplaces. Businesses must concentrate on advancing green innovations as customer awareness of environmental concerns rises. MNCs, in particular, require this to compete in the global market; otherwise, competition would pressurize them to leave the market after a very short amount of time. Schumpeter's writings from the early 20th century served as an inspiration for contemporary views of innovation and economic growth. Based on technical advancements in all ways of thinking and doing, Schumpeter defined innovation in all sorts of profit-making motivations and actions. Fisher defines innovation as one or all of the components of new thinking: new methods for producing and testing connected economic and social activity. According to Schumpeter's theory from 1942, innovative business practices foster a process known as "creative destruction". Economic growth is a form of technological evolution that continually obliterates the previous economic system and builds up the new one, revolutionizing it from within. Businesses who ignore innovation during the production process, transportation, etc. will be driven out of the market.

Theoretically, a rise in entrepreneurs will hasten economic growth. Technological innovation and economic growth have recently been linked in theoretical analysis of Romer and Solow from 1956 (1986), as well as in empirical investigations by Mansfield (1972) and Nadiri (1993). These ideas are well established in the literature on economics. A sizable corpus of empirical investigations concentrated on technical advancement and business innovation. These studies have shown that a company's level of technical innovation affected its financial results and growth in market share. According to Nadiri (1993), innovation impacts ultimate output and productivity through Cobb-Douglas production functions. As an organization grows, produces, and grows, its innovation pace affects its long-term success. The late 1980s saw the beginning of a sizable number of research on green technology in relation to corporate implementation methods. The management and stockholders of firms were prioritized in the majority of these investigations over environmental concerns. But now that customers are demanding improvements, their worries are centerd on anticipated future changes in environmental control requirements. As a result, organizations generally concentrate on finding ways to implement these changes without having a detrimental impact on profits. Regarding individuals operating in free-market economies with a business motive, this idea is reasonable. It is inevitable that more people will become aware of environmental challenges. Typically, new disruptive technologies don't outperform the current technology on established performance measures when they first enter the market, but rather result in breakthrough performance on new value parameters, according to Bower and Christensen's (1995) research. To implement novel products into a market, Moore's (1991) study emphasizes consumer identification and the need to replace old technologies with new, inventive ones in order to adapt to a changing market. Environmental sustainability is expected to provide a sizable worldwide market for cutting-edge sustainability solutions. It will be a significant market, exceeding other important industries like the industrial or automotive industries. According to market predictions, during the next 10 years, energy supply, energy efficiency, transportation, and water technologies will see a rise in demand on average by 5–8%

annually. Focusing on and implementing green technology will be economically advantageous for emerging nations.

2.3 Green Marketing and Its Evolution

Green marketing refers to efforts designed to create and support the commercialization of goods or services to meet human needs while having the least possible negative effects on the environment. Modifications to product and packaging design, manufacturing procedure adjustments, and advertising changes are a few instances in this regard. According to Peattie, "green marketing is an integrated administrative process that is in charge of finding, predicting, and satiating societal and consumer wants in a profitable and sustainable manner". This implies that there is a connection between morality and green marketing. The environment is taken into consideration as one of the moral dilemmas that marketing must examine and discuss. Thus, green marketing is a component of social cost. Business leaders should consider two key factors when developing a green marketing strategy: creating products that meet consumer needs while minimizing negative impacts on the environment and creating consumer perceptions that demonstrate the product's quality and commitment to the environment. Clients and businesses are likely to experience a qualitative shift as a result of a green marketing strategy. If businesses want to satisfy the demands of ecological consumers, they must create both practical and emotional advantages for the product. A proactive approach tries to gain an edge on the competition by deliberately putting items in the minds of its consumers. Green marketing requires the support of all value chain participants. Environmental awareness must be shared by all parties involved.

2.4 Consumer Citizenship Behavior

Consumer citizenship behavior is any voluntary customer action that benefits the business and goes above and beyond the requirements of the typical customer role. Customers engage in civic engagement at their own free will, and this kind of involvement adds enormous value to a business. The idea of corporate citizenship behavior states that customers are more inclined to show their support for a firm, such as a brand community, by engaging in actions like making purchases from that business, recommending it to others, and spreading good word of mouth. Accordingly, some study has shown that consumer citizenship is demonstrated through a range of activities, such as constructive participation in recommending service changes, good WOM communication, and other pleasant and respectful actions.

3. Research Methodology

The study's methodology, which was developed by Tranfield, Denyer, and Smart (2003), is the cornerstone of its systematic literature review technique. This report provides a comprehensive and systematic examination of the relevant scientific literature to examine EVs' societal cost and consumer citizenship behavior order to examine the collective cost and consumer citizenship behavior of EVs. This research pertains to and is based from the papers of Web of Sciences and concentrated on the priorities of the green innovation, only reviews and scientific publications were taken into consideration. The study did not include book chapters or conference papers since they are not precisely connected to research publications. The WoS database is one of the most well-known and well-liked scientific databases.

"Papers were researched focusing on various variations of the topics/keywords 'electric vehicle' and 'social cost'," which resulted in 237 papers being found. Sixty-five publications were chosen for inclusion in the study after being reviewed, starting with the titles and abstracts, in accordance with the technique suggested in. After carefully examining all of those publications, 28 of them were chosen because they fit our focus on evaluating the social costs of EVs. Instead of concentrating on environmental problems that could be linked to socio-economic consequences, we should focus on society and consumer behavior. It was further divided into groups based on how they rated various in the framework designed to explore preferences and needs for possible green breakthroughs, social variables are included of EVs on EV infrastructure: acceptance, perception, social cost, impact, welfare, and consumer citizenship behavior. Papers were arranged according to the year of publication, examined, and put into the aforementioned categories. The results have provided a summary of the findings with a focus on the areas of research that needed to be prioritized for different study years. Furthermore, each article analyzed the Social Cost of Green Innovation Leading to Green Marketing Analyzing Consumer Citizenship Behavior, defining and pinpointing the influence of the desire for a sustainable lifestyle in the developing world.

Primarily, it was determined if the study's objective was to determine the causes, precursors, motivators, hurdles, or intents to embrace the green innovation of EVs for consumer behavior. This criterion has the effect of allowing both conceptual and empirical works to be included. However, studies containing empirical consumer data and outcomes were included in the selection of pertinent publications, with a focus on how research may teach us about real buyers' intents and actions. According to some research models, the demand for EVs may be oil prices and/or different financial incentive conditions; other non-attitudinal and behavioral elements, were used to make the prediction. Such research was disregarded since it provides little insight into customer views and actions, yet being crucial for understanding and simulating broader market conditions. Behavioral reactions to an innovation are defined in academic literature as buying and using it. Many antecedents or predictors have been proposed to explain this behavioral response. The key determinants of the customer's adoption behavior are various forms of purposeful measures of consumer citizenship behavior, such as preparedness and desire to embrace green innovations, which are frequently used in research as a stand-in for consumer citizenship activity. Therefore, our analysis of the literature depicts that the primary methodology used to study public adoption of EVs has been to concentrate on intentions rather than actual adoption. EVs currently hold a minimal share of the market, so this makes sense. Based on our review of the literature, we determine that consumer adoption of EVs has been studied by five different types of theoretical frameworks. To estimate market share, the evidence is generally controlled by rational, economic modeling methodologies that use aggregate or individual, disaggregated national data in an effort to discover the variables that influence consumer automobile buying behavior. However, some of the literature disapproves of such methods or at least adds to them with information from theories that draw on marketing, transitional or social learning models, or environmental or social psychology. New technology like electric cars is unfamiliar to the general public. The examination of customer behavior toward such automobiles is severely hampered by this. A thorough knowledge of consumer behavior towards the adoption of EVs in particular and in general is provided by this systematic evaluation of the available worldwide research. The Consumers and Vehicles subproject of the E-Vehicle Economics and Infrastructure program will use this to support future primary data gathering efforts.

3.1 Understandings

Consumer citizenship feelings like satisfaction and joy, are believed to affect EV purchasing intentions. These feelings were described in the section on consumer innovativeness. Moons and De Pelsmacker (2012) introduced a further layer by hypothesizing customer sentiments regarding EVs. Their claim was supported by Richins (1997) and other consumer behavior theories which suggest that buyers' attitudes toward EVs will be influenced by how they feel about cars, EVs, and driving attitudes about them will be influenced by their feelings towards cars, EVs, and driving in general as well. A consumer's judgment of the style, design, and size of EVs provide the basis for visceral reactions. Moons and De Pelsmacker (2012) propose three emotional processing stages. The feelings consumers experience while using and encountering EVs are related to their behavioral emotions. As far as EV driving is concerned, reflective feelings affect identity and self-image. Moons and De Pelsmacker (2012) assessed customers' level of satisfaction with EVs by asking them about pleasant or negative feelings they experienced. It was not assessed whether consumers felt pleasure, joy, pride, or humiliation, as is the case with Richins (1997)'s theory or Schuitema et al.'s findings. Communicating with consumers about EVs, educating them about them, and developing policies that encourage their acceptance requires an understanding of the emotions that accompany the adoption of EVs. We have discovered that emotions play a significant role in pro-environmental behavior (Bamberg and Möser, 2007; Steg and Vlek, 2009), as well as car choice. However, there has not been much theorization on emotions in relation to pro-environmental behavior (such as adopting EVs), as well as how emotions are related to values, beliefs, and norms (Steg, 2005).

4. Interpretations

A significant transformation in policy, economy, and energy usage is brought about by the newest cutting-edge technology in the transportation industry, EVs. New products affect consumer demand and behavior based on their features, costs, and performance. Pricing, cruise range, performance, environmental impact, durability, and convenience are some of the factors that determine customer demand. However, the significance of EVs will only be clear till there is a sizable market penetration. One of the most important considerations in terms of user convenience is how easily and reliably the battery can be recharged. Because of the technology's ability to fully charge a car in 6–8 hours, overnight charging is the ideal choice for conventional automobiles, which do not have a severe refilling problem due to the abundance of gas stations. Surveying the public after a test drive is one of the most popular ways to see how consumers feel about cutting-edge automobile technology. Researchers observed consumer behavior toward the product with positive results obtained using a well-balanced sample of data and the correlation technique to evaluate the outcomes. Customers generally agree that EVs are a viable choice since they generate less carbon dioxide than traditional vehicles, which benefits the environment. The smooth sailing is another perk that most clients find appealing. These scores indicate that there will be a substantial market demand for EVs. Since EV technology will significantly reduce air and noise pollution, highly populated cities would particularly benefit.

Globally, green marketing is growing more and more important in the modern world. Given its close ties to environmental protection, green advertising is seen as a powerful strategy for the marketing of goods, services, and business concepts. Customers who buy green products are seen as a key driver of consumption in the current economic climate. As consumers become more environmentally conscious, new market opportunities have emerged for the global economy. Ever since the 1980s, environmental marketing and green marketing have taken a keen interest among academics. Since the early 1990s, interest in green merchandise and associated concepts has increased. Since the 2010s, consumer confidence in green retailing operations has grown significantly at a rate that makes them viable and sustainable. According to observations, the industry for green marketing has expanded quickly in the selling of EVs.

5. Discussion and Conclusion

Since there aren't many articles that assess how EVs affect societal costs and consumer behavior, it can be inferred from the study of current research that this is a big problem that needs to be solved. The welfare of green innovation and its commercialization were two more categories where there were not enough articles. Although there is a paucity of research on how EVs are seen in terms of their status as a symbol of green innovation, acceptability and perception are routinely researched. Additionally, those publications seldom ever discussed how the EV will affect society or the necessity of green innovation in general. The study of all three components of sustainability is a tendency that is becoming more and more common in studies analyzing the social cost, which also typically examines economic and environmental issues in addition to social ones. For instance, studies should integrate the previously indicated evaluations of consumer behavior and social costs, assess one perspective—consumer citizenship experience—and societal costs to increase the study's breadth. Furthermore, no articles have evaluated the social issues of EVs by combining all the aspects. Future research on this might face difficulties and be seen as having a gap that needs to be filled. Future research should pay more attention to society's degree of preparation, since this is a topic that has not received as much attention and should be taken into account when determining how to increase EV's adoption in the field of green marketing. The level of society's preparedness is related to both its perception of EVs and its level of approbation. It is intriguing, nonetheless, that authors typically evaluated the environmental and financial aspects of sustainability while attempting to determine the societal cost of EVs. It was discovered that 87.5% of the publications evaluated the social repercussions. This is due to the interconnectedness of the three sustainability-related concepts and the fact that studying sustainability in its whole exhibits the study's extra worth. As more businesses seek to include all three facets of mantainability into their operations, this shows a pattern and points to a possible area for future research on the evaluation of the social component of EVs. As further articles concentrating on EV perception have been published throughout the years, particularly starting in 2018, the relevance of EVs' perception is growing. Although there may be less articles on EV, but perceptivity as the technology develops and as consumers get more knowledgeable about the technology and how it operates through social, business, and policy-related channels will increase.

In contrast to 2016, the number of EV studies on societal cost has grown after 2018. This may be as a result of the yearly growth in EV sales that have exceeded 50% since 2017. As a large number of customers increasingly adopt the technology, businesses work to improve their operations while considering sustainability, this significantly benefits the health of the environment. Countries are enacting legislation focused on sustainability, prompting the mentioned businesses to enhance their sustainability, analyze sustainability metrics, and establish connections between electric vehicle (EV) sustainability and the energy required for charging them. Innovations in batteries and charging systems that place a strong emphasis on the lifecycle analysis of batteries are receiving more attention. However, the social factor should also be considered for EV commercialization to be widespread, regardless of scientific improvements. In most studies, a single country was the primary geographic focus. For a better understanding of EV-related societal concerns, studies with a global scope and a number of nations, as in Knez et al.'s study [19], may be helpful. EV producers, distributors, sales, consumers, and end-of-life companies are working hard to achieve sustainability-related objectives. A number of papers discussed climate action, decent employment, and sustainable cities and communities at the moment. This could be due to the fact that EVs provide an emission-free means of transportation that can help combat climate change. Additionally, they may enhance the sustainability of cities. There is rarely any discussion in academic publications about "good health and well-being", "green innovation and infrastructure", or "affordable and clean energy". Although "climate action" and "cheap and clean energy" might be closely connected, especially in the case of electric cars, it is curious that the papers regularly mentioned "climate action" and seldom discussed the latter as green innovations and how they affect society on a level of social cost that will be further researched on the basis of consumer behavior. More articles on "climate action" and "decent job and economic growth" were published, some of which also examined the social and environmental costs of EVs.

6. Further Research

Future investigations into forecasting of the EV effects on society should concentrate on advancing and commercializing the value of green marketing. More indicators should be included in research that takes into account social effect, green innovation, and good consumer citizenship. Generally speaking, there weren't many indicators used in papers that looked at the social effect. The authors do, however, acknowledge that other studies and analyses of other social characteristics that were not included in this study may have been published. As a conclusion, this research offers fresh perspective and illustrates how the social sustainability research emphasis for EVs was established and organized by looking at articles published in WoS. A lack of articles was identified in the study on forecasting social preparedness and user experience. A further point made was that EVs are regularly assessed for their social and environmental impacts, as well as their economic and environmental effects, thereby making papers more thorough and providing a clever answer. Future studies should include numerical indicators and other variables when assessing the social impact of EVs. In order to achieve operations and strategic goals, it might be taken into account when forecasting business and research operations.

References

[1] WCED-World Commission on Environment and Development. *Our Common Future*. Oxford University Press: Oxford, UK, 1987.

[2] Chaturvedi, U., Sharma, M., Dangayach, G.S., and Sarkar, P. Evolution and adoption of sustainable practices in the pharmaceutical industry: An overview with an Indian perspective. *J. Clean. Prod.*, 168: 1358–1369, 2017. [CrossRef]

[3] Garcia, S., Cordeiro, A., Nääs, I.D.A., and Neto, P.L.D.O.C. The sustainability awareness of Brazilian consumers of cotton clothing. *J. Clean. Prod.*, 215: 1490–1502, 2019.

[4] Currie, G., Truong, L.T., and De Gruyter, C. Regulatory structures and their impact on the sustainability performance of public transport in world cities. *Res. Transp. Econ.*, 69: 494–500, 2018.

[5] European Commission. Transport in the European Union: Current Trends and Issues. 2018. Available online: https://ec.europa. eu/transport/sites/transport/files/2018-transport-in-the-eu-current-trends-and-issues.pdf (accessed on 15 April 2020).

[6] Kumar, R.R., and Alok, K. Adoption of electric vehicle: A literature review and prospects for sustainability. *J. Clean. Prod.*, 253: 119911, 2020.

[7] European Commission. Road Transport: A Change of Gear. 2012. Available online: https://ec.europa.eu/transport/sites/ transport/files/modes/road/doc/broch-road-transport_en.pdf (accessed on 5 December 2020).

[8] Raugei, M., and Winfield, P. Prospective LCA of the production and EoL recycling of a novel type of Li-ion battery for electric vehicles. *J. Clean. Prod.*, 213: 926–932, 2019.

[9] Petrauskienė, K., Skvarnaviˇciute, M., and Dvarioniene, J. Comparative environmental life cycle assessment of electric and conventional vehicles in Lithuania. *J. Clean. Prod.*, 119042, 2020.

[10] Burchart-Korol, D., Jursova, S., Folega, P., Korol, J., Pustejovska, P., and Blaut, A. Environmental life cycle assessment of electric vehicles in Poland and the Czech Republic. *J. Clean. Prod.*, 202: 476–487, 2018.

[11] Del Pero, F., Delogu, M., and Pierini, M. Life Cycle Assessment in the automotive sector: A comparative case study of Internal Combustion Engine (ICE) and electric car. *Procedia Struct. Integr.*, 12: 521–537, 2018.

[12] Onat, N.C., Kucukvar, M., Aboushaqrah, N.N.M., and Jabbar, R. How sustainable is electric mobility? A comprehensive sustainability assessment approach for the case of Qatar. *Appl. Energy*, 250: 461–477, 2019.

[13] Jawahir, I.S., Dillon, O.W., Rouch, K.E., Joshi, K.J., Venkatachalam, A., and Jaafar, I.H. Total life-cycle considerations in product design for sustainability: A framework for comprehensive evaluation. *In: Proceedings of the 10th International Research/Expert Conference "Trends in the Development of Machinery and Associated Technology" TMT*, Barcelona, Spain, 11–15 September 2006.

[14] Pardo-Ferreira, M.C., Rubio-Romero, J.C., Galindo-Reyes, F.C., and Arquillos, A.L. Work-related road safety: The impact of the low noise levels produced by electric vehicles according to experienced drivers. *Saf. Sci.*, 121: pp. 580–588, 2020.

[15] The Methodological Sheets for Sub-Categories in Social Life Cycle Assessment (S-LCA) United Nations Environment Programme and SETAC. Available online: https://www.lifecycleinitiative.org/wp-content/uploads/2013/11/S-LCA_methodological_ sheets_11.11.13.pdf (accessed on 2 January 2020).

[16] United Nations (UN). Transforming our World: The 2030 Agenda for Sustainable Development. 2015. Available online: https://sustainabledevelopment.un.org/content/documents/21252030%20Agenda%20for%20Sustainable%20Development% 20web.pdf (accessed on 30 January 2020).

[17] Tranfield, D., Denyer, D., and Smart, P. Towards a methodology for developing evidence-informed management knowledge by means of systematic review. *Br. J. Manag.*, 14: 207–222, 2003.

[18] Mongeon, P., and Paul-Hus, A. The journal coverage of Web of Science and Scopus: A comparative analysis. *Scientometrics*, 106: 213–228, 2016.

[19] Knez, M., and Obrecht, M. Electric Vehicles infrastructure related drivers' needs and preferences; faculty of logistics: Celje, Slovenia. *World Electric Vehicle Journal*, 12(1): 15, 2015. (In Slovenian). doi: 10.3390/wevj12010015.

[20] Zheng, J., Sun, X., Jia, L., and Zhou, Y. Electric passenger vehicles sales and carbon dioxide emission reduction potential in China's leading markets. *J. Clean. Prod.*, 243: 118607, 2020.

CHAPTER 5

The Evolution of Forecasting Techniques

Traditional Versus Machine Learning Methods

Anna Thomas,[1,*] *Amala Johnson*[2] and *Meby Joseph Manoj*[1]

1. Introduction

Forecasting is used effectively and efficiently to support decision-making for the future. Over time, several methods have been created to conduct forecasting. Finding a forecasting technique with the ability to provide the best estimate of the system being modeled has always been a challenge. The selection and comparison criteria for forecasting methodologies can be organized in a variety of ways. Accurate forecasting has a great demand for various fields like weather prediction, economic condition, business forecasting, demand and supply forecasts and many more. When deciding whether to utilize a certain model to predict future events, accuracy is very important. In every field, machine learning (ML) algorithms are being used to forecast future events. These algorithms can handle more complex data and make predictions that are more accurate. Based on the least values of forecasting errors, forecasters create a model to determine the best strategy for prediction. For centuries, forecasting has been used to assist individuals in making future-related decisions. In the past, forecasts were based on intuition and experience, but as technology has advanced, so have forecasting methods. Currently, advanced ML models and methods

[1] Department of Statistics and Data Science, CHRIST (Deemed to be University), Bengaluru, Email: meby.manoj@res.christuniversity.in

[2] Department of Data Science, CHRIST (Deemed to be University), Pune, Lavasa, Email: amala.johnson@christuniversity.in

* Corresponding author: anna.thomas@res.christuniversity.in

for data analysis are used to provide forecasts. To forecast the future, these models incorporate a range of inputs, including historical data, present trends, and economic indicators. Forecasting is a vital tool for businesses to employ when making future plans. It is used in a wide range of industries, from finance to weather prediction.

Making predictions about the future based on evidence from the past and the present is the process of forecasting. It is used in a variety of industries to help businesses plan for the future and make decisions about investments, operations, and other activities. Forecasting can be used to predict future sales, customer demand, economic trends, and other business-related factors. Forecasting gives organizations an idea of what to expect in the future, enabling them to make plans accordingly. It may also help to make better decisions by providing an estimate of the expected future outcomes. Economic, financial, corporate, engineering, and scientific forecasting can be applied in a wide range of areas. Future sales forecasts, stock market trends, weather patterns, and energy consumption predictions can all be made using it. It can also be used to estimate future demand for goods and services. Moreover, forecasting can be utilized to inform decisions regarding the management of risks, capital expenditures, and resource allocation.

Time-series models, Regression analysis, Prophet Modeling, econometric modeling, ML algorithms are some of the forecasting techniques. The field of forecasting techniques has been significantly impacted by artificial intelligence (AI). The ability of AI-based forecasting algorithms to quickly and accurately assess massive amounts of data has led to their increasing popularity. AI-based forecasting methods can spot patterns and trends in data that a human eye might ignore, leading to more precise predictions. Due to their improved data processing capabilities, AI-based forecasting techniques also enable forecasts that are quicker and more precise than those made using more conventional methodologies. AI-based forecasting methods can also be used to spot future risks and opportunities, enabling better-informed choices. Moreover, AI-based forecasting methods can offer more thorough analyses of the data, facilitating a better decision.

The following section contains the remaining content of the chapter. Section 2 provides an overview about the evolution of forecasting techniques. Section 3 deals with the Traditional forecasting techniques. ML techniques are included in Section 4. Accuracy measures of forecasting models are discussed in Section 5. In Section 6, a comparison between traditional and ML forecasting methods is discussed. Section 7 deals with the advantages and disadvantages of both traditional and ML techniques. Section 8 focuses on the case studies that bring out a comparison between traditional and ML algorithms for forecasting. Section 9 brings out the conclusions obtained from the study.

2. Evolution of Forecasting Techniques: An Overview

Forecasting techniques have evolved significantly over the years. Forecasting techniques have improved along with technology since the days when predictions were made only on the basis of intuition and experience. Early forecasting techniques were based on simple extrapolation of past data, natural phenomena. Observation of natural phenomena such as the phases of the moon, the changing of the such as moving averages, linear regression, and exponential smoothing. However, these methods are limited in their ability to capture complex patterns in data. Early forecasting techniques relied on simple extrapolation of past trends, but modern forecasting methods are more complex and use sophisticated statistical models. These models involve the use of time series data, the application of regression techniques, and the implementation of econometric models. Recently, more precise predictions have been made by employing ML algorithms like artificial neural networks (ANNs) and support vector machines (SVMs). As forecasting techniques continue to evolve, they will become increasingly accurate and reliable.

In early times, forecasting was done using a variety of methods such as astrology, divination, and observation of seasons, and the behavior of animals was also used to make predictions. These methods were often used in combination with each other, and were used by people to make decisions about when to plant crops, when to go to war, and other important decisions. From these the forecasting techniques continue to evolve into time-series analysis, moving averages, linear regression, and exponential smoothing to get more accurate predictions. Predictions of future events can be made by analyzing past data points using time series forecasting. It is commonly used in business and economics to predict future trends and outcomes. A type of statistical modeling used to analyze and predict trends in time series data is known as ARIMA (Autoregressive Integrated Moving Average). Its aim is to recognize patterns in the data and anticipate future values [1]. Regression analysis is a method of statistics used to identify relationships between different variables. It can be utilized to predict future values by examining the correlations between different variables [2]. A method for predicting future values based on historical data points is exponential smoothing. It is applied to a time series to reduce volatility and improve prediction accuracy.

Traditional forecasting practices are usually based on prior data, which may not be reliable in predicting future trends. Furthermore, they are not able to factor in external influences such as changes in the economy, technology, or customer preferences. It also requires a lot of manual effort and also it is time-consuming. This may lead to search more accurate forecasting techniques. The evolution of ML forecasting techniques has been a gradual process over the past few decades. Initially, ML algorithms were used to build simple linear models such as linear regression. As the area of ML progressed, newer and more intricate algorithms, such as SVMs and decision trees, were created. These algorithms allowed for more accurate predictions and better generalization of data. Deep learning (DL) has gained popularity as a forecasting method in recent years. DL algorithms can recognize intricate patterns in data and forecast outcomes more accurately. These algorithms can forecast a wide variety of variables and can work with enormous datasets. Also, the advancement of cloud computing has simplified the process of accessing massive datasets and applying ML techniques to them [3]. This has made it possible for researchers to create more complex models and utilize them to make predictions. Finally, the growth of AI has made it possible to create forecasting models that are more complex. A wide range of variables can be considered by AI-based forecasting models, which can be utilized to produce predictions that are more precise [4].

2.1 Uses of Forecasting Techniques

1. **Planning:** Forecasting gives organizations an idea of what to expect in the future, aiding them in making plans for the upcoming period. This helps organizations to plan their resources and activities accordingly. Forecasting techniques are used to plan for the resources needed to achieve business goals. This helps businesses plan for personnel, equipment, and other resources.
2. **Financial Forecasting:** Forecasting techniques are used to predict future financial performance, such as sales, profits, and cash flow. This helps businesses plan for the future and make decisions about investments and other financial activities.
3. **Risk Management:** Forecasting techniques are used to identify and manage potential risks. This helps businesses plan for potential losses and take steps to mitigate them.
4. **To identify trends:** Forecasting techniques are used to identify trends in the market. This helps businesses stay ahead of the competition and capitalize on emerging opportunities
5. **To anticipate changes in the market:** Forecasting techniques are used to anticipate changes in the market. This helps businesses prepare for potential shifts in customer preferences and make strategic decisions.

3. Traditional Forecasting Techniques

Forecasting techniques are methods used to predict future outcomes based on past data and trends. Beginning with the ancient Greeks, forecasting methods have been applied for a very long time. The Greeks used a form of forecasting called 'divination' to predict the future. This was done by observing the behavior of birds, animals, and the stars. In the 1700s, the first modern forecasting techniques were developed. These techniques used mathematical models to predict future events. This was done by analyzing past data and trends. Time series analysis and Regression analysis are one of the major traditional forecasting techniques.

3.1 Time Series Forecasting Techniques

Time-series analysis is a method of examining data points that have been collected over a certain period of time. It is used to detect trends, seasonal fluctuations, and correlations in the data, as well as to make predictions. This technique can be applied in various fields, including economics, finance, marketing, and engineering. It can be used to analyze stock prices, sales data, weather patterns, and more.

1. **Autoregressive Integrated Moving Average Model (ARIMA):** This approach combines autoregressive and moving average models to predict future values. A linear combination of prior values from its own series makes up an ARIMA model. The AR part of ARIMA indicates that the evolving variable of interest is regressed on its own lagged values and the MA part indicates that the regression error is actually a linear combination of error components, the values of which

happened simultaneously and at distinct points in the past. The I (Integrated) indicate that the data values have been replaced with the difference between their values and the initial values in the series. ARIMA's general form (p, d, q) is depicted as:

$$Y_t = c + \phi_1 Y_{t-1} + \ldots\ldots + \phi_p Y_{t-p} + \theta_1 \eta_{t-1} + \ldots\ldots + \theta_q \eta_{t-q} + \eta_t \quad (1)$$

where, d denotes the order of the non-stationary component, p denotes the order of the AR part, and q denotes the order of the MA part, respectively, $\phi_1, \phi_2, \phi_3, \ldots$ are AR parameters and $\theta_1, \theta_2, \theta_3, \ldots$ are MA parameters. ARIMA models can be used to forecast short-term or long-term time-series data.

2. **Generalized Autoregressive Conditional Heteroskedasticity (GARCH):** It is a statistical model used to describe and predict the volatility of financial time series. It is a particular kind of autoregressive model that captures the time-varying nature of volatility. The volatility clustering that is seen in financial time series is captured using GARCH models. The model consists of two parts: an autoregressive component that captures the persistence of volatility and a moving average component that captures the shocks to the system. The GARCH model is used to forecast future volatility and to quantify the risk associated with a given asset. This model can be used to incorporate the effects of macroeconomic variables. GARCH models are commonly used to predict stock market volatility, as they can capture the nonlinear dynamics of the market. Investors can better manage their risk and make more informed decisions by predicting the volatility of a financial asset [5]. A GARCH model is represented in the following form

$$z_t = \sqrt{h_t \eta_t},\ ht = \alpha_0 + \sum_{i=1}^{p} \alpha i\, Z^2_{t-i} + \sum_{j=1}^{q} Bj\, h_{t-j} \quad (2)$$

where, ε_t is iid random variables, $\alpha_0 > 0$, $\alpha_i \geq 0$, $\beta j \geq 0$ and $\sum_{i=1}^{\max(p,q)} (\alpha i + \beta j) < 1$

3. **Holt exponential smoothing:** Holt exponential smoothing is a technique for predicting future values in a time series by taking into account the weighted average of prior data points. It is possible to apply in a time series data having only trend component. The technique is useful for forecasting short-term trends and can be used to forecast multiple time periods ahead. It helps to clarify the fundamental principles of a time-series construction. Holt exponential smoothing can be used in a variety of applications, including forecasting sales, inventory, and demand, predicting customer behavior, and analyzing financial markets. This technique can be used to recognize trends in time-series data and to make short-term forecasts. It can be used in sales forecasting, financial forecasting, weather forecasting, and demand forecasting [6]. This model can be represented as

$$Z_t = m_t + e_t \quad (3)$$

where, the trend component is represented by m_t, while error component is represented by e_t.

4. **Moving Average Model:** A time series forecasting method referred to as moving average forecasting uses the mean of prior data points to predict future values. This technique is based on the assumption that the future values of a time series will be similar to its past values. It is a simple and effective method for forecasting short-term trends. The moving average forecasting technique can be used to forecast a wide range of time series, including sales, demand, and prices. It is also useful for smoothing out fluctuations in the data and identifying underlying trends. Moving average forecasting technique is simple to understand and apply. It can be used for short-term as well as long-term forecasting. Moving average forecasting technique can be used in stock market prediction, weather forecasting, sales forecasting, production planning, financial forecasting, and demand forecasting [7].
5. **Vector Autoregressive Model (VAR):** A statistical model known as a vector autoregressive model (VAR) is used to find out the linear dependencies between multivariate time series data. It is a generalization of the univariate autoregressive model (AR) to multiple time series. Forecasting, impulse response analysis, and structural analysis can all be performed using VAR models, which are used to determine the dynamic relationship between several variables. VAR can be used to analyze the effects of shocks on multiple variables simultaneously. It is also useful in forecasting the future values of multiple variables based on their past values. VAR can also be used to identify the causal relationships between various variables. This is especially useful in economic analysis, where it is important to understand how different economic variables are related to each other. VAR is also useful in financial analysis, where it can be used to identify the relationships between different financial variables. This can be used to determine the risk variables connected to various assets and to create risk management plans. VAR models can be used to identify causal relationships between variables. VAR models can be used to analyze the impact of one or more exogenous variables on a set of endogenous variables. It can also be used to identify the underlying structure of a system of variables [8]. A VAR (p) model is represented as

$$X_t = \phi_0 + \phi_1 X_{t-1} + \phi_2 X_{t-2} + \cdots\cdots + \phi_p X_{t-p} + a_t. \tag{4}$$

where,
$L^j X_t = X_{t-j}$.
X_t : $p \times 1$ random vector.
ϕ_0 : $p \times 1$ vector of constant terms.
ϕ_i : $p \times p$ autoregressive coefficient matrix.
a_t : $p \times 1$ white noise process
X_t = Multivariate time series

3.2 Regression Analysis

This technique uses data from the past to identify relationships between variables. It can be used to predict future values based on the relationships identified. Regression analysis can be used to predict sales, customer demand, stock prices, and other trends.

It can also be used to identify relationships between different variables and to identify potential causes of a particular outcome. Regression analysis is a powerful forecasting technique that can be used to make informed decisions and to better understand the relationships between different variables [9]. Linear regression, Ridge regression, Logistic regression, Polynomial regression, and Lasso regression are some of the different regression analysis forecasting methodologies. The most fundamental type of regression analysis, known as linear regression, forecasts a continuous outcome variable using one or more predictor variables. Another sort of regression technique used to reduce the effect of multicollinearity in a regression model is ridge regression. Logistic regression is used to predict a binary outcome based on one or more predictor variables. Using a polynomial equation containing one or more predictor variables, polynomial regression is used to forecast a continuous outcome variable. A regression model's predictor variable count can be decreased by using lasso regression.

4. Machine Learning Forecasting Techniques

The defining quality that sets humans apart from other creatures on earth is our capacity for learning from our experiences and studying them. Then they created machines to assist in the works which will work under the instructions of the creator. Later on, as technology moves forward, they started to enable the machine to automatically learn from the past data and experience to improve the performance and predict, just like the human does. Arthur Samuel who introduced the term "Machine Learning" in 1959, while at IBM defined it as "the systematic study that permits computers to acquire knowledge without even being specifically instructed".

Computers can optimize performance criteria via ML by using sample data or existing knowledge. Our model has been constructed to a certain extent, and learning involves the use of a computer programme to alter the model's parameters using training data or prior information. Both descriptive and predictive models can be used to draw conclusions from the data and forecast the future. There are many different types of machines learning algorithms, including unsupervised learning, supervised learning, semi-supervised learning, and reinforcement learning. Many kinds of issues and applications are addressed by various sorts of algorithms. In the past few years, ML has gained popularity, and has also found several uses in fields like image identification, natural language processing, fraud detection, recommendation systems, and many more [10].

In the case of analyzing and forecasting using time series data, ML techniques can be very helpful. The term "time series" refers to a group of data points that have been accumulated over time, such as stock prices, weather reports, or website traffic. Time series data usually show trends, seasonality, and unusual patterns, making it difficult to model using conventional statistical methods. Time series data can be analyzed using ML algorithms to provide predictions based on previous relationships and patterns.

An ML system forecasts the output of the data, whenever it receives a new data, by learning from prior data and building the prediction models using it. The following is a brief explanation of the ML techniques utilized in the study:

1. **Artificial neural networks (ANNs):** ANNs, a type of ML algorithm, are designed to mimic the structure and function of the human brain. ANNs are made up of interconnected processing nodes arranged in layers, each layer carrying out a particular duty during the data analysis process. Prior to being output through the output layer, data is first received at the input layer, where it is processed and changed. To maximize its efficiency on a certain activity, like classification, regression, or prediction, the network learns to modify the strength of connections between the nodes during training.

   A formula, such as the sigmoid function or the rectified linear unit (ReLU) function, determines the output of a neuron. Weights are used to represent the connections between neurons and control how strong the signal is conveyed between them. In order to reduce the cost function, which determines the difference in inaccuracy between the output produced and the output projected, the weights are changed throughout training. An extra set of learning rules use a method called backpropagation that enables the ANN to correct its output results by accounting for errors [11]. Each time the output is identified as an error during the supervised training phase, the data is propagated backward. Each weight is changed based on how much it contributed to the error.

   Due to the accessibility of big datasets and powerful computer resources in recent years, ANNs have gained popularity as a potent tool for tackling challenging ML challenges. However, creating and training ANNs can be difficult and calls for knowledge of statistics, computer science, and mathematics. ANNs have been successfully applied in a number of areas, including image identification, audio recognition, natural language processing, and finance.

2. **Long Short-Term Memory Networks (LSTM):** These kinds of networks are the most commonly used recurrent neural network, where it has the ability to pick up long-term dependencies. It is developed to address the limitations of traditional recurrent neural networks in processing sequential data with long-term dependencies. The "vanishing gradient" problem, which happens when the gradients in the network are very small as they are propagated back through time, is the major problem with conventional recurrent neural networks and they consequently find it difficult to identify long-term dependencies in sequences.

   LSTM networks address this issue by introducing a unique kind of memory cell that can selectively read, write, and forget information for extended periods of time. The numerous gates in this memory cell that control the flow of information include an input gate that controls the flow of fresh data, a forget gate that determines which information to discard, and an output gate that determines which information to output.

   LSTM networks are excellent for categorizing, processing, and making predictions based on time series data since there may be lags of varying lengths between critical events in a time series [12]. LSTM networks are suitable for applications like speech recognition, machine translation, and natural language processing because they can learn and recall long-term dependencies in sequential data by selectively managing the flow of information.

3. **K-Nearest Neighbor regression (KNN):** K-Nearest Neighbor (KNN) regression is a classification and regression tasks can both be accomplished using the same supervised ML method. A continuous output variable (i.e., a numerical value) is to be predicted using KNN regression using a collection of input features. The method finds the K training data points that are the new data point's closest neighbors in order to predict it. A hyperparameter that can be changed to improve performance is often the value of K. The output values of the K nearest neighbors are then averaged or weighted to provide the anticipated value for the new data point.

   It can be used with time-series data with the intention of predicting future values of a time series from historical data. When using KNN to analyze time-series data, there are a few crucial factors to take into account and adjustments that must be performed. The selection of a distance metric is an important factor. The Euclidean distance may not be the best distance metric in KNN regression for time series since it considers all dimensions equally and ignores the temporal nature of the data. The selection of K is another factor. The selection of K, which specifies the number of comparable time series that will be utilized to construct the forecast in time-series data, can have a substantial impact on the prediction's accuracy. When using KNN to analyze time-series data, the window size and step size must also be taken into account.

   Time series data can be used to predict future values using the KNN algorithm, however it is important to carefully examine the distance metric, K selection, window, and step sizes. Also, time-series data with intricate nonlinear correlations or seasonal patterns may not be the best candidates for KNN [13]. KNN regression is often a good approach for predicting continuous values based on input features, especially when there is ambiguity or a complex nonlinear relationship between the input features and the output variable.

4. **Generalized Regression Neural Networks (GRNNs):** A kind of neural network known as Generalized Regression Neural Networks is the process of foretelling a continuous output value from one or more input variables. It is a form of ANN that is frequently used for regression problems. GRNN is a type of feedforward neural network, which propagates input data forward through a number of hidden layers before arriving at an output layer that generates the regression forecast. As opposed to other kinds of neural networks, GRNN uses a single-pass method that can quickly and effectively learn from new data, so it does not need an explicit training step.

   The main idea behind GRNN is that it uses a radial basis function (RBF) to calculate the distance between an input vector and a set of training instances. The output values of training samples are then averaged using this distance, with the weights being set by a Gaussian function that is centerd on the input vector. In comparison to other regression models, GRNN has a number of benefits, including a strong ability to generalize to new data, a rapid learning curve from small datasets, and the capacity to handle nonlinear correlations between input variables and output values [14]. To obtain optimal performance, it might need some parameter adjusting and be sensitive to the RBF kernel selection.

5. **Convolutional Neural Network (CNN):** The CNN is a kind of neural network that is often used in computer vision applications for tasks including image categorization, object recognition, and segmentation. A CNN operates on the input image by applying a number of convolutions and pooling processes. The pooling layers demonstrate the feature maps to minimize their size and improve their resistance to tiny fluctuations, while using filters to extract information from the input image for the convolutional layers.

   CNNs can also be utilized for time-series data, including sensor data or financial market data, even though they are most often linked with image data. The main distinction is that the input data is now processed as a one-dimensional series of data points rather than as a two-dimensional image. Nevertheless, when paired with other DL architectures like RNNs or LSTMs, CNNs can be a potent tool for studying time-series data [15]. However, to guarantee that the network is able to successfully learn useful characteristics from the input data, rigorous preprocessing and architectural design are required.

5. Accuracy Measures

Accuracy metrics are measurements that are used to assess a model's performance. They are used to assess how well a model can forecast the result of a specific data collection. These metrics are employed to evaluate many models and choose the most appropriate one for a certain issue. Common accuracy measures include accuracy, precision, recall, F1 score, ROC AUC, Mean Absolute Percentage Error (MAPE), Root Mean Square Error (RMSE), Root Mean Square Logarithmic Error (RMSLE), and Mean Absolute Square Error (MASE).

1. **Root Mean Square Logarithmic Error (RMSLE):** It is the accuracy measure that compares the logarithms of the expected and actual values to determine how accurately predictions are made. It is employed to evaluate how well a predictive model performs when applied to regression problems. The RMSLE is calculated as the square root of the mean of the squared logarithmic errors.
2. **Mean Absolute Percentage Error (MAPE):** It is the average of the absolute percentage mistakes), which is a statistical measure of forecasting method accuracy. It is determined by averaging the absolute percentage errors between the forecasted and actual numbers. It is common practise to assess the accuracy of various forecasting techniques using the MAPE, a measure of a forecasting method's accuracy.
3. **Mean Absolute Square Error (MASE):** It is a measure of the accuracy, and is of a forecast model. It is calculated by dividing the forecast model's mean absolute error by the naive forecast model's mean absolute error. The result is then multiplied by 100 to give a percentage. MASE is a useful metric for assessing the accuracy of a forecast model, as it allows for comparison between different models and can be used to identify the best model for a given forecasting task.
4. **Root Mean Square Error (RMSE):** The difference between expected and actual values is measured by the RMSE. By taking the square root of the mean of the squared discrepancies between the anticipated values and the actual values,

it is determined. The accuracy of the model's predictions of real values increases with decreasing RMSE.

5. **Precision:** Precision is a measure of how accurately a measurement or calculation is performed. It is the degree of exactness or accuracy with which a measurement or calculation is made. It is usually expressed as a percentage or fraction. It is calculated by dividing the total number of forecasts made by the number of predictions that were correct. By contrasting the results that actually occurred with those that were projected, precision can be utilized to assess how accurate a forecast was. This can be used to identify areas where the forecast could be improved and to identify any potential biases in the forecasting process. By using precision to measure the accuracy of forecasts, businesses can make more informed decisions and improve their forecasting accuracy.
6. **Recall:** Recall is a metric for how well a model can identify each pertinent instance of a given class. Other names for it include sensitivity and true positive rate. It is determined as the sum of the true positives and false negatives divided by the number of true positives.
7. **ROC Curve:** A binary classifier system's performance is graphically depicted using a Receiver Operating Characteristic (ROC) curve as the discrimination threshold is changed. Plotting the true positive rate (TPR) vs the false positive rate (FPR) at various threshold values results in the curve. The proportion of actual positives that are accurately detected is known as the true positive rate, whereas the percentage of negatives that are mistakenly classified as positives is known as the false positive rate. The ROC curve is a helpful tool for a number of tasks, including classifier evaluation and model selection.
8. **F1 Score:** The harmonic mean of the precision and recall scores is used to calculate the F1 score, which serves as a gauge of a model's accuracy. Due to the fact that it strikes a compromise between precision and recall, it is frequently employed in the evaluation of classification models. A higher score on the F1 scale, which spans from 0 to 1, indicates superior performance.

6. Differentiation Between Traditional and Machine Learning Forecasting Methods

Traditional forecasting methods make forecasts based on human judgment and experience. These methods include qualitative approaches, trend analysis, and time-series analysis. ML forecasting techniques are those that use algorithms to make predictions. These methods include neural networks, random forests, and linear regression. Both conventional forecasting techniques and ML techniques have advantages and disadvantages. Traditional methods are often more accurate and reliable, but they can be time-consuming and require a lot of manual work. ML techniques may be quicker and more productive, but they may also be less accurate and require more data to work properly. Traditional forecasting techniques are often used in situations where the data is limited and the goal is to make a reliable prediction. ML techniques are often used in situations where the data is abundant and the goal is to make a more accurate prediction [16]. These studies can help businesses choose the best forecasting method for their needs. Contrarily, ML algorithms frequently employ more complex characteristics and predictive techniques to increase forecast accuracy while reducing a loss function.

As technology advances, corporate applications of AI and ML will continue to change the sector, for instance, due to its application across functions, forecasting using AI and ML is of great interest to most businesses. Companies have historically utilized statistical forecasting techniques like linear regression and Holt's exponential smoothing to help them make decisions. But in many data and analytics programs across organizations and industries, ML-based forecasting has supplanted conventional methodologies. Traditional forecasting techniques are descriptive in nature and focus on the analysis of univariate or multivariate datasets using finite, countable, and understandable factors. The level of disclosure into how these conventional models work is one of their key characteristics. The outcomes that these models produce are simple to follow. Similar to traditional methods, ML forecasting techniques aim to improve forecast accuracy while minimizing a loss function. The strategies used by ML forecasting algorithms usually involve more intricate characteristics and predictive methodologies [17]. The loss function is commonly assumed to be the sum of squares because forecasting and prediction errors are common.

The two forecasting strategies have different minimization techniques. The majority of ML techniques use nonlinear methods to minimize the loss functions, in contrast to the majority of classical procedures, which mostly use explicable linear processes. More computational power is needed for ML approaches than for statistical ones. Traditional forecasting methods frequently require manual data entry and processing, which takes time and labor intensively. Traditional forecasting techniques are often biased towards short-term forecasting and may not be able to accurately predict long-term trends. They frequently cost a lot of money and take a lot of resources to implement. ML forecasting techniques are difficult to understand how the machine learning algorithm is making its predictions, and thereby making it difficult to explain the results. It can take a while to train and fine-tune ML models because they need a lot of trial and error. Compared to conventional forecasting methodologies, ML techniques are more accurate. This is due to the fact that they are more trustworthy because they generate predictions using a lot of data. Finally, there is a place for both conventional and ML forecasting methods in predicting. One might be more appropriate than the other depending on the circumstance.

7. Advantages and Disadvantages of Traditional Forecasting and Machine Learning Forecasting Techniques

7.1 Advantages of Traditional Forecasting Techniques

1. **Easy to understand**: Traditional forecasting techniques are based on historical data and are relatively simple to understand and implement. They are also relatively easy to use, as they require minimal data and do not require complex mathematical calculations.

2. **Cost-effective**: Traditional forecasting techniques are cost-effective because they require minimal data and are relatively easy to use. Examples of traditional forecasting techniques include time-series analysis, linear regression, exponential smoothing, and moving averages, all these are cost-effective and require minimal data.
3. Short-term forecasting: Traditional forecasting techniques are useful for short-term forecasting because they provide a reliable and accurate prediction of future outcomes.

7.2 Disadvantages of Traditional Forecasting Techniques

1. **Time consuming**: Traditional forecasting techniques are often time-consuming and labor-intensive. The data collection, building of models and analysis, and result interpretation are all manual activities that are heavily utilized in traditional forecasting methodologies. As it takes a lot of work to gatherand analyze data, construct models, and interpret outcomes, this can be labour- and time-intensive. Furthermore, the ability of conventional forecasting methods to adapt for environmental changes or incorporate new data sources is sometimes constrained.
2. **They are often inaccurate**, as they do not take into account external factors that could affect the forecast: Traditional forecasting techniques rely on historical data to make predictions, and do not account for changes in the external environment that could affect the forecast. This could include changes in economic conditions, technological advances, or changes in consumer preferences. Moreover, conventional forecasting methods may fail to predict with precision the influence of novel products or services, or the effects of new entrants in the market.
3. **More expensive**: They can be expensive, as they require significant resources to develop and maintain. They also require more data to be collected and analyzed, and often require specialized personnel to interpret the data and develop the forecasts.

7.3 Advantages of Machine Learning Forecasting Techniques

1. **Improved Accuracy**: Compared to traditional forecasting methodologies, ML techniques are more accurate. This is due to the fact that they are more reliable because they generate predictions using a lot of data.
2. **Automation**: ML forecasting methods can be employed without the assistance of a human manual power because they are automated. They are hence significantly quicker and more effective than conventional forecasting techniques.
3. **Reduced Cost**: Because MLforecasting approaches employ less manual work and resources than conventional methods, they are more cost-effective. This makes them more suitable for businesses with limited budgets.

7.4 Disadvantages of Machine Learning Forecasting Techniques

1. **Overfitting**: Overfitting occurs when an ML model is trained improperly, resulting in good performance on the training data but poor performance on new data.
2. **Time-consuming**: Training ML algorithms can be computationally demanding and time-intensive.
3. **Data Quality**: Poor quality data can lead to inaccurate results and poor predictions.

8. Empirical Analysis

In this segment, an empirical analysis was conducted to compare traditional and ML forecasting techniques. The study's main objective was to forecast coffee prices by combining ML methods like KNN and GRNN models with classic time series methods like ARIMA, GARCH models, and Holt's exponential smoothing. The accuracy of these models was assessed using metrics such as RMSLE, MAPE, and MASE, based on in-sample forecasts [18]. The dataset used for the study consisted of 239 monthly observations of coffee prices from November 2002 to September 2022, obtained from the International Coffee Organization, and the data was analysed in R. Figure 1 shows the time plot of coffee prices, revealing that the variable is non-stationary.

In this study, the Augmented Dickey Fuller (ADF) test was utilized to determine the non-stationarity and degree of non-stationarity. The obtained P value of 0.4511 confirms the non-stationarity of the variable. Subsequently, a first differencing approach was used to make the variable stationary, and the ADF test was re-performed, yielding a value of 0.01, confirming the stationarity. The first section of the study focused on traditional forecasting techniques, while the second section explored ML forecasting techniques.

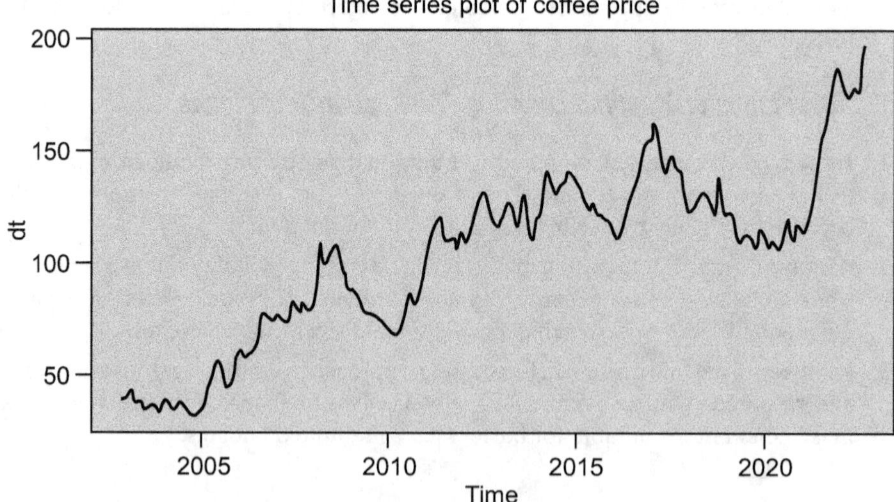

Fig. 1: Fluctuations in Coffee Prices Over Time.

8.1 Traditional Forecasting Techniques

1. **Autoregressive Moving Average Model (ARIMA):** The ARIMA model (0,1,1) was developed to forecast coffee prices, and residual analysis was performed after constructing the model. Normality assumptions were assessed using the Kolmogorov-Smirnov test, revealing that the residuals were normally distributed, with a p-value of 0.94. All assumptions of the residual analysis were satisfied. In-sample forecasting was conducted, and accuracy measures were calculated based on the forecasted values. The ARIMA model generated in-sample forecasted values of 179.8724, 180.4617, and 181.0509 for coffee prices.
2. **Holt's Exponential Smoothing:** The price of coffee is primarily influenced by trends, and to predict its future behavior, the Holts Exponential smoothing method is used. To determine whether the residuals are normally distributed, the Kolmogorov-Smirnov test was conducted, resulting in a p value of 0.91, which indicates normal distribution. The model adequacy was checked, and all assumptions of residual analysis were met. In addition, in-sample forecasting was performed, and the accuracy measures based on in-sample forecasting were calculated. The in sample forecasted values of coffee price using Holt's Exponential Smoothing were determined to be 179.3131, 179.8263, and 180.3394
3. **Generalized Autoregressive Conditional Heteroskedasticity (GARCH) model:** The presence of GARCH effect can be detected by conducting the ARCH LM test. The null hypothesis assumes that there is no heteroscedasticity in the data, while the alternative hypothesis suggests the presence of heteroscedasticity. Since there is an ARCH impact for coffee price and the p-value is less than 0.05, the null hypothesis is rejected. The normality of the residuals was examined using the Kolmogorov-Smirnov test, and the p-value of 0.91 indicates that the residuals are normally distributed. The model adequacy was verified, and all assumptions of residual analysis were met. In addition, in-sample forecasting was conducted, and the accuracy measures based on in-sample forecasting were calculated. The GARCH model was used to forecast the in-sample values of coffee price, and the forecasted values were 179.4, 179.4, and 179.4. Table 1 lists the accuracy measure values for the Holt's Exponential Smoothing, GARCH, and ARIMA models.

From the Table 1, it is observed that the Holt exponential smoothing method has less accuracy compared to ARIMA and GARCH models.

Table 1: Validation Metrics of Holt's Exponential Smoothing, ARIMA, and GARCH models.

Validation Metrics	Autoregressive Integrated Moving Average model	Holt's Exponential Smoothing	Generalized Autoregressive Conditional Heteroskedasticity Model
RMSLE	0.06330309	0.06595173	0.0691726
MAPE	0.05964217	0.05715444	0.05964337
MASE	18.28867	1.02929	1.074437

8.2 Machine Learning Forecasting Techniques

For ML forecasting techniques, we have taken KNN, NNs.

1. **Neural Network (NN):** The NN is utilized as a component in numerous ML algorithms to process complex inputs and make them understandable to computers. The NN model has been constructed and its suitability has been verified through testing. All the residual assumptions have been satisfied and the in-sample forecasting have been done. The in sample forecasted values of coffee price using the NN model are 177.7161, 176.7084, and 175.7699.
2. **K-nearest neighbors (KNN):** The KNN algorithm is a supervized ML algorithm that can be applied to solve classification and regression predictive problems. The KNN model has been constructed, and its suitability has been verified through testing. All residual assumptions have been met, and in-sample forecasting has been conducted. The in sample forecasted values of coffee price using KNN model are 178.6966, 178.5931, and 178.4897
3. **Long Short-Term Memory (LSTM):** It is a recurrent neural network (RNN) architecture that is specifically designed for time series data forecasting. The LSTM model has been constructed, and its suitability has been verified through testing. All residual assumptions have been satisfied, and in-sample forecasting has been performed. The in-sample forecasted values of coffee price using the NN model are 177.161, 176.984, and 177.679.

Table 2: Validation Metrics of KNN, NN, and LSTM models.

Accuracy Measures	KNN	NN	LSTM
RMSLE	0.05240315	0.06595173	0.07270322
MAPE	0.05124303	0.05715444	0.06124503
MASE	1.004936	1.02929	1.105946

By comparing the accuracy measures of both traditional and ML forecasting techniques, we got KNN as the best model having less accuracy measures. Using the KNN model we have forecasted the future values of coffee price for the next 5 months. The forecasted future values of coffee prices from October 2022 to February 2023 are 197.3066, 197.2031, 197.0997, 196.9963, 196.8929. In conclusion, by comparing both ML forecasting techniques and the traditional ones for the study of coffee price we obtained the former as more accurate models. Similar to this there are many case studies existing in the literature proving that ML forecasting techniques give more accurate results compared to the traditional forecasting techniques.

9. Conclusion

The development of forecasting techniques has been a complicated and lengthy process. Although traditional methods such as time series analysis and regression analysis have been employed for many years, they have been largely replaced by more sophisticated ML approaches. ML methods have been demonstrated to be more precise and dependable in forecasting future outcomes. Nonetheless, traditional methods still have their own niche in forecasting and should not be entirely discarded. Ultimately, the selection of a forecasting technique hinges on the specific requirements of the organization and the available data. Traditional methods rely on manual analysis and are restricted in their ability to handle large quantities of data. In contrast, ML methods are capable of rapidly and accurately processing vast amounts of data, resulting in more precise and dependable forecasts. The utilization of ML methods has allowed businesses to make more knowledgeable decisions and has led to more accurate forecasts of future trends. As ML technology evolves, it is probable that it will become the preferred approach for forecasting in many businesses. Traditional versus ML forecasting techniques have been examined in the "Evolution of Forecasting Techniques: Traditional Versus Machine Learning Methods" study, and it was shown that ML techniques have the potential to considerably improve forecasting accuracy. ML methods can detect intricate patterns in data that traditional methods are incapable of detecting, which can result in more precise forecasts. However, ML methods may require a substantial amount of data for training and can be computationally demanding. Consequently, it is essential to consider the trade-offs between traditional and ML methods while selecting a forecasting technique. ML methods are also more adaptable and can be modified to different types of data and forecasting issues. As machine learning techniques continue to develop, they will become increasingly important for forecasting in the future.

References

[1] Kumar, M., and Anand, M. An application of time series ARIMA forecasting model for predicting sugarcane production in India. *Studies in Business and Economics*, 9(1): 81–94, 2014.

[2] Fumo, N., and Biswas, M.R. Regression analysis for prediction of residential energy consumption. *Renewable and Sustainable Energy Reviews*, 47: 332–343, 2015.

[3] Arnab Mitra, Arnav Jain, Avinash Kishore, and Pravin Kumar. A comparative study of demand forecasting models for a multi-channel retail company: A novel hybrid machine learning approach. *Operations Research Forum*, 3: 58, 2022.

[4] Lago, J., De Ridder, F., and De Schutter, B. Forecasting spot electricity prices: Deep learning approaches and empirical comparison of traditional algorithms. *Applied Energy*, 221: 386–405, 2018.

[5] Horv, L., and Kokoszka, P. GARCH processes: structure and estimation. *Bernoulli*, 9(2): 201–227, 2003.

[6] Kalekar, P.S. Time series forecasting using holt-winters exponential smoothing. *Kanwal Rekhi School of Information Technology*, 4329008(13): 1–13, 2004.

[7] Durbin, J. Efficient estimation of parameters in moving-average models. *Biometrika*, 46(3/4): 306–316, 1959.

[8] Nakajima, J. Time-varying parameter VAR model with stochastic volatility: An overview of methodology and empirical applications, 2011.

[9] Draper, N.R., and Smith, H. *Applied Regression Analysis* (Vol. 326). John Wiley & Sons, 1998.

[10] Aggarwal, C.C. Neural networks and deep learning. *Springer*, 10(978): 3, 2018.

[11] Agatonovic-Kustrin, S., and Beresford, R. Basic concepts of artificial neural network (ANN) modeling and its application in pharmaceutical research. *Journal of Pharmaceutical and Biomedical Analysis*, 22(5): 717–727, 2000.

[12] Yu, Y., Si, X., Hu, C., and Zhang, J. A review of recurrent neural networks: LSTM cells and network architectures. *Neural Computation*, 31(7): 1235–1270, 2019.

[13] Guo, G., Wang, H., Bell, D., Bi, Y., and Greer, K. KNN model-based approach in classification. In On The Move to Meaningful Internet Systems 2003: CoopIS, DOA, and ODBASE: OTM

Confederated International Conferences, CoopIS, DOA, and ODBASE 2003, Catania, Sicily, Italy, November 3–7, 2003. *Proceedings*, pp. 986–996, 2003. Springer Berlin Heidelberg.

[14] Bendu, H., Deepak, B.B.V.L., and Murugan, S. Application of GRNN for the prediction of performance and exhaust emissions in HCCI engine using ethanol. *Energy Conversion and Management*, 122: 165–173, 2016.

[15] Chua, L.O., and Roska, T. The CNN paradigm. *IEEE Transactions on Circuits and Systems I: Fundamental Theory and Applications*, 40(3): 147–156, 1993.

[16] Calkoen, F., Luijendijk, A., Rivero, C.R., Kras, E., and Baart, F. Traditional vs. machine-learning methods for forecasting sandy shoreline evolution using historic satellite-derived shorelines. *Remote Sensing*, 13(5): 934, 2021.

[17] Oliveira, T.P., Barbar, J.S., and Soares, A.S. Computer network traffic prediction: A comparison between traditional and deep learning neural networks. *International Journal of Big Data Intelligence*, 3(1): 28–37, 2016.

[18] Thomas, A., and John, N. Analysis and forecasting of crude oil price based on univariate and multivariate time series approaches. *In Data Science and Security: Proceedings of IDSCS 2022*, pp. 137–151, 2022. Singapore: Springer Nature Singapore.

CHAPTER 6

Workforce Forecasting after COVID-19 Pandemics Using Artificial Intelligence

Anjali Mathur

1. Introduction

After the pandemic and lockdowns, most of the jobs turned into "work from home" mode. That creates two classes of employees: (1) one employee with multiple employers, and (2) job with entrepreneurship. It can be considered "Brain Drain" for the organizations or the job providers. Predictions of such employees or the forecasting the behavior of workforce become critical. In the new scenario, it does require to analyze the behavior of employees and do predictions on the basis of their behavior. Due to the uncertain behavior, the workforce can be classified in three classes: (1) want to leave the job, (2) want a salary hike, and (3) seek the changes in assigned work (might be transfer). Employees are very casual to express their feelings either in HR survey or with face-to-face communication or by changing the behavior. In this research work, the analysis of behavior and sentiments are carried out. The research is based on those parameters that are new in fashion, like, e-mail communications, virtual meetings, multiple meetings, virtual official groups (Whatsapp, telegram, etc.), as these are the new methods of communication at the job place. During lockdown, most of the time had been spend on meetings or on virtual platforms. Now, after lockdown, still it is in a culture to arrange online meetings and multiple meetings in a day. The proposed research work has using the records of these virtual meets and other virtual platforms. For this research, the behavior analysis of 249 employees has been observed. Total number of meetings in a day

Associate Professor, VIT Bhopal University, Bhopal, Email: anjalimathur@vitbhopal.ac.in and Total number of meetings in a month (as compared to pre-pandemic era), also changed the behavior of employees.

2. Literature Review

The traditional technique to forecast the workforce is asking the questions in an employee engagement survey. A survey should touch on important components of engagement like employee satisfaction, alignment, and tons more. What does employee satisfaction mean in one's organization, and therefore the way one is ready to measure that? Employee satisfaction survey is generally based on the following points:

a. Employee Engagement Index Survey
b. Employee Communication Survey
c. Involvement Survey Questions
d. Management and Leadership Survey
e. Work Environment Survey
f. Workplace Wellness Survey
g. Employee Recognition Survey
h. Employee Benefits and Compensation Survey
i. Personal and Professional Growth
j. Work-Life Balance Survey
k. Employee Satisfaction Survey
l. Employee Experience Survey
m. Employee Motivation Survey

It's well known what a really satisfied team member looks like, and therefore the way they act: over delivering, eager to come to work, always seeking out their individual role. But dissatisfaction is usually harder to detect or measure. Artificial Intelligence provides methods to do sentiment analysis, prediction and classifications, and decision making. Lots of work is available using these methods for workforce forecasting. Research work is available on Qualitative models [2], Time series models [2], Optimization models [2], Generic mathematical models [2], Simulation models [2]. Machine Learning- based research work is also available for Employee Retention Strategies [1], data mining strategy for predicting attrition that uses Support Vector Mechanism [1], Decision Tree [1], K-NN [1] Logistic Regression[1].

3. Artificial Intelligence and Machine Learning

Machine learning is a branch of artificial intelligence that enables computer systems to automatically improve their performance on a given job without being explicitly programmed. Data is utilized in machine learning to train models, which are subsequently applied to decisions or predictions. The procedure entails choosing and preparing the data, selecting an appropriate method, training the model on the data, and assessing the model's performance. Supervised learning, unsupervised learning, and reinforcement learning are the three primary categories of machine learning. Training a model on labeled data involves supervised learning, where the right result is given for each input. When a model is trained on unlabeled data, it must independently find patterns and relationships. This process is known as unsupervised learning. By trial and error, reinforcement learning trains a model and provides feedback in the form of rewards or penalties based on the model's actions.

3.1 Classification

Using a collection of features or qualities, classification is a sort of supervised learning in machine learning that entails predicting the category or class of an input data item. Building a model that accurately categorizes new, previously unobserved data points into one of several predefined classes or categories is the aim of classification. The training data set for classification is made up of labeled examples, where each example is assigned a class label or category. On the basis of this labeled data, the model is trained to understand how the input attributes relate to the class labels. Once trained, the model can be used to anticipate the class label of upcoming, unobserved data points. Numerous applications, including spam detection, picture recognition, sentiment analysis, medical diagnosis, and credit risk assessment, all use classification. A classification model's performance is assessed using metrics including accuracy, precision, recall, and F1 score.

3.2 Naïve Bayes

A probability based machine learning approach called Naive Bayes is utilized for classification problems. The term 'naïve' refers to the assumption made by Naive Bayes that features are independent of one another. Prior probabilities for each class category are first determined using the frequency of the class in the training data set, which is how the method operates. Using the frequency of the feature value in the training data set for that class, it then determines the likelihood of each feature value for each class. To determine the posterior probability of each class given a brand new, unobserved data point, it finally combines the prior probability and the likelihood. The output is then expected to be the class with the highest posterior probability. Naive Bayes is capable of handling a high number of characteristics and classes while being computationally economical.

A Naïve Bayes calculation's fundamental formula is as follows:

$$P\left(\frac{A}{B}\right) = \frac{P\left(\frac{B}{A}\right) * P(A)}{P(B)} \tag{1}$$

where P(A|B) is the likelihood that event A will occur given that event B occurs, P(B|A) is the likelihood that event B will occur given that event A occurs, P(A) is the likelihood that event A will occur, and P(B) is the likelihood that event B will occur. This equation may be used to determine the likelihood that an event will occur given the probabilities of other occurrences.

3.3 Decision Tree

A machine learning approach called a decision tree is utilized for both classification and regression problems. It is a straightforward but effective algorithm that creates a tree-like model of decisions and potential outcomes. Recursively breaking the data into smaller subsets depending on the values of the input features is how the method operates. This process continues until the subsets are pure or substantially pure with respect to the target variable. The decision tree selects the characteristic at each stage of the algorithm that divides the data into subsets that are the most homogeneous with regard to the target variable. The method measures the quality of the split using a variety of metrics, including information gain, gain ratio, or Gini index. It can handle categorical and continuous data and is simple to grasp and analyze. Outliers and missing data can also be handled by decision trees, and they can also detect nonlinear correlations between the input and output variables.

3.3.1 Formulae for Decision Trees

a. Gini Index: Given a label dataset, the Gini Index measures the overall variation across classes. It is determined by deducting one from the total of the squared probabilities for each class.

$$GINI = 1 - \sum_{i=l}^{c} (P_i)^2$$

where, C = total number of observations,

i = the index of an observation,

P_i = the proportion of observations at the ith index.

b. Entropy: It is a metric for a random variable's degree of uncertainty. It is determined by adding the negative of the product of each class's probability and the class's probability's log.

Entropy, E, is defined as:

$$E = \sum_{i=1}^{C} - P_i log_2 P_i$$

where, $P(x_i)$ is the probability of class i and x_i is a label for the ith class.

For the classification techniques, decision trees and Naive Bayes both have advantages and disadvantages, and the best algorithm depends on the characteristics of the data and the specifications of the classification issue. While decision trees can handle more complicated data structures, they may need more computational resources and are more prone to overfitting. Naive Bayes is typically faster and better suited for small datasets with few characteristics.

4. Methodology

The classification method is used in this work. Employees can be classified in 4 classes: Satisfied, "Will Resign", "Need a salary hike", and "Need a transfer".

Classification is a method to divide the data in a predefined set of boundaries, known as a class. Most common techniques for classification are K-Nearest Neighbors, Naive Bayes, Decision Tree, Support Vector Machines, Regression.

In this research work the data is collected for 249 employees for various behavior parameters. And the employees are classified in four classes. The input features and the target classes are shown in Tables 1 and 2, respectively.

Table 1: Input Features.

Input Features
Attendance on virtual platform
Virtual Mode (Active/Passive)
Behavior while Verbal communication (Aggressive/Arguing/peaceful)
Behavior while written communication (Aggressive/Arguing/peaceful)
Focused points during discussions (Money/Work culture/Job comparisons/business/no comments on these issues)
Dedication towards the assigned work (Completion on time/long pending/delaying the work)
Behavior while Responding (Aggressive/Arguing/peaceful)
Complaint letters (Total numbers)

Table 2: Target classes.

Target classes
Will Resign
Need Salary Hike
Need Transfer
Satisfied

4.1 Python Implementation

Two classification models are used in this research work: (1) Naive Bayes, (2) Decision Tree. The collected data set with input features is shown in Fig. 1.

The general algorithm for performing any classification models on python is as follows:

1. Collection of dataset
2. Feature Engineering
3. Labeling the categorical values
4. Training the dataset
5. Applying the classification model
6. Measuring the accuracy (shown in Figs. 2 and 3, 4).

5. Observations and Result

Once a good accuracy percentage is achieved, the model becomes ready for future predictions. (See Fig. 5)

As there are a total of 8 features with some labels, each label can be input to the model and the model will find out a better class. After applying the label encoder,

	Employee Id	Attendance on virtual platform	Virtual Mode (Active / Passive)	Behavior while Verbal communication	Behavior while written communication	Focused	Dedication towards the assigned Work	Behavior while Responding	Complaint letters tot	Target
0	4001	Yes	Active	Aggressive	Supporting Arguments	Money	Completion on time	Aggressive	1	Will Resign
1	4002	Yes	Active	Arguing	Opposing	Work Culture	Late Completion	peaceful	0	Need Salary Hike
2	4003	No	Passive	peaceful	Aggressive	Work Culture	Not Completion	Arguing	2	Need Transfer
3	4004	No	Passive	Aggressive	Supporting Arguments	Job Comparisions	Completion on time	Aggressive	3	Will Resign
4	4005	Yes	Passive	Arguing	Opposing	Job Comparisions	Late Completion	peaceful	1	Will Resign
...	...	...	...	...	...	...	...	...	...	...
244	4245	Yes	Active	Arguing	Opposing	no_comments	Late Completion	peaceful	1	Will Resign
245	4246	Yes	Active	peaceful	Aggressive	Money	Not Completion	Arguing	0	Need Salary Hike
246	4247	Yes	Active	Aggressive	Supporting Arguments	Work Culture	Completion on time	Aggressive	2	Need Transfer
247	4248	No	Passive	Arguing	Opposing	Work Culture	Late Completion	peaceful	3	Will Resign
248	4249	No	Passive	peaceful	Aggressive	Job Comparisions	Not Completion	Arguing	1	Will Resign

Fig. 1: Dataset.

```
feature_train,feature_test,target_train,target_test = train_test_split(X,y,test_size=0.33,random_state=115)
model = GaussianNB()
model.fit(feature_train,target_train)
pred = model.predict(feature_test)
format(accuracy_score(target_test.astype(int),pred)*100)

'93.97590361445783'
```

Fig. 2: Applying Naive Bayes model.

```
]: feature_train,feature_test,target_train,target_test = train_test_split(X,y,test_size=0.33,random_state=115)
   clf_entropy = tree.DecisionTreeClassifier(criterion='entropy')
   clf_entropy.fit(feature_train.astype(int),target_train.astype(int))
   y_pred_en = clf_entropy.predict(feature_test)
   format(accuracy_score(target_test.astype(int),y_pred_en)*100)

]: '89.1566265060241'
```

Fig. 3: Applying Decision Tree classifier.

```
: [Text(0.3333333333333333, 0.875, 'X[7] <= 0.5\nentropy = 1.498\nsamples = 249\nvalue = [62, 62, 125]'),
   Text(0.16666666666666666, 0.625, 'entropy = 0.0\nsamples = 62\nvalue = [62, 0, 0]'),
   Text(0.5, 0.625, 'X[7] <= 1.5\nentropy = 0.917\nsamples = 187\nvalue = [0, 62, 125]'),
   Text(0.3333333333333333, 0.375, 'entropy = 0.0\nsamples = 63\nvalue = [0, 0, 63]'),
   Text(0.6666666666666666, 0.375, 'X[7] <= 2.5\nentropy = 1.0\nsamples = 124\nvalue = [0, 62, 62]'),
   Text(0.5, 0.125, 'entropy = 0.0\nsamples = 62\nvalue = [0, 62, 0]'),
   Text(0.8333333333333334, 0.125, 'entropy = 0.0\nsamples = 62\nvalue = [0, 0, 62]')]
```

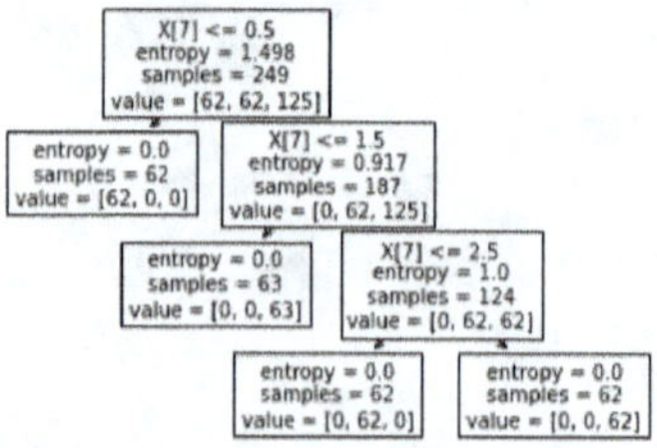

X[7] <= 0.5
entropy = 1.498
samples = 249
value = [62, 62, 125]

entropy = 0.0
samples = 62
value = [62, 0, 0]

X[7] <= 1.5
entropy = 0.917
samples = 187
value = [0, 62, 125]

entropy = 0.0
samples = 63
value = [0, 0, 63]

X[7] <= 2.5
entropy = 1.0
samples = 124
value = [0, 62, 62]

entropy = 0.0
samples = 62
value = [0, 62, 0]

entropy = 0.0
samples = 62
value = [0, 0, 62]

Fig. 4: Decision Tree classification.

```
clf_entropy.predict([[1,0,2,1,5,0,2,1]])

C:\Users\Anjali\anaconda3\lib\site-packages\sklearn\base.py:450: UserWarning: X does not have valid feature names, but Decision
TreeClassifier was fitted with feature names
  warnings.warn(

array([3])

clf_entropy.predict([[1,1,0,0,0,0,0,0]])

C:\Users\Anjali\anaconda3\lib\site-packages\sklearn\base.py:450: UserWarning: X does not have valid feature names, but Decision
TreeClassifier was fitted with feature names
  warnings.warn(

array([2])
```

Fig. 5: Prediction.

the value of encoder for target will be [0,1,2,3] that represents ["Will Resign", "Need Salary Hike", "Satisfied", "Need Transfer"]. The output of the prediction function with the value array ([3]) shows the employee "Need Transfer", as well as the value array ([2]) shows that the employee is 'satisfied' by the job.

6. Conclusion

Workforce forecasting can be done easily by using artificial intelligence, machine learning and data science. Input features and target output are the major role players to find out the desired class or to predict anything.

References

[1] Alshehhi Khaled, Zawbaa Safeya, Abonamah Abdullah, and Tariq, Muhammad Usman. Employee retention prediction in corporate organizations using machine learning methods. *Academy of Entrepreneurship Journal*, 27: Special Issue 2, 2021.

[2] Anahita Safarishahrbijari. Workforce forecasting models: A systematic review. *Journal of Forecasting.* doi: 10.1002/for.2541, 2018.

[3] Simeunovic, N. Nenad, Kamenko Ilija, Bugarski Vladimir, Jovanovic, M., and Lalic Bojan. Improving workforce scheduling using artificial neural networks model. *Advances in Production Engineering & Management*, 12(4): 337–352, December 2017. ISSN 1854-6250.

[4] Marsland, S. *Machine Learning: An Algorithmic Perspective.* (1st Edn.) Chapman and Hall, 2009.

Chapter 7

Gender Disparity in Artificial Intelligence: Creating Awareness of Unconscious Bias

Sugyanta Priyadarshini[1,*] and *Sukanya Priyadarshini*[2]

1. Introduction

The field of Artificial Intelligence (AI) has shown an exponential growth over the past few years [1]. With the growth occurring at a rapid rate, AI has delved deep down into the lives of people, thereby influencing their behavior and opinion [2] at the individual and collective level [3]. Being a general-purpose technology, AI highly mediates the socio-cultural, economic, and political relationships of people [4]. AI holds enormous pot e strides made over decades to achieve gender equality have been stealthily reversed by gender-biased algorithms [2]. Technically AI system is not biased, instead it is shaped by biased people. So, it is the value of creators that is presented in the AI algorithms [7]. If observed closely, the gender biasness in the AI system isn't intentional; it is just the reflection of human nature. As AI is created by humans, it is obvious that it will embody some of the biases upheld by humans themselves. Being in a male-dominated arena, AI relies on the implicit gender-biased inputs [8]; resulting in the promotion of gender inequality [9]. Therefore, the prime reason for gender-biased algorithm is the lack of gender diversity in the AI workforce [10].

[1] Department of Humanities, Kalinga Institute of Industrial Technology Deemed to be University, ORCID: 0000-0001-7660-6162

[2] Department of Humanities, Berhampur University. Email: admin@writerden.com

* Corresponding authors: sugyanta.priyadarshini@kiit.ac.in

This gender disparity poses a threat to women as they might be left behind in the socio-cultural, economic, and political spheres of life. Moreover, it can be said that the unconscious biases of people have led to the unintentional biases in AI system [11]. For instance, as per a new study by the University of London (UCL), AI tools designed to detect liver disease through blood tests are twice as likely to overlook diseases in women as compared to men [12]. Moreover, multiple AI bots and voice assistants also reinforce gender stereotypes [13]. Additionally, Covid-19 pandemic has accelerated the gender gap as new mothers with young children have reduced their working hours by four to five times as compared to the fathers [14]. According to World Economic Forum (WEF) research, only 22% of AI experts worldwide are women, with the remaining 78% being men [15]. Although the participation of women in STEM (science, technology, engineering, mathematics) has increased [16], but the AI algorithm displayed a gender bias by favoring men over women in STEM careers [17]. Moreover, the AI Index-2022 reflects a wide gender gap in the AI workforce. As per the report, within the last 10 years, females account for just 18.3% of all AI PhD and CS PhD graduates. And amongst the 17 top universities, female faculty in AI accounts for just 16.1%. This lack of diversity in academics and workforce has widened the gender gap. Therefore, to minimize this gender-biased algorithm, the representation of women in this field must be encouraged. Since the female section is mostly affected by gender bias, hence their higher representation in the field of AI, will aid to recognize, comprehend, and resolve the bias. AI can behave as a two-edged sword. On the one hand, if developed with a gender-biased algorithm, this may magnify the existing gender inequalities; on the other hand, if designed with a gender-balanced algorithm, it has the potential to mitigate the existing gender inequalities and inequities by challenging the already existing gender stereotypes. Therefore, it is important to have a deeper understanding of gender bias to counteract ethical challenges and make strategic decisions [18].

AI has penetrated deep into daily lives of people, which made both of its faces—positive and negative—to come in to the limelight. On one hand, AI and its related technologies have been acclaimed for bringing forth a marked revolution in digital economy, whereas on the other hand it has been mocked for its systemic unfairness, lack of transparency, and biasness. Considering the negative face of AI, a set of principles and guidelines must be established to make AI algorithms fair and ethical [19]. Hence, more research on fair AI [20,21,22] be encouraged. Fair AI refers to use unbiased probabilistic algorithms [23] that holds the potential to quantify and mitigate the gender disparity. The voyage of women to establish themselves at par with men has been tough. Although the genesis of both men and women are similar, but a large section of women still lags behind in every aspect of life. Even in the 21st century, women are still struggling to establish their rights. In this digital era, AI has turned out to be the main driving force of the fourth Industrial Revolution [24]. Therefore, the gender bias which was already pertinent in the social, political, and economic context has begun to impact the technical context. And with this it has widened the already existing gender gap.

There has been a plethora of papers that highlight the technical ways of using apt algorithms to mitigate gender bias [25,26,27,28,29]. Although multiple fair algorithms have been suggested, but there is no universal solution to shrink the gender gap in AI.

According to [2], generation of unbiased algorithm alone isn't sufficient to uproot gender bias from the AI system. It requires conscious researchers and developers to take into account the inherent gendering of words in order to attenuate gender bias. As previous research papers have already accentuated the technical angle to resolve gender bias in AI, this paper attempts to draw attention towards making an effort to dig deep into the existing framework of AI to gain a conceptual clarity on the factors behind the gender bias. This research paper is an attempt to answer the queries that remain unresolved and unanswered in the earlier research papers of different authors. Given the above, the aim of this paper is to focus on the role of AI in creating gender bias and contribute to the emerging body of research in the diverse field of AI. The objective of this research paper is to (1) investigate the factors that contribute to the widening gender bias in AI, and (2) to suggest approaches that hold the potential to mitigate gender bias in AI systems and bring gender parity.

2. Research Methodology

Literature used in the current analysis is retrieved from diverse scientific database such as Scopus and Web of Science. The search string is deliberately designed to provide a wider approach to the research objectives. The search string is ("Artificial Intelligence" OR "AI" AND "Gender bias" OR "Gender disparity") which is looked upon under "Title, abstract, and keywords" of the documents. The total number of documents published after 2000 recorded were 139. Limiting the documents to conference papers, articles, conference reviews, review articles, book chapters, and books published in English language, 132 documents were retracted. It was observed that majority of documents focused in computer science and social science domain. However, there are certain papers that are not closely associated to the research theme, resulting which articles on energy, neuroscience, material science, physics and astronomy, biochemistry, genetics, and molecular biology are omitted (15 papers excluded). At the end, 117 documents remained in the database. Further, screening of the abstract was initiated to verify their relevance with the research objective. After screening, 42 documents remained at the end. Further, as per suggestion from the research fraternity, 5 research articles and 4 book chapters based on their relevance were taken into consideration. After identification and screening procedure, 51 documents were finally reviewed and prepared for analysis.

There were three stages in the analysis. The selected research documents were initially downloaded and all the gathered information was saved in the excel file (Stage 1). Secondly, based on the abstract's relevance, selected documents were segregated (Stage 2). Finally, 9 suggested papers by other researchers in a similar domain were reviewed and considered (Stage 3). Figure 1 provides a quantitative report of the identification and screening process in detail.

3. Results

3.1 Understanding Gender bias

Gender can be defined as the socially constructed roles, responsibilities, rights, restrictions, identities, behavior, and possibilities that are assigned to men, women,

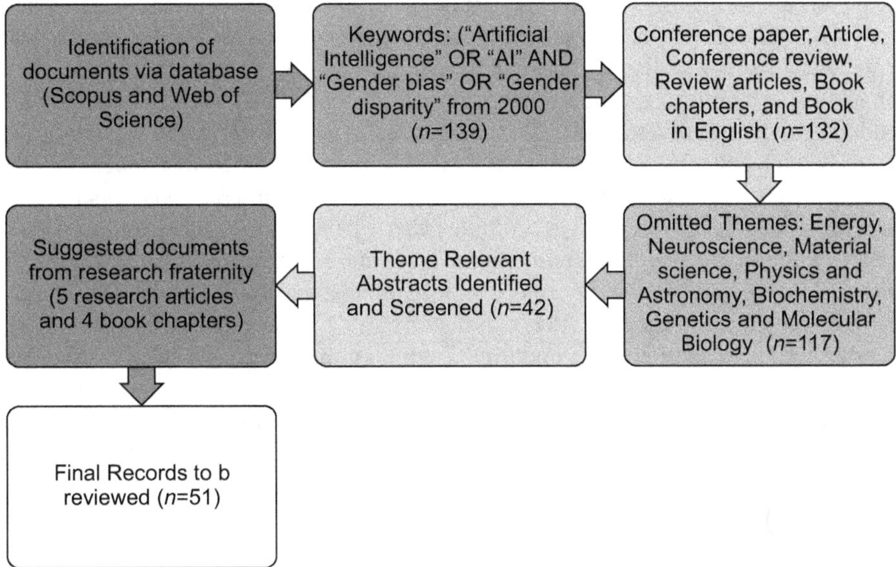

Fig. 1: Quantitative report of the identification and screening process. *Source*: Author's own.

girls and boys; which in turn varies as per time, place, culture, and society [30,31,32]. As it is socially constructed, it is learnt through the process of socialization. Therefore, within a certain setting of the society, gender has its own norms that dictate what is to be expected, permitted, and valued in both and women. Further, these gender norms also decide individual responsibilities, access and control over available resources, freedom of decision-making and specific activities to be undertaken by men and women. Moreover, being a hierarchical concept [30], it accelerates the already existing inequalities in socio-cultural, economic, and political spheres of life. Further, biasness is a human trait that is often backed by unreasoned judgment either consciously or unconsciously. Therefore, it can be said that bias is the tendency or predisposition that undermines objectivity which in turn distorts the general public's accurate assessment of scientific information [33].

Gender bias is quite prominent in academics and especially in the field of science [34]. More stereotypical male than female characteristics are linked to science [35,36]. It is believed by many that women lack the potential and the competitive power to make a successful career in science [37]. With less than 25% of PhD grads [38], full, associate, and assistant professors [39] being women, computer science stands out as a particularly male-dominated STEM field. Moreover, according to the experiment conducted by [40] on gender bias in publication quality and research on Matilda effect [41] in data communication, it was observed that publications by male authors had greater scientific quality, in particular if the topic was male typed [40]. Interestingly, gender bias was also found in scholarly peer reviews. It was observed that women are underrepresented in the peer-review process [42]. There has been a cluster of research papers on gender bias in the field of computer science. Although awareness has been created but nothing much has been achieved in eliminating gender discrimination in the field of computer science and especially AI. Moreover, it was observed that gender bias has resulted into gender imbalance that is a statistical under representation of women, especially in the field of IT [43]. Therefore, a deeper understanding of gender bias was quintessential to understand the reasons behind this existing bias.

The concept of gender discrimination and strenuous efforts to establish gender parity isn't a new-fangled topic of discussion. Many International institutions, organizations, and government of multiple nations have made constant and consistent efforts towards creating rules, regulations, laws, and mechanisms to resolve the age-old issue of gender inequality and establish gender parity. Even though socially progressive laws and measures are undertaken to reduce the gender gap, still existing gender norms have an implicit underlining in the minds of people [40]. To evaluate the profundity of gender bias, many research and experiments have been conducted. According to the experiment conducted by [44] to evaluate the discrimination potential, he observed that men are twice as likely as women to be recruited for a mathematical work if no information of the candidate except their appearance is provided.

3.2 Reasons behind Gender Bias in AI

As the world is transforming fast with digital revolution, and digitalization is becoming a part and parcel of everyone's life, gender gap is also clearly evident in technological context. But the scenario was quite different earlier. In the 1940s, when computers became a reality, an all-women team wrote the first software. Further, in the early 1950s, most of the people believed that women can be good programmers as they have more patience as compared to men and hence there was 50:50 men: women staff. Even in the 1970s, equal number of men and women were interested in coding [45]. But by 1984, there was a setback as a significant decline was marked in number of women and ever since the lack of diversity in this field deepened [46].

3.2.1 Lack of Gender Diversity

The lack of gender diversity in AI and gender biasness in AI are closely inter-wound. It is because of the lack in diversity that we observe biasness in AI algorithms. Several available literatures confirm the existence of gender parity in each and every sector and especially in STEM [15,17,34,35,36]. The lack of gender diversity in STEM has converted computer science into a homogenous industry, where men enjoy hegemony over it [20]. As the whole IT industry has a paucity of gender diversity, the AI development teams suffers from deficiency of varied mind sets, views, opinions, and thought process [27,47,48]. Therefore, we can say that rampant use and application of AI has in turn strengthened the dominance of male over female.

For instance, as per the reports of Reuters, tech giant Amazon created an AI hiring tool on experimental basis. Amazon hoped that by learning from its previous preferences, the resume scanning tool would be able to efficiently identify qualified applicants by comparing their applications to previous hires. But the system began to downgrade the resume of the candidates who had the word women in it. After the engineers discovered this bias, they tried to fix it and make it gender-neutral.

Being unsure of the fact that whether this algorithm is free of biasness or not, they decided to drop the use of this AI tool [49,50,51]. This instance strengthened the fact that data isn't biased; the problem lies with the morals and consciousness of AI designers. Moreover, as per independent research by Anupam Datta from Carnegie Mellon University in Pittsburgh, Google online advertisement system exhibits that as compared to women, a large section of men is getting high paid jobs as they are more suited for executive positions [52]. Further, it was discovered that some Facebook adverts prevent female job seekers from seeing job advertisements that primarily prefers male applicants, and hence contributing to the gender gap [53].

These aforementioned examples suggest that the already existing societal explicit as well as implicit gender stereotyping has made its way into technology. These inequities and biases about gender is shaping AI and its tools. Because of gender biases in AI, it fails to identify the demands, concerns, and achievements of women. This in turn affects not only the self-development of a woman but also her self-worth. Therefore, lack of gender diversity is one of the prime reasons for gender bias in AI.

3.2.2 Societal and Technical Reasons behind Gender bias

Gender bias isn't a contemporary issue. It has existed since the dawn of human civilization. It is in human nature that the dominant group stays unaware and unconscious to the experiences of other existing groups [54]. In the long run, these long-standing inequalities and stereotypes in the society has resulted into gender bias [55,56].

AI a socio-technical setup, that is, a system that involves not only the software and hardware process but also by the personal preconceptions of people who create it and use it. Hence, the AI system is inevitably prone to gender bias. Further, AI systems have the potential to construct and amplify the fed data and thereby reinforcing gender bias and even worse, making them seem natural. Therefore, the problem here isn't biased algorithms. In fact, it is the problem of the AI researchers and developers who breed the existing prejudice ideas that damage the vulnerable women section and support the injustice. Moreover, AI models don't have the potential to make 100% confident predictions. It happens because their comprehension of data is heavily relied on available statistics that has no complete certainty as humans use in practice [57]. Hence, some AI systems let people communicate with them directly in order to accommodate for this inherent unpredictability. As a result, AI systems learns to join the pieces and further reworks and remodels its ideas about the world. But if the AI programmer isn't free from preconceived notion of gender, it will further amplify gender bias.

Being inspired by Dean spade concept of administrative violence [58], a new-fangled term "data violence" [59] has evolved. It suggests that the structural inequalities, rigid attitude and implicit thoughts and behaviour of researchers in designing AI algorithms have led to gendered violence in the data world; which in turn has reinforced gender discrimination. Further, algorithmic bias is referred as civil rights issue of present times. Prior research examined that one possible reason behind the bias is 'technochauvism' [60]. It means that there is a popular belief that the solution to everything lies in data and technology as it is the dynamic technology that has changed how people interact with computers, other people, and their environment. But they don't realize that social fairness and mathematical fairness are completely different. Mathematically, fair solutions aren't adequate to achieve social fairness.

3.2.3 Lack of unbiased Training Datasets

The under and overrepresentation of varied social groups affects the fairness in AI and results into unfair patterns in training datasets. An unfair algorithm will always impart decision that is favourable to a particular section of people and not the entire population [61]. With high popularity and wide use of AI in every field, the fairness in AI algorithm has become a pronounced point at issue. AI along with its inherent biases in employment, advertisements, pay scale, chatbots, face and voice recognition, search engines, language and many more, reflects that unfairness and biasness has already creeped into all AI applications. This raises queries on the reliability, accountability, and investments on AI in different sectors [62].

Moreover, as per the prior research, the entire fault doesn't lie with the developers [52]. The bias in AI isn't programmed by the researchers and developers; in fact, it is the biased training data that is being programmed into the models. Therefore, the bias lies in the data. The rules aren't written down by the programmers. The training data itself has historical biases that can further be absorbed and amplified by the machine learning models. In short, the developers are developing a model that derives relationships from training data automatically [63]. As per the observations of [18] the flaw in the system is because of the fact that datasets are misinterpreted, that is, either their codification is incomplete or incorrect. Moreover, higher the diversity of the data sets, lesser is the biasness. Just collecting a ton of raw data won't be a panacea from biasness; carefully designing the data sets can be a possible solution to remove biasness. Additionally, the machine learning algorithms in AI such as neural networks provide decisions not just by programmed rules but also by the learning patterns from past and keep evolving with new input data. So, the algorithm is susceptible to biases. Moreover, as these learnt algorithms aren't transparent to the users, hence it is difficult to identify and mitigate them [20].

3.3 Strategies to Mitigate Gender Biases in AI

3.3.1 Societal Approaches: Attempts in making Invisible Gender bias in AI Visible

As discussed earlier, lack of awareness about narrowed gender diversity in AI has widened the gender gap in the field of AI [64]. It is suggested that improving transparency, accountability, fairness, and equality can play a major role in diminishing the existing gender gap in AI. Moreover, if the AI algorithms are designed keeping these social attributes in mind, then this may attenuate the issue of gender bias; if not, then it may accelerate gender discrimination and strengthen the male autonomy in AI [64]. Moreover, there is a need to bring forth a change in the narrative, language, bias in texts (e.g., naming of groups, ordering of items in lists, biased descriptions of gender in newspapers, use of gender specific metaphors, etc.) in order to minimize gender bias in AI [2]. There has always been a close relationship between workplaces, discriminatory practices, and discriminatory tools which creates a feedback loop that shapes gender bias in AI and its tools. Therefore, breaking this feedback loop can alter the functioning of AI and its tools; making it a beneficial tool in bridging gender inequality. The researchers believed that the inability of large firms in reflecting gender bias is because of much lesser emphasis on "data pipeline" studies—a term that is used to refer the lack of diversity in hiring pool. Therefore, boosting diversity in the field of AI can be the key to minimize the existing gender bias. Firstly, having a diverse AI team will influence data analytics and AI coding processing [65] and secondly, AI designers of diverse age groups, sexual orientation, and gender will aid to lessen the increasing biasness in AI.

To increase diversity in STEM and AI fields, more funding must be encouraged for women entrepreneurs in the technical field. Better scholarship opportunities for higher education and research must be provided for women in STEM. Further, more funding for corporate social responsibility gender equality initiatives should be incited. The unfair gender policies and practices must be replaced with egalitarian and transparent policies to reduce existing gender inequalities. The organizational policies, budgets, and strategies must be gender transformative. An ecosystem of trust by policymakers is suggested by [66] that ensures protection of both human and consumer rights in AI-based decision-making systems. Additionally, as fairness is an important part in AI system, efforts must be undertaken to promote social responsibility and moral ethics amongst data developers. Moreover, regular training sessions on ethics and morality must be held to inculcate fairness and accountability among AI teams, that will in turn downstream into AI systems. Further, the issue of intersectionality bias which is a kind of destructive discrimination that targets a particular sensitive identity with respect to their gender, race, religion, disability, and sexuality must be addressed in AI in order to provide a fair and inclusive environment for everyone [67].

Further, creative campaigns must be launched to draw attention towards gender inequality in the AI field. The trusted influencers can be hired to spread awareness on this issue. Integrating gender expertise into AI systems can aid AI designers to better understand the areas of gender bias and take necessary action to mitigate it. Online conferences, sessions, and workshops must be organized to interact, converse, and share ideas on mitigating the gender gap in AI. The engagement of entire spectrum of stakeholders at regional, national, and international levels from multiple sectors like academics, policymakers, government, private sector, and disciplines like technology, ethics, civil rights and human rights can together aid to promote deep thinking, awareness, and action on creating gender equality in AI.

3.3.2 Technical Approaches to Mitigate Gender biases in AI

Almost all AI algorithms are driven by data. Hence, the technical approach to mitigate gender bias focuses primarily on ways to provide unbiased and fair data. The biases can be classified into pre-existing, technical, and emergent. The existing human stereotypes produce pre-existing biases in the system. And when the biased AI training data sets develop problematic outcomes from self-learning, it stems out technical biases. The interaction between technology and users further give rise to the emergent biases [68]. Therefore, the AI system can never be completely free of biasness because it is evolving every second. Hence, attempts can be made to make AI system fair, accountable, and transparent with minimal biasness [69].

Fair AI results can only be produced if there is a fair representation of people from different diverse groups [64]. The misreporting of data is because of unethical data practices. Hence, advised that at the design and implementation stage, testing and auditing the AI-based decision-making algorithms must be done [70]. Further, it was observed that removing sensitive characteristics that identify bias doesn't prevent the data training models from learning bias. This is because the other correlated features or redundant encodings may be used as proxies and replicate the bias. And sometimes correlation between neutral features can lead to biases. Hence, to avoid biases, researchers must discover and understand such correlation between casual features that can eventually lead to biases [55,71]. Therefore, identifying and amending the biased data is a continuous process as the AI designer has to always check and recheck for the genesis and replication of new biases from existing data. Further, greater emphasis must be provided on creating context specific algorithms to produce fair and unbiased data [18].

Additionally, [72] proposed that the while designing the decision-making algorithms, data from diverse and vulnerable section of society, with different gender must be accommodated, so that the multidimensional data will aid to neutralize the biases. Moreover, under representation [73] or over representation [74] or misrepresentation [75,76] of data can catapult biases. Sometimes, the AI designed datasets don't perform well. This means the AI designed algorithms don't produce the desired output and hence can lead to biased outcome [77]. Therefore, it is important to cross check the algorithms designed for a specific purpose in order to make sure it is fair and unbiased [70,78]. Further, to minimize the discriminatory results, more attention must be paid on data collected and AI data designers [18]. The hybrid collaborative system [79] that signifies the synergy between man and machines may prove to be a better solution. The superpower of AI and intuition of human can act as a beneficial force to minimize bias in the system. Hence, human involvement in the final decision- making of AI may bridge the gender gap created by AI algorithms [64].

4. Discussion

The AI sector possess the power to revolutionize every sector. It has the potential to create an efficient, less biased, and fair world with positive machine-human interaction. But the field of AI isn't a clean slate. It also has the power to multiply the unconscious stereotypes. Thus, today gender bias in AI has become a subject of prime concern. It is believed by many that technology and women don't gel well and hence the tech ecosystem isn't conducive for women. The technology sector is often considered to be all-boys club. These gender prejudices have resulted in creating unconscious biases in AI and interestingly many are unaware of this existing gender bias. Therefore, this paper has contributed towards creating more awareness on identifying the reasons for gender bias in AI and methodologies that can be employed in bridging the gender gap in the field of AI.

The prime reason for gender bias in AI was lack of diversity. With the predominance of male AI designers, the unconscious gender stereotypes are passed onto the AI training data sets which, in turn, breed more emergent biases. Gender bias in this paper has been considered as a socio-technical problem that is becoming more evident with higher interaction of AI in our daily lives. Moreover, biased data in AI can boost gender prejudices. If sexism is in input data, it is obvious for AI systems to reflect sexist behavior in output. And with a male-dominated tech ecosystem, it is obvious for AI to exhibit gender-biased decisions. Hence, key factors like biased mind sets and datasets along with lack of diversity have shown to create a major setback in bringing gender equity and equality. Hence, a robust framework is needed for AI designers, researchers, software developers, and other stakeholders that will guide them to identify the loopholes and create an effective mechanism to mitigate gender bias in AI. From a technical aspect, the AI algorithms must be consistently evaluated to check the biases. The sensitive characteristics in algorithms must be identified and eliminated from time to time. In this research paper, our studies highlight the methodologies and strategies to create a fair, accountable, and transparent AI data training set. By a fair data set, we suggest an AI training data set that is free of gender stereotypes and that has the potential to reflect an all-inclusive and diverse society. The paper also concentrates on providing education on moral principles and ethics to AI researchers and designers to create awareness of unconscious bias and minimize gender discrimination in AI. It also emphasizes on creating creative campaigns to encourage greater participation of women in STEM and AI. Further, the government as well as private organizations at all levels must stress on the rising concerns of the widening gender gap in AI and implement policies and create strategies to mitigate gender bias. Therefore, this paper aims to identify the possible reasons of gender bias, the mitigation strategies to counter gender bias, and respective research findings, propositions, and proposed framework.

5. Societal Implication

This paper summarizes the reasons and remedies for the existing gender bias in AI. Gender prejudices have been found in every sector. Be it academics, medicine, industry, law, psychology, or sports, gender bias can be found everywhere. So, with bias being embedded deeply in human nature, it is destined to enter into the human-machine interaction. And hence, AI is gender-biased. The biases in AI data are often opaque, but they have the potential to reinforce the already existing gender biases. With the fourth Industrial Revolution being AI driven, gender bias in AI is an area that needs more research and attention. Therefore, this research paper aims to contribute to the society, the definite technical and societal reasons of biased AI. Further, as AI is new and constantly evolving, the ethical issues within it must be identified, observed, and resolved to devise an effective solution. Instead of a superficial and flawed understanding, deeper research is quintessential in this field so that we can dig deep in understanding AI from different aspects. This paper hopes to motivate researchers to identify causes behind loopholes in AI designing and data training and also provide a robust research agenda to mitigate the biases. As the technical landscape is transmogrifying, hence it is necessary for AI researchers, software designers, professionals, and organizations to keep an update on the dynamics and its implications on people and society. This paper contributes towards creating awareness about the flawed training data sets that can breed biasness in the system, which in turn creates imbalance in the society. This research also aims to establish a balance between society and technology as well as technology and ethics. Moreover, it intends to install a foundation for new research work on multidimensional role of AI in creating gender bias. This paper has contributed towards identifying the socio-technical biases in AI data sets and potential mitigation approaches. Lastly, the research paper provides wide scope for evolving the proposed theoretical framework to mitigate gender bias in AI. Hence, this paper will be a great addition to the socio-technical context of gender bias in AI.

6. Conclusion

Invention has always been the key to survival and progress for human. Humankind has always looked out for easy, smarter, effective, efficient, faster, and convenient ways of doing work. That is the reason we keep experimenting and inventing new technologies. In the 21st century, the interaction between human and technology has created a revolutionary and progressive AI which possesses both the positive and negative outcomes. Although there are many positive effects of AI, but ethical and fairness challenges faced by AI have been a hot topic of discussion and research. The biases in AI reflect that our society is still far behind in achieving gender equality and equity. The women are still a marginalized section in technology sector and have poor representation in this field. No doubt, the civilization has marched a long way in achieving higher trajectory in technology, but still gender biases exist in every domain. Since, humans are the creators of AI, the system has biasness. So, the onus falls on AI designers to create fairness and an ethical framework around AI. Identifying the traits of gender bias in AI is a difficult and complex work that requires acute attention. Hence, more research must be encouraged in this field with women at the core of it. AI technology is a boon and a bane. With multiple benefits of AI in every field, it can definitely outweigh the risks associated with it. And, to achieve it, combined efforts of AI technicians, social scientists, tech lawyers, ethics, experts on technology, along with higher gender diversity and robust ethical framework is a prerequisite to maintain a vigilance on evolving AI algorithms from time to time to counteract biasness.

References

[1] Aggarwal, K., Mijwil, M., Garg, S., Al-Mistarehi, A.-H., Alomari, S., Gök, M., et al. Has the future started? the current growth of artificial intelligence, machine learning, and deep learning. *Iraqi Journal for Computer Science and Mathematics*, 3: 115–123, 2022. https://doi.org/10.52866/ijcsm.2022.01.01.013.

[2] Leavy, S. Gender bias in artificial intelligence: the need for diversity and gender theory in machine learning. *Proceedings of IEEE/ACM International Workshop on Gender Equality in Software Engineering*, pp. 14–16, 2018. Sweden. https://doi.org/10.1145/3195570.3195580.

[3] Wilson, H., Rauwolf, P., and Bryson, J. Evolutionary psychology and artificial intelligence: the impact of artificial intelligence on human behaviour. *In*: *The SAGE Handbook of Evolutionary Psychology: Applications of Evolutionary Psychology*, pp. 333–351, 2020.

[4] Klinger, J., Mateos-Garcia, J.C., and Stathoulopoulos, K. *Deep Learning, Deep Change? Mapping the Development of the Artificial Intelligence General Purpose Technology* (SSRN Scholarly Paper No. 3233463), 2018. https://doi.org/10.2139/ssrn.3233463.

[5] Makridakis, S. The forthcoming Artificial Intelligence (AI) revolution: Its impact on society and firms. *Futures*, 90: 46–60, 2017. https://doi.org/10.1016/j.futures.2017.03.006.

[6] Ali, O., Flaounas, I., Bie, T.D., Mosdell, N., Lewis, J., and Cristianini, N. Automating news content analysis: an application to gender bias and readability. *Proceedings of the First Workshop on Applications of Pattern Analysis*, pp. 36–43, 2010. https://proceedings.mlr.press/v11/ali10a.html.

[7] Crawford, K. Opinion | Artificial Intelligence's White Guy Problem. *The New York Times*. 2016, June 25. https://www.nytimes.com/2016/06/26/opinion/sunday/artificial-intelligences-white-guy-problem.html.

[8] Manyika, J., Silberg, J., and Presten, B. What do we do about the biases in AI? *Harvard Business Review*. 2019, October 25. https://hbr.org/2019/10/what-do-we-do-about-the-biases-in-ai.

[9] Nuseir, M.T., Al Kurdi, B.H., Alshurideh, M.T., and Alzoubi, H.M. Gender discrimination at workplace: Do Artificial Intelligence (AI) and Machine Learning (ML) Have opinions about it. pp. 301–316. *In*: Hassanien, A.E., Haqiq, A., Tonellato, P.J., Bellatreche, L., Goundar, S., Azar, A.T., et al. (Eds.). *Proceedings of the International Conference on Artificial Intelligence and Computer Vision (AICV2021)*, 2021. Springer International Publishing. https://doi.org/10.1007/978-3-030-76346-6_28.

[10] Zhang, H., and McCombe, I. *The Role of AI in Mitigating Bias to Enhance Diversity and Inclusion*. 2019. https://www.ibm.com/downloads/cas/2DZELQ4O#:~:text=AI%20holds%20great%20promise%20for,outputs%20which%20can%20damage%20D%26I.

[11] Shacklett, M. *AI can be Unintentionally Biased: Data Cleaning and Awareness can Help Prevent the Problem*. TechRepublic. 2021, March 18. https://www.techrepublic.com/article/ai-can-be-unintentionally-biased-data-cleaning-and-awareness-can-help-prevent-the-problem/

[12] UCL. *Gender Bias Revealed in AI Tools Screening for Liver Disease. UCL News*. 2022, July 11. https://www.ucl.ac.uk/news/2022/jul/gender-bias-revealed-ai-tools-screening-liver-disease.

[13] Robison, C.C., and M. How AI bots and voice assistants reinforce gender bias. Brookings Institute. 2020, November 23. https://www.brookings.edu/research/how-ai-bots-and-voice-assistants-reinforce-gender-bias/.

[14] Collins, C., Landivar, L.C., Ruppanner, L., and Scarborough, W.J. COVID-19 and the gender gap in work hours. *Gender, Work & Organization*, 28(S1): 101–112, 2021. https://doi.org/10.1111/gwao.12506.

[15] World Economic Forum. Gender bias is a threat to future Artificial Intelligence (AI) applications: Opinion. *AI for Good*. 2019, September 7. https://aiforgood.itu.int/gender-bias-is-a-threat-to-future-artificial-intelligence-ai-applications-opinion/

[16] Davila Dos Santos, E., Albahari, A., Díaz, S., and De Freitas, E.C. Science and Technology as Feminine: Raising awareness about and reducing the gender gap in STEM careers. *Journal of Gender Studies*, 31(4): 505–518, 2022. https://doi.org/10.1080/09589236.2021.1922272.

[17] Lambrecht, A., and Tucker, C. Algorithmic bias? An empirical study of apparent gender-based discrimination in the display of STEM Career Ads. *Management Science*, 65, 2019. https://doi.org/10.1287/mnsc.2018.3093.

[18] Marabelli, M., Newell, S., and Handunge, V. The lifecycle of algorithmic decision-making systems: Organizational choices and ethical challenges. *The Journal of Strategic Information Systems*, 30(3), 2021. https://doi.org/10.1016/j.jsis.2021.101683.

[19] Jobin, A., Ienca, M., and Vayena, E. The global landscape of AI ethics guidelines. *Nature Machine Intelligence*, 1(9), Article 9, 2019. https://doi.org/10.1038/s42256-019-0088-2.

[20] Feuerriegel, S., Dolata, M., and Schwabe, G. Fair AI: Challenges and opportunities. *Business & Information Systems Engineering*, 62, 2020. https://doi.org/10.1007/s12599-020-00650-3.

[21] Richardson, B., and Gilbert, J.E. *A Framework for Fairness: A Systematic Review of Existing Fair AI Solutions* (arXiv:2112.05700) 2021. arXiv. https://doi.org/10.48550/arXiv.2112.05700.

[22] Bellamy, R.K.E., Dey, K., Hind, M., Hoffman, S.C., Houde, S., Kannan, K., et al. AI Fairness 360: An extensible toolkit for detecting and mitigating algorithmic bias. *IBM Journal of Research and Development*, 63(4/5): 4:1–4:15, 2019. https://doi.org/10.1147/JRD.2019.2942287.

[23] Ghahramani, Z. Probabilistic machine learning and artificial intelligence. *Nature*, 521: 452–459, 2015. https://doi.org/10.1038/nature14541.

[24] World Economic Forum. *The Fourth Industrial Revolution: What It Means and How to Respond.* 2016. https://www.weforum.org/agenda/2016/01/the-fourth-industrial-revolution-what-it-means-and-how-to-respond/

[25] Celis, L.E., Huang, L., Keswani, V., and Vishnoi, N.K. *Classification with Fairness Constraints: A Meta-Algorithm with Provable Guarantees* (arXiv:1806.06055). 2020. arXiv. https://doi.org/10.48550/arXiv.1806.06055.

[26] Zhao, J., Wang, T., Yatskar, M., Ordonez, V., and Chang, K.-W. *Men Also Like Shopping: Reducing Gender Bias Amplification Using Corpus-level Constraints* (arXiv:1707.09457). 2017. arXiv. https://doi.org/10.48550/arXiv.1707.09457.

[27] Wang, T., Zhao, J., Yatskar, M., Chang, K.-W., and Ordonez, V. *Balanced Datasets Are Not Enough: Estimating and Mitigating Gender Bias in Deep Image Representations.* pp. 5310–5319, 2019. https://openaccess.thecvf.com/content_ICCV_2019/html/Wang_Balanced_Datasets_Are_Not_Enough_Estimating_and_Mitigating_Gender_Bias_ICCV_2019_paper.html.

[28] Deshpande, K.V., Pan, S., and Foulds, J.R. Mitigating demographic bias in AI-based resume filtering. *Adjunct Publication of the 28th ACM Conference on User Modeling, Adaptation and Personalization*, pp. 268–275, 2020. https://doi.org/10.1145/3386392.3399569.

[29] Castaneda, J., Jover, A., Calvet, L., Yanes, S., Juan, A.A., and Sainz, M. Dealing with gender bias issues in data-algorithmic processes: a social-statistical perspective. *Algorithms*, 15(9), Article 9, 2022. https://doi.org/10.3390/a15090303.

[30] WHO. *Gender and Health*. 2019. https://www.who.int/health-topics/gender.

[31] Lindqvist, A., Sendén, M.G., and Renström, E.A. What is gender, anyway: A review of the options for operationalising gender. *Psychology & Sexuality*, 12(4): 332–344, 2021. https://doi.org/10.1080/19419899.2020.1729844.

[32] Heidari, S., Babor, T.F., De Castro, P., Tort, S., and Curno, M. Sex and gender equity in research: rationale for the SAGER guidelines and recommended use. *Research Integrity and Peer Review*, 1(1): 2, 2016. https://doi.org/10.1186/s41073-016-0007-6.

[33] Handley, I.M., Brown, E.R., Moss-Racusin, C.A., and Smith, J.L. Quality of evidence revealing subtle gender biases in science is in the eye of the beholder. *Proceedings of the National Academy of Sciences*, 112(43): 13201–13206, 2015. https://doi.org/10.1073/pnas.1510649112.

[34] Larivière, V., Ni, C., Gingras, Y., Cronin, B., and Sugimoto, C.R. Bibliometrics: Global gender disparities in science. *Nature*, 504(7479), Article 7479, 2013. https://doi.org/10.1038/504211a.

[35] Carli, L.L., Alawa, L., Lee, Y., Zhao, B., and Kim, E. Stereotypes about gender and science: women ≠ scientists. *Psychology of Women Quarterly*, 40(2): 244–260, 2016. https://doi.org/10.1177/0361684315622645.

[36] De Gioannis, E. The conundrum of gender-science stereotypes: A review and discussion of measurements. *Quality & Quantity*. 2022. https://doi.org/10.1007/s11135-022-01512-8.

[37] Eaton, A.A., Saunders, J.F., Jacobson, R.K., and West, K. How gender and race stereotypes impact the advancement of scholars in STEM: Professors' biased evaluations of physics and biology post-doctoral candidates. *Sex Roles*, 82(3): 127–141, 2020. https://doi.org/10.1007/s11199-019-01052-w.

[38] NCSES. *Women, Minorities, and Persons with Disabilities in Science and Engineering: 2021 | NSF–National Science Foundation*. (2021. https://ncses.nsf.gov/pubs/nsf21321/report.

[39] NCSES. *Survey of Doctorate Recipients, 2019 | NSF –National Science Foundation*. 2019. https://ncses.nsf.gov/pubs/nsf21320.

[40] Knobloch-Westerwick, S., Glynn, C.J., and Huge, M. The matilda effect in science communication: an experiment on gender bias in publication quality perceptions and collaboration interest. *Science Communication*, 35(5): 603–625, 2013. https://doi.org/10.1177/1075547012472684.

[41] Rossiter, M.W. The Matthew Matilda Effect in Science. *Social Studies of Science*, 23(2): 325–341. 1993. https://doi.org/10.1177/030631293023002004.

[42] Helmer, M., Schottdorf, M., Neef, A., and Battaglia, D. *Research: Gender Bias in Scholarly Peer Review*. ELife; eLife Sciences Publications Limited. 2017, March 21. https://doi.org/10.7554/eLife.21718.

[43] Masiero, S., and Aaltonen, A. Gender bias in information systems research: A literature review. *Proceedings of the 41st International Conference on Information Systems Hyderabad.* 2020, December 13.

[44] Reuben, E., Sapienza, P., and Zingales, L. How stereotypes impair women's careers in science. *Proceedings of the National Academy of Sciences of the United States of America*, 111(12): 4403–4408, 2014. https://doi.org/10.1073/pnas.1314788111.

[45] Kristiansen. *Nei, det er ikke biologi som gjør at vi har få kvinner i IT-bransjen.* Digi. No. 2020, October 13. https://www.digi.no/artikler/debatt-hvorfor-sa-fa-kvinner-i-it-bransjen/500823.

[46] Thompson, C. The secret history of women in coding. *The New York Times.* 2019, February 13. https://www.nytimes.com/2019/02/13/magazine/women-coding-computer-programming.html.

[47] Turner Lee, N. Detecting racial bias in algorithms and machine learning. *Journal of Information, Communication and Ethics in Society*, 16(3): 252–260, 2018. https://doi.org/10.1108/JICES-06-2018-0056.

[48] Sun, T., Gaut, A., Tang, S., Huang, Y., ElSherief, M., Zhao, J., et al. *Mitigating Gender Bias in Natural Language Processing: Literature Review* (arXiv:1906.08976). 2019. arXiv. https://doi.org/10.48550/arXiv.1906.08976.

[49] Lauret, J. *Amazon's Sexist AI Recruiting Tool: How did it go so wrong?* Medium. 2019, August 16. https://becominghuman.ai/amazons-sexist-ai-recruiting-tool-how-did-it-go-so-wrong-e3d14816d98e.

[50] Lavanchy, M. *Amazon's Sexist Hiring Algorithm Could Still Be Better Than A Human.* 2022, August 19. https://www.imd.org/research-knowledge/articles/amazons-sexist-hiring-algorithm-could-still-be-better-than-a-human/.

[51] Dastin, J. Amazon scraps secret AI recruiting tool that showed bias against women. *Reuters.* 2018, October 10 https://www.reuters.com/article/us-amazon-com-jobs-automation-insight-idUSKCN1MK08G.

[52] Datta, A., Sen, S., and Zick. A. Algorithmic transparency via quantitative input influence: Theory and experiments with learning systems. *IEEE Symposium on Security and Privacy (SP)*, San Jose, CA, USA. pp. 598–617, 2016. doi: 10.1109/SP.2016.42.

[53] Teodorescu, M., Morse, L., Awwad, Y., and Kane, G. Failures of fairness in automation require a deeper understanding of human-ML augmentation. *MIS Quarterly*, 45: 1483–1500, 2021. https://doi.org/10.25300/MISQ/2021/16535.

[54] Fei-Fei Li. Unmasking A.I.'s Bias Problem. *Fortune.* 2018. https://fortune.com/longform/ai-bias-problem/.

[55] Ntoutsi, E., Fafalios, P., Gadiraju, U., Iosifidis, V., Nejdl, W., Vidal, M.-E., et al. *Bias in Data-driven AI Systems—An Introductory Survey* (arXiv:2001.09762). 2020. arXiv. http://arxiv.org/abs/2001.09762.

[56] Chauhan, P., and Kaur, G. Gender bias and artificial intelligence: A challenge within the periphery of human rights. *Hasanuddin Law Review*, 8: 46, 2022. https://doi.org/10.20956/halrev.v8i1.3569.

[57] Friedrich, S., Antes, G., Behr, S., Binder, H., Brannath, W., Dumpert, F., et al. Is there a role for statistics in artificial intelligence? *Advances in Data Analysis and Classification*, 16(4): 823–846, 2022. https://doi.org/10.1007/s11634-021-00455-6.

[58] Spade, D. *Normal Life: Administrative Violence, Critical Trans Politics, and the Limits of Law.* Duke University Press, 2015. https://doi.org/10.2307/j.ctv123x7qx.

[59] Hoffmann, A.L. Data violence and how bad engineering choices can damage society. *Medium*, (2018, April 30). https://medium.com/s/story/data-violence-and-how-bad-engineering-choices-can-damage-society-39e44150e1d4.

[60] Broussard, M. *Artificial Unintelligence*. MIT Press. 2018. https://mitpress.mit.edu/9780262537018/artificial-unintelligence/.

[61] Mehrabi, N., Morstatter, F., Saxena, N., Lerman, K., and Galstyan, A. *A Survey on Bias and Fairness in Machine Learning* (arXiv:1908.09635). 2022. arXiv. http://arxiv.org/abs/1908.09635

[62] Fahse, T., Huber, V., and van Giffen, B. *Managing Bias in Machine Learning Projects.* pp. 94–109. 2021. https://doi.org/10.1007/978-3-030-86797-3_7.

[63] Waters, J.K. *Controversies at Google Raise the Question: What is Ethical AI?: Pure AI.* 2021. https://pureai.com/articles/2021/03/03/ethical-ai.aspx.

[64] Hayes, P., Poel, I. van de, and Steen, M. Algorithms and values in justice and security. *AI and Society*, 35(3): 533–555, 2020. https://doi.org/10.1007/s00146-019-00932-9.

[65] West, S.M., Whittaker, M., and Crawford, K. *Discriminating Systems*. 2019. https://ainowinstitute.org/discriminatingsystems.pdf.

[66] Cirillo, D., Catuara-Solarz, S., Morey, C., Guney, E., Subirats, L., Mellino, S., et al. Sex and gender differences and biases in artificial intelligence for biomedicine and healthcare. *Npj Digital Medicine*, 3(1), Article 1, 2020. https://doi.org/10.1038/s41746-020-0288-5.

[67] Ghai, B., Hoque, M.N., and Mueller, K. *WordBias: An Interactive Visual Tool for Discovering Intersectional Biases Encoded in Word Embeddings* (arXiv:2103.03598). 2021. arXiv. http://arxiv.org/abs/2103.03598.

[68] Friedman, B., and Nissenbaum, H. bias in computer systems. *In*: *Computer Ethics*. pp. 215–232, 2007. Routledge. https://doi.org/10.4324/9781315259697-23.

[69] Roger Von Laugenberg. *Bias and Discrimination in Algorithms – Where Do They Go Wrong?* VICESSE. 2020, June 29. https://www.vicesse.eu/blog/2020/6/29/bias-and-discrimination-in-algorithms-where-do-they-go-wrong.

[70] Miron, M., Tolan, S., Gómez, E., and Castillo, C. *Evaluating Causes of Algorithmic Bias in Juvenile Criminal Recidivism*. 2020. SpringerLink. https://link.springer.com/article/10.1007/s10506-020-09268-y.

[71] Loftus, J.R., Russell, C., Kusner, M.J., and Silva, R. *Causal Reasoning for Algorithmic Fairness* (arXiv:1805.05859). arXiv. 2018. https://doi.org/10.48550/arXiv.1805.05859.

[72] Clifton, J., Glasmeier, A., and Gray, M. When machines think for us: The consequences for work and place. *Cambridge Journal of Regions, Economy and Society*, 13(1): 3–23, 2020. https://doi.org/10.1093/cjres/rsaa004.

[73] O'Neil, C. *Weapons of Math Destruction: How Big Data Increases Inequality and Threatens Democracy* (1st Edn.). Crown, 2016.

[74] Boyd, Danah, and Crawford, K. Critical questions for big data. *Information, Communication & Society*, 15(5): 662–679, 2012. https://doi.org/10.1080/1369118X.2012.678878.

[75] Barocas, S., and Selbst, A.D. *Big Data's Disparate Impact* (SSRN Scholarly Paper No. 2477899), 2016. https://doi.org/10.2139/ssrn.2477899.

[76] Calders, T., and Žliobaitė, I. Why unbiased computational processes can lead to discriminative decision procedures. pp. 43–57. *In*: Custers, B., Calders, T., Schermer, B., and Zarsky, T. (Eds.). *Discrimination and Privacy in the Information Society: Data Mining and Profiling in Large Databases*, 2013. Springer. https://doi.org/10.1007/978-3-642-30487-3_3.

[77] Hardt, M., Price, E., Price, E., and Srebro, N. Equality of opportunity in supervised learning. *Advances in Neural Information Processing Systems*, 29, 2016. https://papers.nips.cc/paper/2016/hash/9d2682367c3935defcb1f9e247a97c0d-Abstract.html.

[78] Grari, V., Ruf, B., Lamprier, S., and Detyniecki, M. 2020. Achieving fairness with decision trees: An Adversarial Approach. *Data Science and Engineering*, 5(2): 99–110, 2020. https://doi.org/10.1007/s41019-020-00124-2.

[79] Grossman, G. This is how we'll merge with AI. VentureBeat, (2020, November 23). https://venturebeat.com/ai/t his-is-how-well-merge-with-ai/.

Chapter 8

Customer Perspective through Artificial Intelligence

Forecasting Green Products' Sustainable Development

Jobin Jacob,[1,*] *Jessia Saji*[1] and *Arun C Antony*[3]

1. Introduction

1.1 The Theory of Planned Behavior

The idea of planned behavior was developed in 1980 as a philosophy of deliberate action to interpret human behavior. The primary element of this theory is an individual's purpose, which is impacted by the attitude of expecting that the behavior will result in the desired result. This theory has helped in determining certain characteristics of an individual that includes smoking, drinking, services, and so on. The theory states that the behaviors are achieved through motivation and control. These characteristics developed are completely voluntary which can sometimes help in the betterment and improvement in any field. The name of the theory itself gives us a clarity that it is a well-planned formation of behaviors different from his or her normative and preconceived beliefs and norms.

The theory of planned behavior regards human actions that reflects the intentional and purposeful decisions. It represents effort taking by an individual to engage in a particular act. There are three factors that determine the intention of attaining

[1] School of Business and Management, CHRIST (Deemed to be University), Bengaluru, Karnataka, India, Email: jessia.saji@christuniversity.in

[2] Christ College, Vadgaon Sheri, Off Pune Nagar Road, Pune, Maharashtra, India. Email: fr.arun@christcollegepune.org

* Corresponding author: jobin.jacob@christuniversity.in

certain characteristics: the behavioral attitude, standardized norms, and control of the preconceived behavior. Behavioral attitude, which determines the worth of an action, is a person's favorable or negative evaluation of a behavior's results. Every person has his or her social expectation on a behavior and this is reflected through the subjective norm. The theory of planned behavior (TPB), developed by [23], contends that an individual's choice to engage in or refrain from engaging in a particular action, such as gambling, can be influenced by their intention to do so. The motivating variables that drive an action are thought to be captured by intentions, which also serve as indicators of how much effort a person is prepared to put forth to carry out the act. In general, an activity is more likely to be performed if there is a stronger intention to engage in it. According to the idea of planned behavior, having a high level of subjective norm and high intention is correlated with high levels of perceived conduct control, which in turn predicts the degree of behavior involvement. Through the notion of planned behavior, which will promote customer loyalty and cooperative reputation, this research aims to explain the citizenship behavior of consumers and organizations. The consumers are the heart of any brand because they can either promote or hamper its name. Customer citizenship behavior includes brand recommendation, assister of other buyers, and giving feedback about the products they use.

1.2 Consumer Behavior Theory

Every human action, emotion, or thinking is categorized under human behavior. This suggests that every daily decision, purpose, feeling, and idea can be categorized as human conduct. Human and consumer behavior are related, according to (1990: 91), who describe consumer actions as the study of human behavior in a consumer position. Consumer behavior entails specific types of individual acts, specifically those related to the purchase of goods and services from marketing organizations [24]. Consumer conduct is described by Schiffman and Kanuk (1997) as the behavior that consumers display in searching for, buying, using, evaluating and disposing items, services, and ideas. Studying how people decide how to spend their available resources (time, money, and effort) on things linked to consumption is known as consumer behavior, according to [25]. It mainly looks into the who, what, when, where, how frequently, and how consumers use the goods they purchase. It also encompasses all the actions that customers take when they search for, purchase, use, assess, and reject products and services that they anticipate will live up to their expectations.

With the rise of the so-called marketing notion, consumer behavior most likely developed into a significant area of research from a marketing standpoint. According to Assael [2], who emphasizes the importance of the marketing idea on retailing, marketers must first identify the benefits that customers seek in the marketplace before developing the strategies to meet those needs. Despite seeming obvious, the selling strategy wasn't developed by marketers until the 1950s. Assael [2] identified two barriers to marketers' earlier adoption of the idea. The first is that the marketing concept was not properly adopted by institutions until the 1950s. The tactics for mass production and mass marketing benefited from distribution and advertising.

1.3 Artificial Intelligence and Customer Engagement

A fundamental strategy for service organizations to gain competitive advantages is to deliver a great customer experience (Berry, 1995). There are numerous consumer touchpoints during the service experience. These contact points consist of how clients interact with various service cues. These services are divided into humane, functional, and mechanical hints by Berry et al. (2006). These service experiences are the terms used to describe these three types of hints and enhance how customers connect with the business. The way employees treat customers makes them feel more human. Functional indicators are critical to the technical quality of the service providing since they reflect the service's dependability and competence. The mechanical experience is the result of the customer interacting with the sensory components of the service, such as sights, sounds, and other ambient aspects. Using taxonomy of service experience, AI-powered services might be referred to as "functional experiences". Numerous studies have shown that most clients are unhappy or frustrated with AI-powered facility and would much rather speak to a live person. This study provides a fresh viewpoint on the interactions between consumers and the service supplied by AI and employees and how such interactions affect their connection with the organization. Given the significant impact staff help has on customer satisfaction and the popularity of AI-powered service in enterprises, customer loyalty is measured by customers' interactions with the firm and their loyal behavior. This analysis will support research on customer engagement and offer novel suggestions for raising customer engagement with the brand.

1.4 Citizenship Behavior and Consumer Satisfaction

Citizenship behavior is not a necessary requirement in any organization, it is a voluntary characteristic which will help in the better functioning of the company. Organizational citizenship behaviors (OCBs) are actions that go above and beyond the minimal responsibilities of a job and improve the social and psychological climate of the workplace. If a customer is dissatisfied, they have two options: either they quit utilizing the firm's services, or they can let the company know how they feel. It's crucial to prioritize customer pleasure because when a consumer is unhappy, they frequently express it by engaging in an antisocial customer behavior. When clients act in a way that endangers the service provider and/or its employees, this occurs. When a customer is loyal to a company, he or she may occasionally pay more for a higher-quality commodity or service. (2013) discovered that direct contact employees who were content with their jobs were more likely to create close ties with the organization's customers when they investigated the influence of employee job satisfaction on consumer repurchase intentions. Customer happiness and loyalty, and overall a company's success frequently depend on employee satisfactions. Keeping front-line staff happy results in greater customer service since they are the faces of the business.

Consumer desire for goods with less adverse repercussions on the environment, people, and animal welfare has increased, bringing ethical consumption into sharper focus. This trend has resulted in the availability of non-animal-tested cosmetics and cruelty-free goods. For instance, shampoo with the label 'cruelty-free' has a higher purchasing intent than its unlabeled competitors. But label type (certified vs. uncertified) did not show any significant variations. Taking into account the social, environmental, and economic aspects to bring a change and positive impact is the purpose of sustainability. People these days look for more organic products with less side effects, for example, people tend to choose shampoos with the label, "No Parabens" because many studies have proved it has side effects like the discovery of traces of parabens in the tissue samples of breast cancer patients in a British scientist's, Philippa Darbre [26], research paper. There is rising concern about the health concerns related with the use of non-green artificial items that react with the human body. Numerous skin and hair conditions have been caused by this. People now understand how important it is to be natural and 'organic' as a result of the extreme environmental changes. The use of renewable techniques is essential for sustainable development Consumers today think significantly differently than they did in the past, which is consistent with changes taking place in the business world. Nowadays, the purchase decision for cosmetics illustrates how consumers are changing into more conscientious ones. Cosmetics are a part of necessity, which fosters confidence in your uniqueness. They work as a proactive agent to improve one's outward look. They are products that enhance the appearance of the skin. They are often chemical compounds that are combined to enhance beauty. Presently, physical appearance has evolved to be vital to the consumers. Some claim that the cosmetics business is an important one that has undergone a significant change in this century.

According to [27], there are numerous instances of pollution and a hazardous environment, which has increased young consumers' concern for their health and raised their standard of living. They now aim to be healthy, thus they have started purchasing green cosmetics. There are much cheaper chemical cosmetics but they are quite hazardous to the skin, but currently the respondent's willingness to spend extra for green products has increased dramatically and has given significance to one's own health and sustainability. They view businesses that invest in environmental development favorably. The interests of young consumers in green cosmetics have developed dramatically over time as they have become more aware of the advantages of green cosmetics for both health and the environment. They are willing to pay more for products that are chemical-free and friendly to the environment. Young customers have recently become more concerned with purchasing eco-friendly and healthful products. The relevance of the theory of consumer behavior is evident when the consumers' views and their decision-making in purchasing of chemical products and environmental green products are dealt with.

2. Literature Review

Emuesiri Oduran, in his article *Understanding the Role of Corporate Social Responsibility in Brand Image, Brand Loyalty, and Buyer Purchase Intention in the American Cosmetic Skincare Industry,* talks about how ever since the concept of brand loyalty was first developed in 1923, numerous marketing studies have focused on it (Copeland, 1923). [28] defined brand loyalty as patronage of a branded product by consumers. Three methods to brand loyalty exist, according to Lichtlé and Plichon

(2008): behavioral, attitudinal, and hybrid. The behavioral approach evaluates consumer loyalty based on their likelihood of making subsequent purchases, purchase sequence, and retention rate. The attitudinal method measures brand loyalty in terms of customer referrals and word-of-mouth advertising by customers. The hybrid approach assesses brand loyalty in terms of repeat purchase behavior along with positive brand perceptions among consumers. According to Organ (1988), Organization citizenship behaviours (OCB) are significant and positive behaviors in workplaces that improve the quality of communication and encourage an active environment among workers that quickly react to environmental changes. Supervisor support, organizational justice, role clarification, and commitment were some of the key antecedents of OCBs identified by researchers. The positive effects of OCBs were also demonstrated by increased individual and group performance, decreased turnover, improved service quality, and increased customer satisfaction.

In the article, *Effective Factors on Customer Satisfaction and Customer Loyalty in FMCGs*, Niloufar Shahlaei Namini views customer satisfaction as the consumers' perceived and expected value. The customer will be satisfied if the perceived value matches the intended value of the customer (Kotler and Armstrong, 2004). According to Anderson, et al. (1994), the benefits of having satisfied consumers include decreased price elasticity, increased loyalty, decreased future transaction costs, lower expenses to acquire new customers, and improved brand reputation. There are two techniques to distinguish the consumer satisfaction; total number of transactions and specific client satisfaction. According to Tauber (1972), each shopper has a different motif for going shopping, whether it is to "learn about new things" or simply enjoy making a purchase. It indicates that people are content with various aspects of shopping (Clottey et al., 2008). Additionally, Clottey et al. (2008) noted a direct correlation between client loyalty and the quality of the product and service. According to Cronin and Taylor (1992), customer satisfaction is viewed as a transaction-specific value that focuses on the client's experience during a particular service interaction. Some individuals believe that an overall assessment of the customer experience is what determines whether a consumer is completely satisfied (Jones and Suh, 2000). They lend credence to the assumption that customer satisfaction depends on both the quality of the service and the service provider's expertise. Customer service is regarded as an attitude (Yi, 1990). In Ghana's telecommunications sector, for instance, there is a relationship between the service provider and the customer, and the evaluation of several interactions between the two will determine the customer's level of satisfaction.

Caroline Pülm, in her article, stressed upon the importance of green and cruelty-free products. The term 'cruelty-free' refers to products and substances that have not been tested on animals at any stage of production. The selling of cosmetics that have undergone animal research is prohibited in the European Union as of March 2013. However many companies, nevertheless, do it clandestinely without alerting clients. Products with the label 'cruelty-free' are offered for sale even when they are not. There are two basic reasons for making these false claims. Sales of goods rise as a result of two factors: first, the term 'cruelty-free' has an advantageous connotation, and second, goods with this designation appeal to buyers' moral heuristics. Independent third parties have the right to accredit products as cruelty-free, giving each item that hasn't been subjected to animal testing a distinct label. These certifications are intended to assist customers in making better informed purchase decisions. These approved labels serve as shopping aids and assist customers in choosing products that are consistent with their moral and personal values. Brand trust, which may be seen as a result of consumer behavior on the market, is the first mediating dimension. Perceived product quality is the second mediating variable, and it significantly affects both future consumer behavior and company performance. To get better marketing outcomes, advertising would be seen as a long-term investment. It is evident that advertising significantly contributes to the growth of ethical business practises and client loyalty as organizations aim for high levels of market competition. Advertising is essential for creating brand trust and affecting perceived quality.

In the article, *Artificial Intelligence, Firms and Consumer Behavior: A Survey*, Abrardi et al. [1] talks about how to improve operational systems, information technology (IT) and how researchers have tried to combine human and AI from a technological standpoint. This research adopts an analysis from a new perspective how customers' emotional intelligence influences their interaction with AI, staff, and how engaged they become as a result. To better understand how human and machine intelligences might work together in the commercial arena, this project expands research on emotional intelligence into consumer research. In addition to altering how competition is conducted and how effective it is, he contends that AI is having a substantial impact on how well existing markets are functioning and is garnering the attention of anti-trust and privacy authorities. In actuality, AI systems pose a challenge to the way that the market mechanisms and consumer decision-making processes are currently structured. The growing use of data, extremely strong scale and scope economies, and extensive network effects lead to a significant incumbent advantage. Additionally, because of these reasons, there are fewer, extremely dominant enterprises in marketplaces that are highly concentrated.

Building recommendation engines that can access many data sources, such as client privacy data, external corporate data, and websites, is a need for machine learning, according to Nikhil Patel and Sandeep Trivedi in their study. Regardless of a customer's past purchasing habits or preferences, businesses may build a relevant, customized experience for each one of them thanks to recommendation engines. Retail behemoths are employing recommendation algorithms more and more to provide customized shopping experiences for particular customers. These recommendations are generally well-received by customers, but millennial customers—who are more likely to purchase products they see advertised in publications or on television—find them particularly effective. The companies involved are using machine learning to create a recommender system that increases the ideas that are provided as a result of aggregating the majority of all previously obtained external data and combining it with analytical and natural language processing (NLP) approaches.

3. Research Methodology

The current literature is examined in the first stage of the research. The appropriate journals are searched for publications relevant to the various aspects of this study (theory of planned behavior, citizenship behavior, brand type, cruelty-free labels).

After reviewing the available literature and determining any gaps that needed to be addressed, the study questions were developed. The findings of past studies have also used as a foundation for collecting primary data. The researcher is attempting to undertsand the citizenship behavior of the customers and how it can lead to customer or brand loyalty. This study is executed through qualitative research using the theory of planned behavior and consumer behavior which can help in encapsulating the behavioral undestanding of the stakeholders and consumers. Under the qualitative research method, a case study method of research is chosen by the researcher to do an in-depth study on the theories taken. A case study helps in an elaborated understanding of a particular area through in-depth analysis of the problem. Here, the area studied would be the customer's role in coorporate reputation and how citizenship behavior is connected to customer loyalty. The case study in this research will be done the same problem mentioned above. It will focus on behavioral differences in the customers and how can their actions be influenced. The resaercher will focus and study about the green cosmetics, that is, the organic makeup and how it promotes in the sustainable development and aims at protecting the environment. The significance of IA is also dealt in this research to undertsand how it helps in forecasting customer loyalty and affects citizenship behavior. Since, this research comprises of multiple concepts and each concept is dealt in a deeper level, the sudy can be considered as an inter-disciplinary approach.

Applying these theories will help us understand how ones behavior can affect the selling and buying process. Planned Behavior theory helps in comprehending how an individual plans and develops certain characteristics, which may be different from his or her normal understanding or actions. So, applying this theory will aid in recognizing the consumer citizenship behavior on another level. The theory of consumer behavior helps us identify an organization, individual, or group and their behavior while choosing, buying, utilizing the resources or goods available. It studies the consumers' attitude while purchasing and the reason behind this behavior and this is significant because it gives a crystal clear idea on their habits, perceptions, and likes and dislikes on a brand. In fact, knowing these factors would help in finding solutions and promote a brand efficiently, respecting the customers' needs and preferences. Ordinal utility is another factor that greatly determines the loyalty and satifaction of the customer. Although it cannot be quantified numerically, the degree of satisfaction a customer feels after the consumption of different commodities can be ranked in order of choice. Through the research study, the concept of ordinal utility is also undertaken as it can influence other customers and their choice with examples. This area of research is particularly taken to study the importance of the consumer-behavior in promoting a brand. Here, the researcher has taken the cosmetic industry to understand the nuances of green cosmetics and chemical cosmetics and what can attract buyers the most. An in-depth case study is done on each of the components to help figure out the concept of customer or brand loyalty. This study aims to answer the question of how customers' behavior affects the corporate reputation, the solutions, and what kinds of products attract them and what do not. The reason behind choosing the green cosmetics is to understand the customers' preferences by promoting sustainability.

4. Impactful Factors Influencing AI Forecasting

The concept of OCB, which signifies "this person's behavior is discretionary, not immediately or explicitly diagnosed by using the formal praise machine and that, in the aggregate, promote the powerful functioning of the enterprise", gave rise to the concept of customer citizenship behavior (CCB) (Organ, 1988). CCB is typically defined as a person's willingness to engage in uninvited, constructive, and positive behaviors toward other clients and a brand. The theory of consumer behavior is concerned with the attitudes and ideas that consumers have when they buy and use products. Attitudes are essential in the context of comprehending consumer behavior since they have a major influence on behavior toward a specific product or brand. In addition to the previously identified factors, attitudes also influence future decisions and are difficult to change. When a consumer's behavior is internalized using this approach, the behavior is better understood. The firm or stakeholder can make products that are more sought and well-liked by the public by understanding, taking into account the preferences of the customer. Some people think that total customer satisfaction is dependent on the overall evaluation of service experience, whereas customer satisfaction is viewed as being transaction-specific value centerd on the experience of the customer on a specific service encounter (Cronin and Taylor, 1992) which supports the idea that customer happiness depends on both the service outcome and the experience of the service provider.

Advertising is among the most important aspect that might affect consumer brand loyalty. According to Chioveanu (2008), advertising persuades consumers to buy specific brands. Additionally, it encourages patron loyalty. Unquestionably, it has a substantial impact on consumer behavior. It is seen as a long-term investment for establishing effective business performance since brand trust is defined as the consumer's belief that a specific brand will fulfill the promised features. In conclusion, it is evident that advertising significantly contributes to the growth of ethical business practises and client loyalty as organizations aim for high levels of market competition. It is a type of commercial communication that disseminates information about products and services (Kotler, 2016). It makes natural that this information might influence how consumers assess the value of a good or service. Repeated product purchases are an indication of brand loyalty. Brand associations that are positively influenced by perceived product quality are, in turn, positively influenced by brand associations. In research, brand loyalty is frequently perceived as a single dimension, but it may also be seen as an attitudinal and behavioral loyalty.

Customer service is one of several strategies that contribute to the success of any business (Sandhu et al., 2013). For all client groups to receive high-quality services that are tailored to their unique needs, customer satisfaction is a requirement. According to Eshghi et al. (2008), the consumer defines service quality as the whole value of a service. Five factors make up service quality: responsiveness, certainty, dependability, empathy, and tangibility (Parasuraman, 1988). This means that to get the intended result, effective Service Quality is required in order to ensure client loyalty (Ahmed, 2010). The ability of service providers to satisfy customers' expectations and desires would determine the quality of the service. The service industry is aware of the significance of service quality as a safeguard for customer retention.

Many well-known drugstores and luxury cosmetic brands are cruelty-free, although their products typically contain a large percentage of synthetic components. By definition, an organic component is natural, and vegan cosmetics must be free of animal testing. By 2022, the size of the global Ayurvedic market is anticipated to nearly triple, rising from $3.4 billion in 2015. Demand of Ayurvedic products is rising in India as a result of growing consumer awareness of their advantages, government attempts to promote their use, and rising purchasing power. Given that people have begun to place a high value on appearance and a positive self-image, the cosmetics business is the one that is most in demand right now. Many new goods have hit the market; however, the majority of them contain harmful ingredients that have damaged users' skin and general health. Customers are now more vigilant about the products they purchase. As a result, natural products that are excellent for the skin and have fewer negative effects are now available. When compared to purchasing inexpensive chemical products that could someday harm them, consumers are willing to spend extra for environmentally friendly goods. Many businesses are currently manufacturing these products while taking into account the needs and preferences of the consumers. Products that respect social, environmental, or ethical concerns have come under increasing scrutiny in recent years (Luchs et al., 2010). Customers are growing more mindful of how their purchases affect the environment and the circumstances under which the goods were made (Zander and Hamm, 2012). Through 2024, the market for cruelty-free cosmetics is expected to grow by 6%, opening up a wide range of opportunities for companies in the cosmetics industry. More than ever, consumers are aware of the methods of manufacturing, the environmental impact of products, and other details, such as how product testing affects the welfare of animals. Due to the latter, new product alternatives have been created and promoted that combine necessary product characteristics with environmentally friendly and moral qualities. A crucial part of sustainable development is the application of green chemistry concepts to the creation of new procedures or components as well as the evaluation of existing procedures and ingredients. With the help of the recently discovered insights, businesses will be better able to comprehend their customers and come to informed decisions on how best address to these factors in their marketing strategies. As many customers now look for green cosmetics, it is the duty of the company to plan and produce them to develop customer and brand loyalty and this will eventually result in the corporate's reputation.

Social networking has become more popular, making it easier than ever to connect with customers. As a result, to create the greatest software possible, it is crucial to conduct research and use analytics rather than relying on generalizations. Businesses may receive both positive and negative feedback from their customers thanks to the direct messaging features on applications like Instagram and Facebook. Businesses who use the appropriate social media channels for their industry may find it quite helpful to learn what their clients need. Not all customers are the same, and not all incentives will result in them coming back. Understanding what clients desire is crucial when developing a loyalty programme. Businesses can modify the program to fulfill those objectives and ensure its success by understanding what customers desire.

The relative high cost of green cosmetics compared to chemical products remains one of the key obstacles to their growth, despite shifting consumer views. Increasing numbers of households with two incomes, shifting views, and changes in household size and lifestyles are just a few of the factors that have changed how consumers purchase. They are willing to spend more because their top objective is to maintain a great, youthful, healthy, and beautiful appearance without endangering their country. More and more green cosmetics brands are available each year. Even though Indian customers are aware of environmental issues and think that expensive, high-end green cosmetics are only for people in higher socioeconomic classes, they are still sceptical. Lack of reliable information on the ingredients in cosmetics is another hindrance to their slow development. Even though companies are working hard to research natural components and provide more clinically verified, secure products with a distinctive marketing proposition, raising awareness is essential (Khan, 2013). Customers who buy green cosmetics are dubious since there are no market controls. As a result, companies must explain to customers how green cosmetics are better for the environment and for people's health.

The service experience is a journey of emotions. Although some interactions can be pleasant and memorable, others are not. For instance, auto-messaging services, a form of AI, are useful for customers who require assistance after office hours and are used by companies like hotels and airlines. This AI tool, however, leaves very little area for writing personalized messages for certain clients, which can make them dissatisfied with the service. Strong emotional intelligence may lead to customers who are more receptive to the services provided by AI systems. However, the experiences of the laborers with employee service are not always positive. Their negative attitudes could affect how clients perceive and engage with the employee. While some customers may express their displeasure, others may display greater sympathy for the employee. Their emotional intelligence, sometimes referred to as emotional abilities, reflects this response. Emotional intelligence has received a lot of attention over the past three decades in the context of organizations as a crucial factor in both individual and organizational success (Prentice, 2019). The impact of a customer's emotional intelligence on their relationship with the service provider and how they use the service has not yet been studied. A particular area of study is how AI of this kind might be combined to enhance organizational effectiveness. In an effort to enhance operational systems, IT experts have tried to combine human and machine intelligence. This research analyzes from a new perspective on how customers' emotional intelligence influences their interaction with AI, staff, and how engaged they become as a result. As proposed earlier in the chapter, most consumers are aware of the environmental issues around them and are finding ways to come back to more chemical-free, sustainable, green products. AI plays a significant role in understanding the customers better and predicting the possibility of bringing this change to the society along with emotional intelligence.

5. Conclusion

The researcher has addressed various fundamental issues that have influenced consumer citizenship behavior and helped to discover consumer attitudes toward particular cosmetic products. The importance of green and cruelty-free cosmetics has been highlighted throughout the chapter, as well as how people value sustainable development by attempting to steer clear of chemical cosmetics. The study focuses on the importance of AI in forecasting brand loyalty and customer satisfaction along with emotional intelligence. AI is a reality in the present. It is widely utilized to implement operational marketing, which involves risk identification and target of clients, brand promotion, and pricing to maximize revenue. It has a tremendous impact on, and transforms marketing communication in the digital age. Academics and business professionals have suggested that AI is gaining traction with the use of big data analysis, machine learning, social media analysis, and algorithms. Decision-making, simulation modeling, and other methods are employed to increase brand recognition in the worldwide market. Consequently, AI is significantly altering brand customer attitudes, marketing tactics, and preferences.

It is the responsibility of companies to identify their target audience and understand their concerns and requirements. The products must be modified in accordance with consumer needs. Cosmetic firms can cultivate loyal consumers who engage in repeat purchase behaviors by prioritizing Corporate Social Responsibility (CSR) activities that lead to a positive image. With the concern of green cosmetics, the companies have begun to manufacture Ayurvedic and cruelty-free products for the customers. Through the study of the theory of both planned and consumer behavior, the attitudinal changes of the buyers while purchasing and using the products are studied in depth. We are all consumers of some products and certain brands are imprinted in us which force us to repeatedly purchase them no matter how many other similar products arrive in the market. That is the power of marketing and this can only be attained through recognizing the target audience along with their growing concerns and need. Green cosmetics are particularly mentioned in the chapter because it is a recent addition on to the market replacing the chemical cosmetics. This has happened because the customers have become more aware and concerned about their skin and health along with that for the environment. This change in the attitude must be first understood by the companies that manufacture cosmetics. This research has also argued the importance of AI that helps in promoting customer satisfaction by understanding the needs of the consumers. It is clear that advertising has a crucial role in motivating consumers to make purchases, which leads to repeated endorsement and brand loyalty.

References

[1] Abrardi, Laura. Artificial intelligence, firms and consumer behavior: A Survey. *Journal of Economic Surveys*, 26 July 2021. 10.1111/joes.12455.

[2] Assael, Henry. *Marketing Principles & Strategy. Study Guide*. Fort Worth, Tex. Dryden Press, 1993.

[3] Bakator, Mihalj. Influence of Advertising on Consumer-Based Brand Loyalty. *Journal of Engineering Management and Competitiveness*, 7(2): 75–83. 2017. 10.5937/jemc1702075b. Accessed 5 Mar. 2020.

[4] Bernazzani, Sophia. What is customer satisfaction? *Hubspot.com*, 2018. blog.hubspot.com/service/what-is-customer-satisfaction.

[5] Buja, Vrullim. The impact of leadership styles on service quality and customer satisfaction: A comparative analysis between foreign and domestic capital banks in Kosovo. *Innovative Marketing*, 18(3): 181–196. 29 Sept. 2022. 10.21511/im.18(3).2022.16. Accessed 10 Oct. 2022.

[6] Cherry, Kendra. What is a case study? *Verywell Mind,* Verywell Mind, 7 Nov. 2022. www.verywellmind.com/how-to-write-a-psychology-case-study-2795722.

[7] Customer Satisfaction a Higher Priority in Marketing Strategy than Branding or Innovation. *Www.b2bmarketing.net,* www.b2bmarketing.net/en-gb/resources/news/customer-satisfaction-higher-priority-marketing-strategy-branding-or-innovation. Accessed 9 Nov. 2022.

[8] Trawnih, Ali. Understanding Artificial Intelligence Experience: A Customer Perspective. *International Journal of Data and Network Science*, 6(4): 1471–1484. 10.5267/j.ijdns.2022.5.004. Accessed 27 Aug. 2022.

[9] Iqbal, Talha. *The Impact of Leadership Styles on Organizational Effectiveness*. GRIN Verlag, 19 Oct. 2011.

[10] Kim, Si Hyun. A social exchange perspective: the mediating effect of customers' perceived overall justice and affect in the relationship between employee performance and customer satisfaction. *Open Access Theses & Dissertations*, 1 Jan. 2014. scholarworks.utep.edu/open_etd/1656/. Accessed 8 Nov. 2022.

[11] Kiss, Christine, and Martin Bichler. Leveraging network effects for predictive modelling in customer relationship management. *SSRN Electronic Journal*, 2005. 10.2139/ssrn.883588. Accessed 3 June 2019.

[12] LaMorte, Wayne. The Theory of Planned Behavior. *Behavioral Change Models,* Boston University School of Public Health, 2019. sphweb.bumc.bu.edu/otlt/MPH-Modules/SB/BehavioralChangeTheories/BehavioralChangeTheories3.html.

[13] Ma, Jingjing. *Linking Observation of Coworkers' Organizational Citizenship Behavior with One's Own Organization Citizenship Behavior and Psychological Strain*. 2017.

[14] *Marketing and Consumer Behavior: Concepts, Methodologies, Tools, and Applications*. Hershey, PA, Business Science Reference, 2015.

[15] Prentice, Catherine, and Mai Nguyen. Engaging and retaining customers with AI and Employee Service. *Journal of Retailing and Consumer Services*, 56(56): 102186. Sept. 2020. 10.1016/j.jretconser.2020.102186.

[16] Putra, Halim. What is the meaning customer citizenship behavior? A Review. *International Journal of Science and Research*, 2018, pp. 910–911. ISBN: 2319-7964.

[17] Singh, Anurupa. Green Cosmetics –Changing Young Consumer Preference and Reforming Cosmetic Industry. *International Journal of Recent Technology and Engineering*, 8(4): 12932–12939. 30 Nov. 2019. 10.35940/ijrte. d6927.118419.

[18] Theory of Consumer Behaviour. *GeeksforGeeks*, 6 June 2022. www.geeksforgeeks.org/theory-of-consumer-behaviour/. Accessed 8 Nov. 2022.

[19] Varsha P.S. The Impact of artificial intelligence on branding. *Journal of Global Information Management*, 29(4): 221–246. July 2021. 10.4018/jgim. 20210701.oa10.

[20] Wuisan, Elisa Christianti, and Agustinus Februadi. Consumers' attitude towards the cruelty-free label on cosmetic and skincare products and its influence on purchase intention. *Journal of Marketing Innovation (JMI)*, 2(2), 28 Sept. 2022. 10.35313/jmi. v2i2.35. Accessed 17 Oct. 2022.

[21] Yazid, Ahmad Shukri. The mediating effect of customer satisfaction on the relationship between Corporate Social Responsibility and Customer Loyalty. *Journal of Advanced Research in Dynamical and Control Systems*, 24(4): 299–309. 31 Mar. 2020. 10.5373/jardcs/v12i4/20201444. Accessed 7 May 2020.

[22] Yi, Youjae. The impact of other customers on customer citizenship behavior. *Psychology & Marketing*, 30(4): 341–356. 25 Feb. 2013. 10.1002/mar.20610. Accessed 2 Apr. 2019.

Chapter 9

Weather Forecasting and Climate Behavioral Analysis Using Artificial Intelligence

Murali Karuppusamy,[1,*] *Gurumoorthy Chelladurai,*[2] *Sambath Krishnamoorthy*[3] and *Sudharshan Rathinavel*[3]

1. Introduction

Human beings forecast specific short-term weather conditions with the known average long-term climatic conditions. The rapid climatic change due to various anthropogenic sources made the weather forecast more complex. As far as weather is concerned, it is very important to everyone including authorities of government and non-government sectors, water users, industrial sector, and other group of entities [7]. To avoid risk in agriculture and food security, it is necessary to predict the weather and climate as suggested by the Intergovernmental Panel on Climate Change (IPCC). The major challenge is to increase crop production by efficient use of available resources. Ensuring availability of food resources to the fast-growing population lies on efficient weather and climate prediction systems to provide early warning to assist farmers about risks in agriculture based on climate changes.

The various sources of weather and climate related risks in agriculture include scarcity and depletion of water resources, drought conditions due to varying rainfall patterns, and so on. The various seasonal hazards cause reduced crop production.

1 Professor, Department of Civil Engineering, Dr. Mahalingam College of Engineering and Technology, Pollachi, Coimbatore, Tamil Nadu, India,

2 Assistant Professor, Department of Civil Engineering, National Engineering College, Kovilpatti, Tamil Nadu, India, Email: gurumoorthycivil2010@gmail.com

3 Assistant Professor, Department of Civil Engineering, Sri Ramakrishna Institute of Technology, Coimbatore, Tamil Nadu, India, Emails: sambathk19@gmail.com; sudharshan.ce@srit.org

* Corresponding author: murali.vlb@gmail.com

It also affects other allied sectors such as fisheries, forestry, and the environment. Efficient weather forecast systems by using Artificial Neural Networks (ANNs) help to promote efficient usage of available water resources, conservation, and optimal utilization of natural resources, and avoid pollution due to agricultural chemicals and other external agents.

According to the Sustainable Development Goals (SDGs) it is necessary to ensure affordable, clean, and reliable energy for the growing population as well as to mitigate climatic change to improve the nation's economy. Which is mainly focused on agricultural products and practices which rely on weather. Apart from agriculture, weather prediction contributes to economic growth by means of analyzing wind and tides paving the way for electricity generation from alternative natural sources. Weather forecasting is a highly complex and highly quantitative atmospheric research activity [3]. National economic losses can be minimized by weather prediction which develops preparedness among general people against seasonal disaster and agricultural management in achieving SDG 2.

2. Artificial Intelligence for Decision Support

The application of Artificial Intelligence (AI) techniques along with understanding of the physical environment helps to predict high impact weather [8]. ANNs are one among the Artificial Intelligence (AI) techniques with interconnected networks of weighted nonlinear functions. ANNs been adopted for various meteorological applications since 1980s. AI techniques can be effectively applied for modern weather forecasts. They will be more helpful to improve the ability of using big data to extract perceptions and acts as a regulator for weather forecasts and the decision-making process.

The various advantageous features of AI techniques providing decision support include the following:

1. Easy handling of large datasets
2. Development of numerical models reducing cognitive overload
3. Post-model bias corrections done with ease.

The AI techniques adopted for decision support include,

1. Artificial Neural Networks
2. Fuzzy Logic
3. Principal Component Analysis
4. Numerical Weather Prediction
5. Geostationary Operational Environmental Satellites R Series
6. Support Vector Machines
7. Decision Tree-based methods.

3. Artificial Intelligence in Weather Forecasting

Weather forecasting can be done using various methods [3] as shown below:

1. Current weather forecasting
2. Numerical methods and
3. Statistical methods.

ANN is a type of adaptive system that varies based on internal or external information that flows through the network [3]. It is an efficient tool for capturing and analyzing complex relationships between input and output data [7].

4. Artificial Intelligence in Climate Behavioral Analysis

Weather forecasting using ANN utilizes various input data as shown below:

1. Month of the year
2. Day of the month
3. Mean sea level pressure
4. Dry-bulb and wet-bulb temperature
5. Mean dew point
6. Maximum, minimum, and mean values of temperature
7. Mean relative humidity
8. Rainfall
9. Wind direction and its speed
10. Mean cloudiness.

The output data will be like that of the input data considered by the network.

5. Case Study

A case study on weather forecast using AI has been conducted in Bangladesh which has been selected for analysis since it is a zone of rapid seasonal alterations with six seasons [18]. In addition to seasonal variations, the country is prone to risks caused by natural calamities occurring frequently. The natural hazards that are mainly caused as a result of inadequate weather prediction systems include tropical cyclones, storm surges, floods, heavy rain, and tornadoes. The various seasonal hazards cause reduced crop production. Due to non-inear characteristics of weather pattern as discussed earlier, the prediction of weather pattern in Bangladesh becomes more complex and highly quantitative in nature.

Using ANN, a study on prediction of heavy rainfall was been conducted to prepare reliable monthly weather forecast. Due to heavy rainfall during June 2015, Barisal city, Bangladesh had been flooded. It was observed that root causes included lack of flood prediction and no proper drainage system. The disaster had been taken into consideration for the study and weather data throughout the year had been

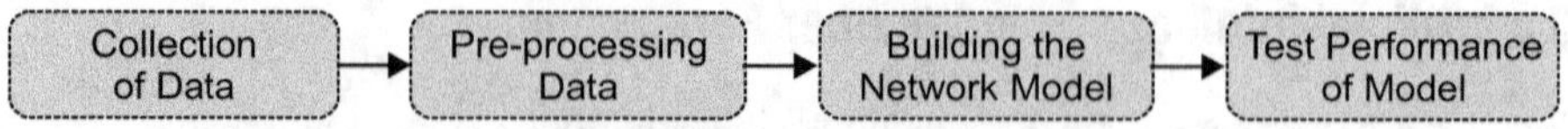

Fig. 1: Steps involved forming Artificial Neural Network model.

collected. The major input factors considered included temperature, humidity, dew point, and pressure.

Based on the inputs, the target value of rainfall percentage of the city was planned to be determined.

Based on the collected data, normalization of those data has been done using MATLAB application. In general, normalization is done to avoid uncertainty due to mixing algorithm criteria for the learning algorithm in MATLAB [18]. After normalization, ANNs will detect the behavior of nonlinear data sets and weather conditions will be predicted.

Among the collected data, major data (70%) is used for training, for validation 15% was used, and for testing another 15% was utilized. Based on regression analysis of 8 months weather data, rainfall prediction has been done. Mean Squared Error (MSE) has been utilized to study the comparison between the real precipitation and the forecasted values based on ANN. Based on comparison, the target value shows that the trained network prepared gives similar result of precipitation data with respect to the measured rainfall. Using the ANN model included with data normalization, reliable seasonal disaster prediction can be done.

The implementation of the ANN model developed a monthly based weathering monitoring model which is more reliable. The monthly rainfall prediction has shown good performance by providing reasonably accurate weather predictions.

6. Conclusion

AI and related data science methods have a proven record to work with big data across a variety of disciplines. The AI method continues to support enhanced weather prediction and to understand more weather-related phenomena. Based on various research about weather forecasting systems, it has been made possible to obtain simpler and reliable weather monitoring model. AI techniques, specifically ANN has been utilized to evolve such a model with only specific weather parameters avoiding abundant data sets and its complex processing methods. The various research works conclude that disaster prediction using weather forecast systems evolved on application of AI techniques are more reliable.

Acknowledgments

Our thanks go to the editors of the volume, Professor Sachi Nandan Mohanty, Professor Tejaswini Kar, and Professor Preethi Nanjundan for giving me the chance to write a chapter on this area of AI for Forecasting: Tools and Techniques and making helpful suggestions.

References

[1] Abdul Manan Ahmad, and Chia Su Chuan. Weather prediction using Artificial Neural Network. *International Technical Conference on Circuits/Systems, Computers and Communications*, pp. 262–264, 2002.

[2] Santhosh Baboo, S., and Kadar Shereef, I. An efficient weather forecasting system using artificial neural network. *International Journal of Environmental Science and Development*, 2010.

[3] Abhishek Saxena, Neeta Verma, and Tripati K.C. A review study on weather forecasting using artificial neural network approach. *International Journal of Engineering Research & Technology*, pp. 2029–2035, 2013.

[4] Gyanesh Shrivastava, Sanjeev Karmakar, Manoj Kumar Kowar, and Pulak Guhatha Kurta. Application of artificial neural networks in weather forecasting: a comprehensive literature review. *International Journal of Computer Applications*, 51: 17–29, 2012.

[5] Govind Kumar Rahul, Saumya Singh, and Saumya Dubey. Weather forecasting using artificial neural network. *International Conference on Reliability, Infocom Technologies and Optimization*, pp. 21–26, 2020.

[6] Kumar Abhishek, Singh, M.P., Saswata Ghosh, and Abhishek Anand. Weather forecasting model using artificial neural network. *4th Procedia Technology*, pp. 311–318, 2012.

[7] Klent Gomez Abistado, Catherine N. Arellano, and Elmer A. Maravillas. Weather Forecasting using Artificial Neural Network and Bayesian Network. https://doi.org/10.20965/jaciii.2014.p0812.

[8] Amy McGovern, and Kimberly L. Elmore. Using Artificial Intelligence to improve real-time decision making for high impact weather. *American Meteorological Society*, pp. 2073–2090, 2017.

[9] Anandharajan, T.R.V., and Abhishek Hariharan. Weather monitoring using artificial intelligence. *International Conference on Computational Intelligence and Networks*, pp. 106–111, 2016.

[10] Mohammad Monfared, Hasan Rastegar, and Hossein Madadi Kojabadi. A new strategy for wind speed forecasting using artificial intelligent methods. *Renewable Energy*, 34: 845 –848, 2009.

[11] Lorenzo Donadio, Jiannong Fang, and Fernando Porte-Agel. Numerical weather prediction and artificial neural network coupling for wind energy forecast. *Energies*, 14: 338, 2021.

[12] Florian Dupuy, Gert-Jan Duine, et al. Valley winds at the local scale: correcting routine weather forecast using artificial neural networks. *Atmosphere*, 12: 128, 2021.

[13] Reza Sabzehgar, Diba Zia Amirhosseini, and Mohammad Rasouli. Solar power forecast for a residential smart microgrid based on numerical weather predictions using artificial intelligence methods. *Journal of Building Engineering*, 32, 2020. https://doi.org/10.1016/j.jobe.2020.101629.

[14] Fermin Rodriguez, Alice Fleetwood, Ainhoa Galarza, and Luis Fontan. Predicting solar energy generation through artificial neural networks using weather forecasts for microgrid control. *Renewable Energy*, 126: 855 –864, 2018 (October).

[15] Gustavo Araujo, Fabio A.A., and Andrade. Post-processing air temperature weather forecast using artificial neural networks with measurements from meteorological stations. *Applied Sciences*, 12: 7131, 2022.

[16] Robert, J., Kuligowski, and Ana, P. Barros. Experiments in short-term precipitation forecasting using artificial neural networks. *Monthly Weather Review*, pp. 470–482, 1998.

[17] Robert, J., Kuligowski, and Ana P. Barros. Localized precipitation forecasts from a numerical weather prediction model using artificial neural networks. *Weather and Forecasting*, pp. 1194–1204, 1998.

[18] Taohidul Islam, Sajal Saha, Ali Ahmed Evan, Nabonita Halder, and Shakti Chandra Dey. Monthly weather forecasting through ANN Model: A case study in Barisal, Bangladesh. *International Journal of Advanced Research in Computer and Communication Engineering*, 5(6): 1–6, 2016.

CHAPTER 10

Assessing Climate Change through Artificial Intelligence

An Ethico-Legal Study

*Achyutananda Mishra, Ananya Pandey** and *Smruti Mohanty*

1. Introduction

IPCC (The Intergovernmental Panel on climate change) [1], in the 6th Assessment Report released in 2022, reports that the net anthropogenic GHGs (greenhouse gases) continued to rise during the period 2010–19. It shows that GHG emission in the last decade is the highest in human history. According to the *World Inequality Report, 2022*, carbon dioxide concentration level in the atmosphere across the globe is the highest in millions of years. Consistent rise in the global emission level leading to alarming rise in atmospheric temperature has been a cause of concern for mankind. Rising atmospheric temperature leading to climate change has severely affected weather patterns; led to melting of glaciers; caused natural disaster and extinction of species, and severely impacted the ground water table. It has put the human race at a crossroads and thrown open an existential question for the world. Attempts have been made, both international and national, to reverse the impact of the rising scenario concerning climate change but have yet to be successful. The technological revolutions arising in recent times, especially in the domain of Artificial Intelligence (AI), offer hope to give a new shape to human civilization. With the aid of human intelligence, AI can perform assessment and predictive work as well which may help in mitigating the effect of adversely affecting climate change and help improve the environment. As per UNESCO (United Nations Educational, Scientific and Cultural

School of Law, Christ University, Bangalore, Karnataka, India.
Emails: achyuta.mishra@christuniversity.in; smruti.mohanty@res.christuniversity.in
* Corresponding author: ananya.pandey@res.christuniversity.in

Organisation), AI can perform assessment and prediction of climate change, which may assist in the protection of the environment. The Council identifies three priority areas relating to use of AI which includes improved understanding and predictions of climate change and geohazards [2]. This chapter aims at exploring the contribution of AI in assessing the behavioral pattern of climate change and the ethico-legal challenges involved therein.

2. Climate Change—A Brief Introduction

Climate change can be referred to as the change in the climate driven on by human activity, which alters the atmosphere on a global scale, besides the changing climate variance seen throughout the same time period. It has adverse impacts such as forest fires, floods, rise in extreme sea level, melting of glaciers, changing weather events, etc.

Nature has endowed man with an environment to sustain and nurture its physical, moral, spiritual, intellectual and social growth. Nonetheless, man has gained power as a result of tremendous advances in science and developing technology. For generations protection, preservation, and development of the environment has been a major issue affecting the well-being of the people across the globe. The interaction of human society with the nature/environment, so far, has been a debatable issue.

Recent anthropogenic activities have led to significant climate change raising questions about the future survival and existence of humankind; it has become a matter of serious concern. It has a catastrophic and cataclysmic effect on human society. Climate change is not only affecting the environment but also is causing a powerful impact on several human rights and the country's economy. Climate change is an existential problem which requires a collective and robust response.

Numerous attempts have been made internationally to curb, mitigate, and regulate the dreadful problem of climate change and its impacts and for the protection of the environment but with little success. Environmental concerns relating to the development of human society dates back to Stockholm Declaration, 1972, which is called the 'Magna Carta of Environment', where the world leaders sat to frame a few principles for the future of mankind [3]. After this, International action towards it gained momentum.

In 1992, the international community expressed its determination to fight climate change and framed the United Nations Framework Convention on Climate Change (UNFCCC) [4], (also known as Earth Summit) acknowledging the contribution of human activities towards substantial increase in GHGs concentration in the atmosphere [5]. Before the adoption of Earth Summit, several instances, inclusive of Vienna Convention (1985) for the protection of ozone layer depletion, a protocol concerning Transboundary Air Pollution Convention, adoption of the first sitting to IPCC for the examination of rise in the global temperature climate change was taken into account. In the second conference of World Climate in 1990, taking into consideration the notion of the urgency towards the protection of environment and climate change, the General Assembly decided to congregate the 1992 [6] United Nations Conference on Environment and Development. The Rio Declaration and Agenda 21, which represented a worldwide consensus on development and environmental cooperation, built a new mechanism for negotiating international accords to safeguard ecological integrity.

3. AI in the Contemporary World

AI in the 21st century is considered to be one of the most fascinating technologies of time as it would help transform the world, inclusive of law, ethics, and philosophy [7]. Oxford Online According to Oxford Dictionary, "AI is a theory of development of computer systems which make it capable of performing tasks equal to that of human intellect like visual processing, speech synthesis, determining choices, and translation software."

It basically functions by large volumes of data combined with quick, repeated processing and algorithms, allowing the programme to learn spontaneously from correlations or characteristics in the input [8]. It can conduct computations, make interpretations, and judgments that generally need human understanding, whereas the evolution of AI comprises the ability to learn from previous experiences, adjust to new conditions, and deal with abstract concepts. AI portrays human intelligence. AI is of utmost significance in the today's world as it has a variety of applications such as it is used in web search, cars, cyber securities, astronomy, education, finance, gaming, healthcare, digital assistants through chat bots, online shopping and advertising, agriculture and farming, weather forecast, and much more [9].

According to a research conducted, AI can assist in reducing GHG emissions by 5%–10% of a company's carbon footprint, or 2.6 to 5.3 gigatons of CO2e globally [10]. AI may assist in alleviating climate change in various ways, including increased energy efficiency and reduced exudations from industry, horticulture, and transportation [11]. By assisting in the identification of risk factors and the creation of mitigation and adaptation strategies, AI can help in detecting emissions at macro and micro levels, reducing their impact, and eliminating existing toxins from the environment. Furthermore, the Carbon Disclosure Project's Global Supply Chain 2021 report found that just 38% of firms are collaborating with manufacturers on environmental issues, despite the fact that supply networks emit 11.4 periods as much as direct operations. Using AI to automate portions of this evaluation process would speed up the process and make it obtainable to more organizations and consumers. AI can be focused on five key uses, i.e., forecasting, energy efficiency, agriculture, industries, and reduction of emissions from transportation by using smart technology, where its technical prowess meets the requirements of governmental decision-makers. AI serves as a tool to aid decision-makers and individuals in addressing climate concerns rather than serving as the actual answer [12].

To identify and access water quality, the UNESCO global water quality site employs machine learning algorithms on satellite image. It is an interesting technique for monitoring freshwater quality by satellite "earth observation" (EO). It facilitates open data access and sharing of the same on water quality, thereby contributing to raising awareness about water quality. This results in fresh data that researchers may use to, among other things, analyze how climate change is affecting water customer preferences and business operations [13].

A few organizations are working to accelerate and simplify effective climate action by independently measuring GHG emissions, such as the inventory developed by 'Climate TRACE' a non-profit Organization that employs AI intelligence, collective data science knowledge, and satellite images and other remote sensing techniques to track human-caused GHG emissions. Similarly, Blue Sky Analytics has the ability to calculate emissions specifically from fires. Pachama which was established in the year 2018 as an organization that blends cutting-edge technology with the most recent developments in ecology while drawing inspiration from California's redwood forests [14]. Their objective changed to utilizing market processes to create a sustainable economy that could address climate change. The goal of Pachama, which uses the most recent developments in satellite data and machine learning to guarantee the integrity of nature-based carbon credits, is to restore nature in order to combat climate change. Pachama uses satellite vision and AI to monitor and track the carbon stored in forests over time, locating high-quality carbon credits. Organizations may optimize their emissions at scale with the help of these AI platforms. Industries such as oil and gas, biopharmaceuticals, automotive, and consumer goods, should use this ready-to-deploy software to account their carbon emissions [15]. This will help them estimate not only the more challenging-to-measure indirect emissions created throughout the company's complete value chain as well as measuring emissions produced directly by the company's own operations.

4. Application of AI

4.1 Introduction

The application of AI in the contemporary world dates back from the use of steam engines and now AI is the most strategic technology in the current civilization and is used in diverse fields. It can be determined that AI is one of the significant studies in the field of technology duplicated as human intelligence and is used to carry forward exhaustive work with a grade of sovereignty to accomplish the goals set under UNSDGs [16]. In contemporary society AI is discharging a significant role in the field of law, trade, business, commerce, banking, healthcare, finance, and education.

Following are the most concerned areas of application of AI in assessing climate change:

4.2 AI and Weather Forecast

The most common scientific method traditionally used for predicting and forecasting weather was empirical focus on the various happenings of nature such as rise in sea-level, melting of glaciers, rise in global temperature, amount of release of carbon dioxide in the atmosphere. But in recent times AI is being used to predict the weather pattern. Several techniques have been developed through the use of AI for forecasting weather such as analyzing the historical data of the climate pattern the future pattern of the weather predicted, use of AI-powered code-driven technical network or a low/no-code organization, informed expertise decision, data from predictive analytics for operations, use of global accounting firms and agencies, real-time predictions, using automated time series, use of the source information and the values, use of algorithms that are accessible and exclusive [17].

Following are several areas connected to the efficient and practical use of AI in climate change mitigation:

(1) AI for mitigation

The framework for applying AI to tackle climate change includes macro and micro-level monitoring, reduction (lower intensity of GHG emissions), and adaptation, lowering emission-generating activities, reduced GHG emissions), removal of GHG, hazard forecasting (long-term forecasting of localized trends), developing early warning system, handling crisis, building infrastructure, safeguarding population, conserving biodiversity and withdrawal (environmental and digital annihilation).

(a) Macro level

The enhanced quantity of environmental emissions is a critical component of models used to forecast future temperatures. AI may assist such models by, for example, improving measurements or detecting the satellite remote-sensing data for extra analysis [18].

(b) Micro-level

Generators could use this level of emissions assessments to emancipate their products' carbon footprints, track their advancement in the direction of environmental, social, and governance (ESG) goals, and identify emission-cutting opportunities [19].

(c) Reduction

As a result of the global climate emergency, efforts to cut associated GHG emissions must be accelerated. Urgent and comprehensive mitigation actions to avoid the most disastrous effects of climate change. The reduction process has three components:

(i) **Reduced intensity of GHG emissions:** AI solutions can aid in the transition to new energy sources. Solar energy supply forecasts can assist us in identifying regions where there is potential for increased solar energy consumption, hence lowering GHG emissions.

(ii) **Reduced greenhouse gas emissions:** If policymakers adopt geoengineering to alleviate the effects of climate change, AI will be an important mechanism for expediting chemical research and assisting in the development of novel materials and processes that result in a decrease in emissions of greenhouse gases. Promotion of behavioral pattern and the related changes can also help to minimize energy use and emissions.

(iii) **Lowering Emission-Generating Activities:** AI may help to reduce emissions via enhanced supply chains through more accurate demand forecasting (to counteract overproduction) or more efficient product transportation (such as shorter delivery times and lower energy utilization). This might be accomplished by collecting data to develop models that estimate demand or improve transportation routes.

(d) Removal of GHGs

There are two forms of carbon dioxide removal: (1) Environmental Cleanup and (2) Technological Extraction. Natural ecosystems like as forests, algae, and wetlands play a significant role in reducing atmospheric carbon. Monitoring these ecosystems demands the gathering and processing of vast amounts of data, a situation in which AI thrives. Environmental cleanup can be supplemented by industrial processes, although these technologies are still in their infancy and have scalability challenges. AI would be a valuable tool in resolving these challenges as soon as feasible.

(e) Hazard Forecasting

Long-term forecasting of localized trends: To foresee localized long-term patterns and to anticipate the possible implications of climate change. For example, how likely is it that a major drought will occur at a certain place during the next decade and what are the possible consequences of the drought for the farming industry, the availability of water, and the wellness of people?

(f) Developing early warning systems

AI can assist us construct mechanisms for prior notification that can deliver timely notifications about imminent occurrences in addition to anticipating long-term trends. AI, for example, may assist us in identifying circumstances conducive to catastrophic weather occurrences such as flooding by evaluating data from weather stations, satellite photos, and sensor networks. AI can aid in wildfire prevention by utilizing data sources such as satellite images, real-time meteorological data, and social media posts to develop better fire detection and spread algorithms [20]. To create a dynamic wildfire risk map with an interactive fire spread simulation, a smart framework that integrates all of these technologies is required.

(g) Handling crisis

If a severe weather event occurs, AI can assist in managing the situation by providing decision-support tools. For example, AI might be used to identify those who are likely to be impacted by the incident and connect them with the services they require. AI may also be used to monitor the situation in real time and provide information on people's locations, infrastructure condition, and the status of rescue efforts.

(h) Building infrastructure

Drought may be mitigated with intelligent irrigation systems that use meteorological data and plant sensors to optimize watering schedules. AI-enabled flood defences that use real-time data on rainfall, river levels, and land elevation can help guard against flooding. Smart cities that control heating, cooling, and ventilation using sensor data can also help.

(i) Safeguarding populations

Substantial mobility makes up one of the likely effects of climate change. By providing decision-making tools for managing refugee camps, observing migration, and coordinating humanitarian efforts, AI can help us deal with all of this.

AI may also be used to track events in instantaneously and provide information on where individuals are located, the state of the infrastructure, and the progress of rescue efforts.

(j) Conserving biodiversity

Substantial climate change is anticipated to have an impact on migration. By offering instruments for making decisions for refugee camp oversight, observing migration, and coordinating humanitarian efforts, AI can help us deal with all of this. AI may also be used to track events in instantaneously and provide information on where individuals are located, the state of the infrastructure, and the progress of rescue efforts.

4.3 AI in Mitigating Agricultural Emissions

A significant contributor to climate change is agriculture, and agriculture is responsible for 10%–15% of global GHG emissions. Precision agriculture refers to maximizing agricultural inputs (such as fertilizer, water, and pesticides) on a focused, local basis, and occasionally even plant by plant, as opposed to uniformly treating all plants across large fields and farms [21]. It will be crucial in improving agriculture's sustainability in the coming years. The new and modern agricultural system has been very consuming and resource-intensive. Tons of fertilizers are used excessively, and are majorly responsible for GHG emission. Additionally, the GHG produced by fertilizers, nitrous oxide, is especially harmful since it heats the atmosphere roughly 300 times more than carbon dioxide. The World Economic Forum estimates that "if 15% to 25% of farms used precision agricultural methods, greenhouse gas emissions could be cut by 10%, water use could be cut by 20%, and crop yields could be increased by 15%". One way for providing farmers with real-time information about how to deploy resources on their fields efficiently is to employ computer vision on aerial data. This will show them where to apply more or less fertilizer, where to mend leaking irrigation systems, and other such information. This business model's attractiveness is that it is software-based, making it scalable and capital-efficient.

4.4 AI in Mitigating Electro Power Emission

The electro power domain is among the most significant patrons of GHGs. About 30 gigatons of CO2 were emitted globally in 2010 (Gt). When fossil fuels such as coal, oil, and natural gas are used to provide the heat needed to power steam-driven turbines, the electricity production industry accounts for roughly 12 Gt (or 40%) of global GHG emissions [22]. Most importantly, as more people switch to driving electric vehicles, there will be a rise in the need for electricity. If no action is made to decarbonize the electric grid, such increased electricity demand will result in higher GHG emissions. Nevertheless, the electricity sector is likely to shifting their dependency to renewable sources. Hydropower, solar, and wind energy are examples of renewable energy sources. Al can be used to improve the predictability of renewable energy sources. In addition to its powers in climate research, Al can be extremely helpful in attempts to cut emissions in the electric power industry.

For example, Google and DeepMind used machine learning to anticipate wind power output in the United States "36 hours ahead of actual generation", resulting in optimal hourly day-ahead delivery obligations [23]. Al, for instance, can be used to modify the propellers of wind farms to keep up with shifting wind directions, so reducing the intermittent problem with wind turbines. The most notable aspect of Al's capabilities is that it can increase the storage of renewable energy. Large-scale data analysis is required for such efficiency maximization, making Al participation suitable [24].

5. AI and Climate Change: Emerging Trends

5.1 International Perspective

Climate change poses a serious threat to the existence of the world and now AI is being used to address the problems of climate change. Although regulation of AI is in its inception stage, "Organisation for Economic Co-operation and development (OECD)" has developed a few principles for regulation of AI, which signals States´ unanimity on key principles.

The report outlines five basic emerging principles based on the concepts concerning artificial intelligence stewardship, namely:

(i) AI should help people and the earth by promoting inclusive growth, sustainable development, addressing climate change, and increasing wealth;

(ii) To create a just society, AI systems should be built to respect the rule of law, human rights, democratic principles, and diversity, and should incorporate appropriate protections, such as allowing for human intervention if necessary;

(iii) AI should be publicized in a transparent and responsible manner.

(iv) AI systems must be dependable and safe throughout their life cycles, and possible hazards must be regularly reviewed and handled.

(v) Organizations and individuals responsible for designing, deploying, or running AI systems must be held accountable for their correct functioning in accordance with the aforementioned principles.

In terms of agreements, focusing specifically on AI and associated climate change, the Global Partnership on Artificial Intelligence (A group of 24 countries and the European Union currently working on global ethical AI adoption plan for climate action and biodiversity preservation), in collaboration with the "Centre for AI & Climate", released the *Climate Change and AI: Recommendations for Government report in November 2021* which provides 48 particular suggestions for how authorities could possibly both support and address environmental issues using AI change.

Also, the study, the situation was exhibited at the Glasgow Climate Change Conference under the aegis of the UN, urges the countries to do the following:

(i) Enhance data ecosystems in sectors crucial to climate change, such as energy and transportation, by developing digital twins.

(ii) Improve research, innovation, and implementation assistance.

(iii) To shape the responsible development of AI as a whole, make climate change a major concern in AI strategy.

(iv) Promote expanded international collaboration.

Governments should create and execute rules and best practices for ethical AI practice and participative design in climate conditions. Certain guidelines might be developed as part of bigger efforts on ethical and responsible AI within international organizations, professional organizations, or other contexts. It is proposed that each record from climate-critical agencies be made freely accessible in terms of the production of information and open data standards be available without jeopardizing privacy, security, financial, or governance interests. Governments should also develop rigoros standards for assessing effect at the national and international levels.

This involves creating a rule to assist determine how to evaluate various sorts of systems and application areås. Universally agreed ethical norms and regulations may also influence the regulation of AI.

Further, the UNESCO Guideline on the Ethics of AI proposes to establish criteria for the governance of AI in UN Member States [25]. The recommendations aim at, among other things, creating a worldwide set of values, principles, and actions to assist governments in creating laws, policies, or other AI-related mechanisms that are in compliance with the laws of other nations.

Certain suggestions including fundamental principles such as equality for all people, human rights and basic freedoms preservation and promotion, environmental and ecological thriving, safeguarding variety and inclusivity, and living in peaceful, just, and interdependent communities, all of which should be promoted rather than jeopardized by AI technologies. It also includes the following principles that can be used to guide the design of AI systems:

(i) proportionality and do no harm;

(ii) safety and security;

(iii) fairness and non-discrimination;

(iv) sustainability;

(v) right to privacy and data protection;

(vi) human oversight and determination;

(vii) 67 transparency and explainability;

(viii) responsibility and accountability;

(ix) awareness and literacy; and

(x) multi-stakeholder and adaptive governance and collaboration.

5.2 National Perspective

In Japan, AI is utilized to transmit tsunami and floods notifications. AI systems and technology have been heavily used in China to construct smart cities or in Brazil to track the clearance activity in the Amazon.

(1) China

The country has approached the lens of AI in assessing the stability of environment and environment related concerns in a well legalized manner. Certain regulations and plans have been adopted by the country so as to encourage the growth of the AI sector. In recent years, China has also accelerated the adoption of specialized rules to oversee AI, including industrial ethics and algorithms.

The regulatory framework is comprised on the following elements:

(i) **Trustworthy AI White Paper:** This white paper framework came into existence on July 2021, established by the China Academy of Information and Communication Technology discusses the significance and difficulty of strengthening the 'reliability' of AI systems. Policy proposals in the white paper include crafting additional Chinese legislation on trustworthy AI, adopting commercial AI insurance plans, and taking a careful approach to research on artificial general intelligence (AGI) [26].

(ii) **Ethical Norms for New Generation AI:** These guidelines were established in September 2021 by the Ministry of Science and Technology to build up ethical norms while using the AI in China. The guidelines comprise of the information security, human control and accountability for AI, as well as the prevention of AI-related cartels.

(iii) **Internet Information Service Algorithmic Recommendation Management Provisions:** In March 2022, Cyberspace Administration of China (CAC) used AI algorithms to warn consumers when an AI algorithm is used to choose which content to display to them and to provide them with the opportunity to opt out of being targeted [27]. Also, data security was the crucial point of concern in the established **provisions.**

(iv) **Guiding Opinions on Strengthening Ethical Governance of Science and Technology:** This was established by the Ministry of Science and Technology aiming to improvise system of technology, improve governance, effectively avoid and regulate ethical issues, and generate social benefits.

(2) Japan

Digital ministers are anticipated to debate the human-centric approach to AI at the G7 meeting in Japan in 2023, which may include regulatory or non-regulatory policy measures. Since the host country, Japan's attitude to AI legislation may have a significant impact on world leaders' ability to reach an agreement.

The Japanese government published the Social Principles of Human-centric AI in 2019 as the guiding principles for deploying AI in society. The guiding principles' fundamental beliefs are sustainability, human-centeed dignity, and variety. Also, Ministry of Economy, Trade, and Industry (METI's) Governance Guidelines for AI Principles Implementation highlights the action goals for adopting the Social Principles and demonstrates how to achieve them with particular examples. It describes how to create and upgrade an AI governance structure in conjunction with stakeholders using an agile governance paradigm.

(3) Canada

AI and Data Act, June 2022, in the legislation, risk mitigation and information disclosure for high-impact AI systems will be required. The federal government submitted Bill C-27, titled The Digital Charter Implementation Act, 2022, on June 16, 2022. The bill's introduction of the Artificial Intelligence and Data Act is one of its three cornerstones (AIDA). If passed, the AIDA would be Canada's first law governing AI.

(4) Brazil

Brazil's House of Representatives adopted the Brazilian Artificial Intelligence Law (Bill No. 21/2020) on September 29, 2021, sending it to the Federal Senate for further consideration. If passed, the bill will set principles, obligations, and rules for creating and deploying artificial intelligence (AI) in Brazil. The measure would also bring Brazil's major legislation, such as the Brazilian Data Protection Law (LGPD) and the Brazilian Consumer Protection Code (CDC), into line. According to the legislation, an "AI system" is one that is based on a computer process that can learn, sense, and understand the external world by processing data and information from a set of human-defined objectives.

(5) European Union (EU)

In EU legislation, the environmental issues of AI are connected to:

(i) European green dal movement;
(ii) suggestions for a new legal framework for artificial intelligence and data stewardship;
(iii) basic elements of EU environmental legislation, such as 'sustainability' and 'integration'; and
(iv) the intersection of environmental law concepts and laws with AI legislation and other sets of standards that relate to AI, such as the general data protection regulation [28].

The Artificial Intelligence Act (AI Act) is a law proposed by the European Commission on April 21, 2021, with the goal of establishing a single regulatory and legal framework for AI. The scope of the Act includes all industries (except the military) and all forms of AI. The proposal, as a component of product regulation, does not impose rights on individuals, but rather controls suppliers of AI systems and companies that employ them in a professional capacity. Other very recent AI projects, along with the European Commission's AI Act (AIA), have continued to insist on a human-centric approach, despite the fact that it appears clear that the challenges, including those triggered by AI, should be addressed from an ontocentric, rather than anthropocentric, perspective [29].

6. Ethico-Legal Challenges of AI in Climate Study

AI has emerged as a significant tool to mitigate/address a few challenges of climate change but it is fraught with few ethical and legal issues. AI plays a dual role in assessing climate change, i.e., as a contributor to climate change [30], and as a mitigator and adaptor to the climate change. The ethical issues of both the roles of AI needs to be taken into account [31]. For example, AI contributes to climate change due to the increase in the carbon footprint by use of fossil fuels for electricity production, on the other hand it provides means and measures to combat climate change in mitigating and adapting the changes necessary for society. All the ethico-legal concerns need to be addressed so as to impart transparency and accountability in the application of AI. Following are the ethico-legal concerns encountered while using AI in the assessment of climate change.

(1) Accuracy

Accurate data collection needs to be filled in the machines. The accuracy required for an AI system is closely related to data, training time, effort, and computing power required, and it is an important component of interpretation and representation.

(2) Automation Bias

Private entities and corporations hold the authority to feed the data, whatever is feasible for them. Many independent researchers have developed AI tools and techniques to forecast weather patterns, having a deep connection with climate change. The ethico-legal issue is that such firms use their insights to constrain individuals' areas of activity, decreasing the common citizen's ability to make autonomous decisions. If the vested interest continues to be implemented by the private entities, large implications shall follow.

(3) Data Privacy

The personal data of the consumers get compromised with access to advanced technology. Data privacy is at threat in case of AI as it uses a huge set of data that can be a significant cause for raising concerns related to privacy and information security. Even though no personal data is processed in AI, it is a significant threat in detecting the usual pattern of the users. Data protection and privacy laws are well established in every country, still the use of AI has posed a challenge on the privacy laws as well, where new and significant ethical concerns can be raised in the due process of time [32]. AI systems may be vulnerable to new sorts of security flaws, such as model feeding attacks. The reliability issue is self-evident in the context of information security and the deployment of AI. If the AI system affects the output value of the desired result, it can be counted one of the ethical concerns in the use of AI.

(4) Sustainability/Sustainable measure/use

First and foremost, sustainability should be non-anthropocentric, which means that both humans and nonhumans should be preserved for their own sake. This is prompted by the scientific fact that humans are members of the animal world and part of nature, and that nonhumans frequently exhibit ethically relevant traits, such as the ability to experience pleasure and suffering or to live a life worth living. Second, long-term sustainability should be prioritized, even the astronomically distant future. The notion of temporal equality drives this: everything should be regarded equally regardless of time period. Finally, a significant amount of work should be directed on sustainability, with optimization taking precedence over sustainability.

Initially, AI should be employed to increase long-term sustainability and optimization. This involves tackling acute risks to the sustainability of global civilisation, such as global warming and pandemics, for existing and near-term forms of AI.

Second, emphasis should be directed to long-term forms of AI, which have the potential to be very important for long-term sustainability and optimization. Long-term AI is rarely explored in connection to sustainability, but the study argues that these problems are a better focus for AI and sustainability research. Long-term AI may help to solve issues such as global warming, but it may also offer new risks, particularly in runaway AI situations. Additionally, it may play a significant part in space expansion, which is critical.

(5) Lack of transparency of AI tools

'Transparency' is a contentious phrase in and of itself, but a lack of openness raises concerns about accountability. The lack of openness makes it more difficult to identify and remedy prejudice and discrimination [33].

(6) Unemployment

Use of AI undoubtedly leads to unemployment. It replaces the jobs and human intelligence by creating a new wave of automation thereby reducing the employment opportunities. Loss of job is not just a financial issue, but also a social and psychological one. The question of new job opportunities inclusive of fair distribution and availability of the employment can be one of the significant concerns in the ethical issues arising out of use of AI system.

(7) Justice

Climate change injustices have acquired a lot of attention lately. Notwithstanding the fact that high-income countries are responsible for the bulk of GHG emissions, low-income countries are expected to suffer more than high-income countries. When it comes to mitigation and adaptation measures, the UNFCCC emphasizes that equality requires high-income countries to take on more commitments than low-income ones.

7. Need for Regulatory and Policy Framework

As the use of AI in combating the menace of climate change raises many ethico-legal concerns affecting rights of individuals and society, there is dire need to regulate its use. For an effective regulation, there should be a commission/committee formed to perform an impact assessment on the use of trustworthy AI in the sphere of climate change, which shall enable the law-making bodies to come up with an effective policy and a regulatory framework [34]. The assessment should take into account the ethical usage and implementation of AI, particularly in the realm of climate change. The policies that should thereafter be made would be evaluated against the ethical, legal, and societal impacts, focusing on mitigation and adaptation of climate change. For high-risk AI systems, data, documentation, and accountability, information supply and openness, human oversight, and robustness, and accuracy will be necessary [35]. The policy framework should mandate to achieve the objectives and goals in promoting AI usage in the climate change field. There is the need of regulating the entire process of feeding and usage of data into AI while conducting any assessment or prediction of climate change. The regulatory mechanism governing the use of AI and climate change should be guided by the three Ms, i.e., management, monitoring, and modeling of climate action and environmental systems [36].

As discussed above, concerns are related to ethics for the usage of AI revolves around data ownership, autonomy, privacy and security, foreseeability, causation, control, and opacity. Any violation of above stated rights or principles should pass the test of few legal requirements and liability. Remedial measures should be provided for any violation of the above fixing liability on the perpetrator which will ensure effective and effective use of AI while using for the purpose of mitigation of climate change.

8. Conclusion and Suggestions

Future is uncertain and unpredictable, but it can be assessed by scientific study of the past and present. Human race is passing through a very crucial stage as climate change has presented an existential question for the world. Climate change has cross-border implications; it affects more than one country or region. Emphasizing on the safety and security of the future of the human race, United Nation Sustainable Development Goal (UNSDG) 13 aims at immediate action to fight climate change and its consequences. The application of AI for mitigation significantly aids in accomplishing the UNSDGs. By adopting and integrating AI for the assessment of climate change leading to adaptation measures, the human race will benefit immensely. As it is filled with few ethico-legal concerns, in the absence of clearly articulated regulatory framework it will be perilous to proceed with such a technology/tool. Hence, there is a need to frame policies and law for its regulation for the greater benefit of the society. The policy approach should establish a clear, practical, and reasonable framework to guarantee that AI is employed in ways that will uphold people's rights and gain their trust, preparing the world for the digital era and facilitating combating climate change.

References

[1] IPCC is a United Nations organization that assesses and reports the science relating to climate change.
[2] UNESCO. Artificial Intelligence for Sustainable Development: Challenges and opportunities for UNESCO's science and engineering programmes. 50 pages. August 2019. Online: https://unesdoc.unesco.org/ark:/48223/pf0000368028.
[3] The Stockholm Declaration of 1972, also known as the Declaration of the United Nations Conference on the Human Environment, was the first global environmental declaration issued by the United Nations.
[4] The United Nations Framework Convention on Climate Change created an international environmental agreement to counteract "destructive human interference with the climate system", partly by stabilizing atmospheric greenhouse gas concentrations.

[5] UNFCCC. United Nations Framework Convention On Climate Change. United Nations, FCCC/INFORMAL/84 GE. 05-62220 (E) 200705, Secretariat of the United Nations Framework Convention on Climate Change, Bonn, Germany. 24 pages. Online: http: unfccc.int/resource/docs/convkp/conveng - Google Search. 2019.

[6] Rio Declaration on Environment and Development, UN Doc. A/CONF.151/26 (Vol. I), 31 ILM 874 (1992), Rio de Janeiro. 14 June 1992. Online: https://www.un.org/en/development/desa/population/migration/generalassembly/docs/globalcompact/A_CONF.151_26_Vol.I_Declaration.pdf. Accessed on 19 February 2023.

[7] Coeckelbergh, M. Artificial Intelligence: Some ethical issues and regulatory challenges. pp. 31–34. *In*: *Technology and Regulation*. Online: https://doi.org/10.26116/techreg.2019.003. 2019.

[8] Boden, M.A. *Artificial Intelligence: A Very Short Introduction*. Reprint Edn. 3. Oxford University Press, Oxford. 2018.

[9] Gailhofer, P., Ankeherold, Jan Peter Schemmel, Cara-Sophie Scherf, Cristina Urrutia, Andreas R. Köhler, and Sibylle Braungardt. The Role of Artificial Intelligence in the European Green Deal, Study for the special committee on Artificial Intelligence in a Digital Age (AIDA), Policy Department for Economic, Scientific and Quality of Life Policies. European Parliament, Luxembourg. 2021.

[10] Fortune. https://fortune.com/2022/09/19/climate-week-artificial-intelligence/. Accessed 18 February 2023.

[11] Robinson, D. AI in Climate Change: Machine Learning Helps Identify Methane Leaks in Gas Wells. *NS Business*. 2018.

[12] Huang, J. Exploring Climate Framework Laws and the Future of Climate Action, Pace. pp. 6206. 2021.

[13] Naik, V., et al. Chapter 6: Short-lived climate forcers. pp. 6–7, 6–8. *In*: Masson-Delmotte, V. et al. (Eds.). *Climate Change 2021: The Physical Science Basis*. Contribution of Working Group, I to the *Sixth Assessment Report of the Intergovernmental Panel on Climate Change* (Eds.). 2021.

[14] Marco lansiti and Paul, D. The simple economics of Artificial Intelligence; Karim Lakhani, competing in the Age of AI. *Harv. Bus. Rev*. pp. 3–4. 2019.

[15] Dreyfus, G., Xu, Y., Shindell, D., Zaelke, D., and Ramanathan, V. Mitigating climate disruption in time. *International Journal of Law and Humanities*, pp. 223, 235–236. 2019.

[16] Collins, H.M. *Artificial Experts: Social Knowledge and Intelligent Systems*. MIT Press, Cambridge, MA. 1990.

[17] Vihul, L. International Legal Regulation of Autonomous Technologies, Centre for International Governance Innovation. Online: https://www.cigionline.org/articles/international-legal-regulation-autonomous-technologies/. 2019.

[18] Ibid.

[19] Ibid.

[20] Beer, David. 2017. The social power of algorithms. *Information, Communication & Society*, 20(1): 1–13. https://doi.org/10.1080/1369118X.2016.1216147.

[21] Ibid.

[22] Lehr and amp, Paul, Ohm. Playing with the Data: What Legal Scholars Should Learn About Machine Learning. 51 *U.C. Davis L. Rev*. pp. 653, 669–702. 2017.

[23] Klučka, J. General overview of the Artificial Intelligence and international law. *In*: Klučka, J., Bakošová, L., and Sisák, L. (Eds.). 2019.

[24] Sands, P. *Principles of International Environmental Law*. (2nd Edn.), Cambridge: Cambridge University Press. 1246 pages. ISBN: 978-0-51-181351-1. 2003.

[25] Van de Poel, I., Fahlquist, J.N., Doorn, N., Zwart, S., and Royakkers L. The problem of many hands: Climate change as an example. *Science and Engineering Ethics*, 18: 49–67, 2012.

[26] More than Digital. China and Europe lead the way in AI. https://morethandigital.info/en/china-and-europe-lead-the-way-in-regulating-artificial-intelligence-ai/. Accessed on 21 February 2023.

[27] China Briefing. AI and regulatory measures in China. https://www.china-briefing.com/news/ai-in-china-regulatory-updates-investment-opportunities-and-challenges/. Accessed on 21 February 2023.

[28] Alonso, E., Sherman, A.M., Wallington, T.J., Everson, M.P., Field, F.R., Roth, R., et al. Evaluating rare-earth element availability: A case with revolutionary demand from clean technologies. *Environmental Science and Technology*, 46(6): 3406–3414, 2012. doi: 10.1021/ es203518d.

[29] Ibid.

[30] Nordgren, A. Artificial intelligence and climate change: Ethical issues. *Journal of Information, Communication and Ethics in Society*, 21(1): 1– 15. https://doi.org/10.1108/JICES-11-2021-0106. 2023.

[31] UNESCO. Recommendation on the Ethics of Artificial Intelligence. https://www.unesco.org/en/artificial-intelligence/recommendation-ethics. Accessed on 27 February 2023.

[32] Praha, Nakladatelství Leges. Artificial Intelligence from the Perspective of Law and Ethics: Contemporary Issues, Perspectives and Challenges. pp. 13–24. ISBN: 978-80-7502-590-6. 2021.33. World Commission on Environment and Development. The Report of the World Commission on Environment and Development: Our Common Future, Geneva. Para. 27, A/42/427. Online: https://sustainabledevelopment.un.org/content/documents/5987our-common-Future.pdf. 1987.

[34] UN General Assembly. Transforming our world: The 2030 Agenda for Sustainable Development. pp. 45, 47–52, A/RES/70/1. Online: https://www.refworld.org/docid/57b6e3e44.html. 2018.

[35] Herbert, A. Simon. Rational choice and the structure of the environment. 63 *Psychol. Rev*., pp. 129, 136. 1956.

[36] Stein, Amy L. Artificial Intelligence and Climate Change. *Yale Journal on Regulation*, 37(3): 890–939. 2020.

CHAPTER 11

Workforce Forecasting Using Artificial Intelligence

Alwin Joseph,[1,*] *Amala Johnson*[2] and *Aashiek Cheriyan*[3]

1. Introduction

Workforce forecasting predicts an organization's future demand and supply of the workforce. Each organization has its strategies to manage and track the appropriate workforce. The adequate forecasting technique for the workforce involves data analysis and pattern mining from various data points. Some of the critical attributes considered for the analysis and the forecasting of the workforce requirement include the data such as demographics, economic trends, and labor market conditions; these help in calculating informed predictions about future workforce requirements [1,2]. The primary aim of workforce forecasting is to ensure that an organization has suitable employees with the appropriate skills to meet its business needs by helping organizations make informed decisions about staffing levels, employee training, and other workforce management strategies.

Artificial intelligence (AI) can play a significant role in workforce forecasting. Tech giants like Amazon, Google, and Meta are using AI-based tools to forecast the demand of users and allocate their physical resources to deliver their products and services flawlessly. Similarly, AI-powered tools can help organizations analyze large amounts of data, identify patterns and trends, and make more accurate predictions about future workforce requirements. AI-based prediction systems can automate repetitive tasks, reducing the time and effort required for manual analysis and reducing the risk of human errors [3,4]. The tools provide accurate results when provided with

[1,2] Assistant Professor, Department of Data Science.
[1] Email: amala.johnson@christuniversity.in
[3] Assistant Professor, School of Business and Management (BBA).
[1,2,3] CHRIST (Deemed to be University) Pune Lavasa Campus, India.
Eamil: aashiek.cheriyan@christuniversity.in
* Corresponding author: alwin.joseph@christuniversity.in

more attributes. At the same time, humans find it very difficult to monitor the various factors that can directly or indirectly affect workforce requirements.

AI-powered workforce forecasting techniques are categorized into multiple groups; these include Predictive Analytics, Natural Language Processing (NLP), Machine Learning (ML), and Neural Networks (NNs). Predictive analytics algorithms analyze historical data to predict future workforce needs and trends. NLP algorithms can help organizations analyze large amounts of unstructured data, such as job postings and resumes, to identify workforce trends and skills requirements. ML algorithms analzse historical data and predict future workforce demand and supply. At the same time, NN is an ML algorithm that forecasts workforce demand and supply by analysing complex relationships between variables [5,6]. In most use cases of the prediction techniques, they parallelly employ multiple algorithms or systems to validate and recheck to assure confidence; even human assistance is made with AI predictions to validate the predictions and take appropriate actions based on the results. AI-based forecasts can also help plan training programs for the existing employees and help them get trained in the new business needs.

Applying AI techniques in the organization's daily operations has been a common practice. The AI process has been applied to various use cases by organizations to make more informed decisions about workforce strategies and improve their ability to meet business needs [7]. AI is increasingly being used in the recruiting process. Many AI-based approaches are devised to streamline and automate various tasks to identify the best and most talented workforce. These processes include Resume Screening, Job Matching, Interview Scheduling, Candidate Engagement, and Diversity and Inclusion. AI adoption will help improve efficiency, accuracy, candidate experience, everyday operations, and plans and predict various needs of customers, employees, and more [8,9,10]. However, ensuring that AI-powered tools are used ethically and in compliance with privacy and laws is essential. Various decisions made with the supervision of AI techniques sometimes need more ethical aspects and may affect the employees or the overall functioning of the organization's work culture.

2. Workforce Forecasting

Workforce forecasting is crucial for each organization; several data points and scenarios must be handled when developing AI-based forecasting models. Workforce forecasting is aimed at helping in the efficient planning and organisation of the organization's daily activities without the impact of employee retention and other factors. Considering all the elements and planning the future is crucial for any organization. Forecasting future needs, adapting, and deciding to meet future needs make companies more profitable. The forecasting can be affected by the heavy attrition rate that persists [11]. Several data concerning the organization help build predictive models to predict future needs. Some of the commonly used data points for designing a workforce forecasting model are represented in Fig. 1. The data points frequently relate to the organization, society, and other social and economic factors, which are more related to determining an organization's future. It is hard to define and identify all the factors that will significantly impact workforce forecasting; however, the primary attributes that can influence the forecasting include those mentioned.

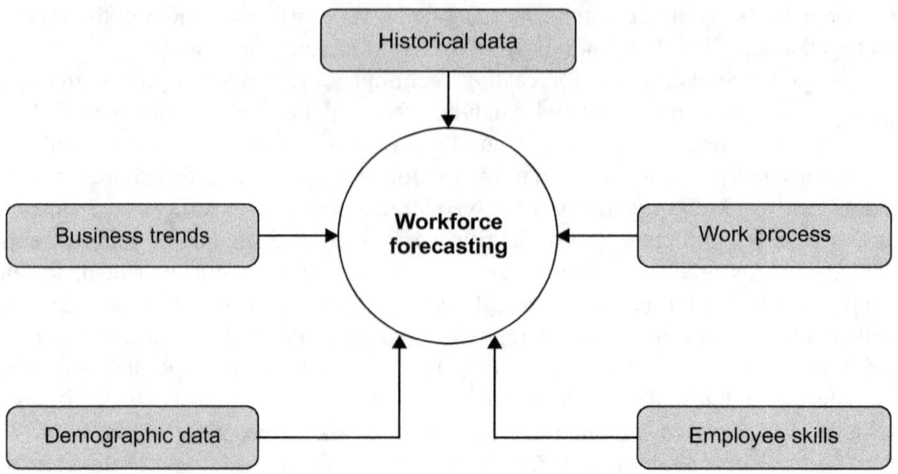

Fig. 1: Type of data used for workforce forecasting models.

The model's accuracy concerns the data it can consume when the predictions are happening, so it is essential to have the maximum available data points to create the models required for the forecasting technique. Workforce forecasting involves a lot of assumptions, data processing, and clubbing of multiple AI models and techniques, and constant work is required to monitor and maintain the efficiency and accuracy of these techniques.

Gathering and pre-processing all the data related to the various factors facilitates the validity of the AI model, making the predictions more accurate. Some critical data points commonly used for forecasting the workforce needs of an organization include the following:

1. **Historical data:** Organizations often consider historical data, such as employee data, sales data, weather patterns, and demographic data, when developing AI-based workforce forecasting models. This data helps to identify patterns and trends in employee demand. Usage of historical data helps identify similar behaviors that had happened before; it can be demanded in more workforce or helps reduce employee attrition and make other informed predictions. By analyzing past workforce data, organizations can identify factors influencing workforce demand and supply, such as changes in industry trends, economic conditions, and technological advancements. Even though historical data plays a crucial role in predicting the future, past trends and patterns may only sometimes be a reliable indicator of future demand or supply, mainly if there are significant changes in the industry or economic conditions. Organizations should also consider other factors influencing workforce demand and supply, such as technological changes, demographics, and government policies.
2. **Business trends:** Business trends such as changes in the economy, new product launches, and fluctuations in customer demand plays a huge role in developing AI-based workforce forecasting models. These trends can impact employee demand, and organizations use AI to identify these impacts and make accurate forecasts. Business trends data can be gathered from various sources, including

the daily news, surveys, purchase patterns of certain items, and other socio-economic factors. Some other key business trends that must be considered while forecasting the workforce include the amount of data-driven decision-making, automation and AI usage, workforce diversity and inclusion, and agile workforce strategies that can give a much clearer idea of the need for the future workforce.

3. **Demographic data:** Demographic data, such as population trends and changes in the workforce composition, retirement and turnover, and diversity inclusion, can also impact workforce needs. Organizations use AI to analyze demographic data and incorporate it into their workforce forecasting models to identify the trend concerning demographic choices; commonly considered demographic attributes are age, gender, ethnicity, educational qualifications, organization's geographical location, nationality, and regionality of the employee and experience. It provides valuable insights into the workforce's size, composition, and characteristics. Organizations can anticipate future workforce needs by analyzing demographic data and developing strategies to address potential labor shortages or surpluses, recruit and retain a diverse and skilled workforce, and track progress on diversity and inclusion initiatives.
4. **Employee skills:** Employee skills can be considered another factor that plays a huge role in planning the workforce while developing AI-based models; it has to be taken care. The data on employee skills help identify the high-demand skills and match employee supply with demand. Skill mapping, succession planning, training and development, and talent acquisition are everyday tasks related to employee skill requirements. Analyzing employee skills also helps to identify potential skill gaps or shortages and develop strategies to address them. Organizations can build a more resilient and adaptable workforce to meet future challenges better.
5. **Work processes:** Work processes, such as the number of employees required for a particular task, are also considered when developing AI-based workforce forecasting models. Organizations use AI to optimize their processes, reducing the number of employees needed for a job and improving efficiency. Work process helps in workload analysis, process improvement, formulation of business strategies, and workforce flexibility. Organizations can build a more efficient and adaptable workforce to better meet future challenges by analysing the work process and developing strategies to adjust staffing levels or redistributing work to better match demand.

These are some prominent attributes considered while developing AI-based workforce forecasting models [6,12]. The models created with these factors are more accurate, efficient, and flexible, enabling organizations to match employee supply with customer demand and improve workforce planning processes. Forecasting the workforce is always a crucial task for organizations that are directly linked to the success and profitability of the organization by employing an adequate workforce in the required time. Advanced planning and forecasting can be done with the help of more data points from these categories to identify the efficient time for each employee to work and then employ them at the right time and many more insights on maximizing the organizational productivity and profitability.

2.1 Need for Workforce Forecasting

Workforce forecasting identifies the essential workforce for any organization; it is critical and heavily depends on determining its efficiency concerning growth. Forecasting helps to plan and meet future workforce needs [8,13,14]. Some key reasons organizations need to engage in workforce forecasting include the following.

1. **Business growth:** As an organization grows, its workforce needs may change. The organization requires hiring new employees or reassigning the roles of existing employees to manage them better. Forecasting and AI tools engagement help organizations plan for and manage these changes effectively.
2. **Resource allocation:** Workforce forecasting can help organizations determine the number of employees and the skills they need to meet their business goals. This information can be used to allocate resources and prioritize workforce development initiatives. Foreseeing the requirements and applying appropriate resources to projects/work will help organizations to complete the required work on time.
3. **Talent management:** Organizations can use workforce forecasting to identify future talent needs and develop strategies to attract, train, retain, and develop employees. AI tools and forecasting help automate the process and make it more streamlined and effective.
4. **Budgeting:** Workforce forecasting can help organizations estimate the cost of hiring and retaining employees, which is essential for budgeting and financial planning. The forecasting techniques can tell the cost requirements to retain, hire or any other process. helping in planning the organization's future activities accordingly.
5. **Compliance:** Certain industries, such as healthcare and education, are subject to regulations that require a minimum number of employees to meet the needs of their customers. Workforce forecasting can help organisations ensure they comply with these regulations.
6. **Competitive advantage:** Organizations that accurately forecast their workforce needs can be better positioned to compete in their industry, as they can quickly respond to changing business conditions and workforce demands. In the current competitive world, competitive advantage is an essential factor; the first come will always have the upper hand, and the needs and trends can be analysed and identified with the help of AI tools that helps organizations to plan their future business goals and set the workforce requirements to achieve their goals.
7. **Anticipating future labor needs:** Workforce forecasting helps organizations anticipate and ensure adequate employees with the right skills and competencies to meet current and future business demands. By analysing the current market trends, and business needs, organizations can identify future needs and concentrate on identifying the best workforce for the conditions.
8. **Identifying potential skill gaps or shortages:** Workforce forecasting helps organizations identify and develop strategies to address them, such as training and development programs or targeted recruitment strategies. Identifying the

current workforce's shortcomings and helping them achieve the skill gaps and shortages is a cost-saving approach to maintaining and strategizing the workforce.

9. **Reducing labor costs:** Workforce forecasting can help organizations reduce labor costs by identifying opportunities to improve efficiency and productivity, such as process improvements or automation. The application of new technologies in the work environment helps organizations to adjust their workforce to focus more on essential job roles by humans, and the frequently done, repeated jobs can be replaced with less number of workforce, thus reducing the overall expense in labour; such factors can be analyzed and identified with the help of new technologies.
10. **Improving workforce diversity and inclusion:** Workforce forecasting can help organisations identify areas where they need to improve diversity and inclusion, such as recruiting more employees from underrepresented groups or developing targeted training programs for diverse employees. Having a diverse workforce helps the organisation and employees to work together better; the synergy of these can be analysed with the help of AI and foresee the implications of the groups, and changes can be made accordingly to maintain the growth of the employees and the organization.
11. **Succession planning:** Workforce forecasting helps organizations identify and develop employees who have the potential to fill key leadership positions in the future, ensuring continuity and stability in the organization. When an employee leaves the organization, an alternative for the employee must be identified to help the company move forward. Identifying exact replacements play a significant role, and AI can help analyze the existing workforce to determine the best replacement.
12. **Adapting to changing business needs:** Workforce forecasting helps organizations adapt to changing business needs, such as shifts in demand for specific products or services or changes in market conditions. With the updates in business needs and plans, the workforce must be adaptable to the organization's changes, or else there will be a high chance that the employees will leave the organization.

All these needs signify the requirement for proper workforce forecasting and identifying an adequate workforce for the organization's future growth. Figure 2 shows that it is essential for the organization to do appropriate workforce forecasting. Workforce forecasting must also include the company to identify its required profit shares. Organizations work to generate profit; to make a profit, they require the needed workforce that works towards the reason and gain profit. The factors discussed above are some of the motives that help the organization helps in achieving this goal.

Overall, workforce forecasting is a critical component of workforce management. It can help organizations make informed decisions about workforce strategies and ensure they have the necessary resources to meet their business goals. Identifying the best workforce helps the organization to achieve and excel in its business goals. Organizations use several AI techniques in planning and forecasting their workforce.

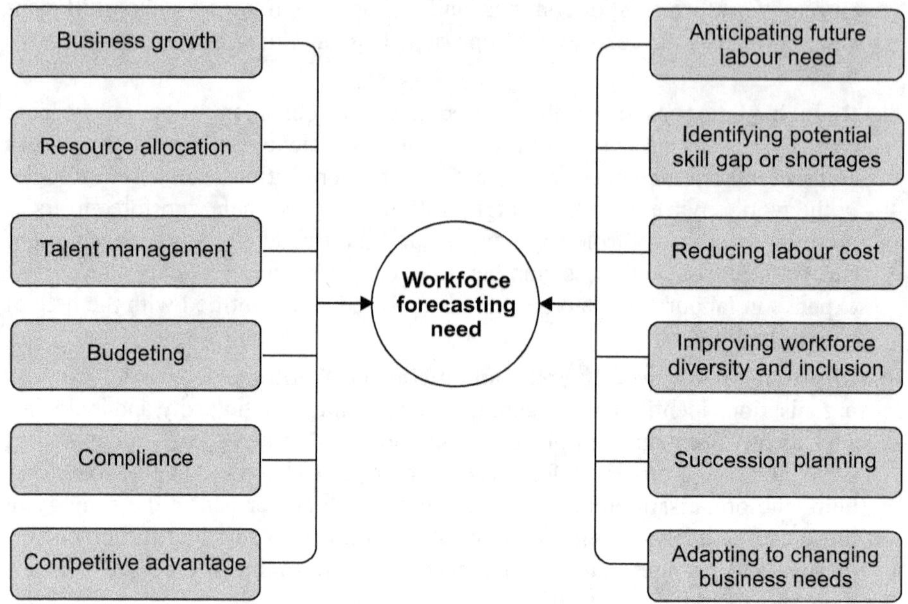

Fig. 2: Need for workforce forecasting.

The success of AI techniques cannot always be guaranteed as it is directly dependent on the data provided to the system. The decisions made with the support of AI techniques are more accurate since the predictions include a lot of attributes, and the study is based on learning trends from the data and analyzing whether the trends will repeat. Most AI techniques focus on understanding data trends, which upholds the data's importance. The more relevant data passed into the AI, the better the insights it can provide regarding the results. The most commonly applied AI techniques for workforce prediction and analysis include:

1. **Predictive Analytics:** Predictive analytics uses ML algorithms to analyse historical data and predict future workforce demand. Predictive analytics models can forecast employee turnover, workforce skills gaps, and other essential workforce metrics. It can forecast future labor demand and supply based on historical workforce trends, industry trends, and other factors. It is a commonly used technique to process data and make informed decisions. A lot of analysis can be done with the help of these analyzing techniques, including customer churn, which is an essential factor to be taken care of when planning the workforce.
2. **Machine learning:** ML algorithms can analyse large amounts of data to identify patterns and trends that may be missed by human analysis. ML algorithms can be used to develop more accurate workforce forecasting models. Clustering algorithms are used to group workers based on skills, experience, and performance, making it easier to identify workforce clusters; combing the current trends and needs helps organizations plan. The decision tree algorithm can analyze data and predict workforce demand based on input variables like location, industry, and job type.

3. **Natural Language Processing (NLP):** NLP analyses unstructured data sources, such as employee feedback and performance reviews, to gain insights into workforce trends and needs. Employee feedback to identify themes and trends. This can provide insights into employee satisfaction and engagement, which can help organizations develop more effective workforce retention strategies. Applying intelligent agents to understand the employees helps reduce biases in decisions.
4. **Chatbots:** Chatbots are used to automate routine tasks, such as scheduling interviews and answering candidate questions, freeing up time for recruiters to focus on more strategic tasks. Chatbots can also automate everyday HR tasks, such as onboarding and benefits enrolment, freeing HR staff to focus on strategic workforce planning. Apart from these, the data coming into the chatbot help in analyzing and identifying trends and issues. Workforce engagement can also be done with the help of chatbots.
5. **Image processing:** Image recognition algorithms can analyze employee photos or video footage to monitor engagement and identify potential workforce issues. Surveillance cameras help to understand the problems of the employees and give special care to the employees, which will help improve employee satisfaction and retention. It will also help in workforce planning to group similar employees who can contribute together, enhancing the efficiency of teams.
6. **Simulation:** Simulation tools can be used to model different workforce scenarios such as changes in market demand or the adoption of new technologies. This can help organizations identify potential labour shortages or skill gaps and develop strategies to address them. Simulations are one of the influential factors in validating the AI tools' predictions; the simulations' results help validate the impact of the decision changes. Simulations also can demonstrate the effects of changes proposed to bring into the dynamics of an environment, they can also show the impact of the various environmental factors, and the changes can be monitored and assessed easily in the employees' working environment.

Organizations use various AI techniques to support workforce planning and forecasting, which is represented in Fig. 3, which includes predictive analytics, NLP, chatbots, decision trees, clustering algorithms and much more [12,15–19]. By using AI, organizations can make more accurate and data-driven decisions about their workforce needs, improving competitiveness and reducing labor costs. Planning and forecasting the needs help them to meet the customer demand and serve the customers, reduce the costs and adjust the workforce demands based on the demand in the market.

2.2 Challenges in Workforce Forecasting

Workforce forecasting and planning are complex and challenging. The challenges are mainly due to the style of the organization's work, unpredicted socio-economic factors, unseen trends that arise in the market, and so on. Organizations need help with predicting and managing their workforce needs. The projections made by AI techniques primarily concern the data we can identify and pass to the many data

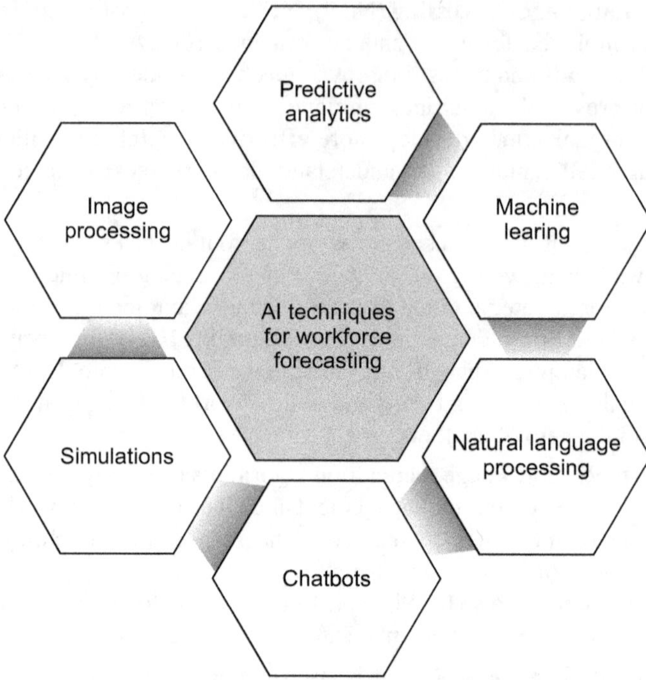

Fig. 3: AI Techniques for Workforce Forecasting.

points that will go unnoticed, significantly impacting the workforce. It may not be able to foresee many factors, but with the help of available data, the organization could get some insights from the current status. It can estimate what will be future requirements and challenges. Some of the most common challenges included while predicting and analyzing future trends are described and depicted in Fig. 4. The challenges are mainly due to the impact of external factors and internal management decisions and policies. The organization should be proactive and oriented on having a long-term plan, training, and developing the workforce to meet their future perspectives and approaches.

1. **Data availability and accuracy:** Accurate data is crucial for effective workforce planning, but organizations often need help to obtain and validate the data they need. This can make it difficult to make accurate predictions about future workforce needs. Structuring the available data, clearing, pre-processing and organizing the data are some of the challenges in making data available. Data privacy and ethical issues also play a significant role when using the organization's specific data for such analysis. The organizations may also need help to gather all the required data and consider building an effective model due to the significant number of data points. It becomes challenging to construct a complete model for 100% accurate forecasting.
2. **Rapid changes in the labor market:** The labor market is constantly changing, and it can be challenging to keep up with new trends and developments. Organizations must be able to adapt quickly to these changes to make informed

workforce planning decisions. Many changes arise due to technical and technological advancements; there are also a lot of unexpected and unpredictable changes in the market trends. The employees may or may not adjust to the changes, making predictions challenging in such situations. However, the rapid changes will demand quick rework on the AI predictions to model the impacts of the changes on the organisation.

3. **Resistance to change:** Employees and managers may resist changes to the workforce, especially if they are perceived as potentially leading to job losses or changes; this can make it difficult for organizations to implement workforce planning initiatives and make necessary changes. The changes the AI model predicts may not have the human touch sometimes, which is unacceptable to the stakeholders, and the AI sees things concerning data. It will only be able to identify and incorporate the proper solutions occasionally.
4. **Budget constraints:** Organizations often face budget constraints that limit their ability to invest in workforce planning initiatives and implement changes to their workforce. The constraints include spending much time setting up the AI model and prediction tools. Also, some of the challenges identified may need to be able to bridge at the time due to the organization's issues and challenges. Even if the AI tells the management to include more workforce for the coming quarter, sometimes the business might need to go according to the expectations, and the organization will not be able to invest extensively in further resources.
5. **Lack of management buy-in:** Effective workforce planning requires buy-in and support from senior management. Workforce planning will be challenging to achieve if they do not see the value of these initiatives and need to understand the complexities of the workforce planning process. The management should also be actively involved in the decision-making process of workforce planning, supporting, or opposing the decisions when it is proposed. Once the process is started, the planning process, when questioned, will create severe setbacks within the organization. The active involvement, understanding, agreement, acceptance, responsibility, credibility, commitment, and participation of management is required even if the AI tools have proposed workforce planning.
6. **Short-term focus:** Some organizations may focus too much on short-term goals and fail to plan for longer-term workforce needs. This can lead to labor shortages or skill gaps in the future. Some of how short-term focus can affect workforce forecasting include limited investment in employee training and development, reactive hiring decisions and failure to consider changing market conditions. Organizations should focus on long-term strategies and perspectives to overcome these challenges. Long-term plans should include investing in employee training and development to ensure that the employees have the skills and capabilities needed for future success, making proactive hiring decisions that align with the long-term goals, and regularly assessing and adapting workforce strategies in response to changing market conditions.
7. **Uncertainty:** The future is inherently uncertain, making workforce planning difficult. Economic or business strategy changes can impact future labour demand and supply. Many uncertainties and market fluctuations are happening

concerning various socio-economic factors. The uncertainties concerning the socio-economic factors may not be accounted for in the AI tools and models; thus, the predictions' validations are required for the success of AI-based forecasting. The models can be trained and improved later when new data of uncertainties arise, leading to a complex system that might even lead to unstable predictions. Handling uncertainties is a much more required and challenging task when automating the forecasting of the workforce required for an organization.

8. **Changing workforce demographics:** Demographic shifts, such as an ageing workforce or changes in the labor force participation rate, can make predicting future labor supply and demand difficult. Demographic factors such as gender, age, education, department, tenure, ethnic group, personality, cognitive style, organizational role, etc., are essential demographic variables that depend on an employee. Demographic variables play a significant role in predicting employee churn, a crucial factor that must be considered when planning for the future needs of the workforce. Analyzing the current workforce and understanding how many of the current employees will continue in the organization and how many of the employees will quit is an essential factor that has to be considered and taken care of. Thus, demographic variables are important factors leading to employee churn, and the forecasting model has to take this as input for planning the requirement for future employee requirements.
9. **Technology disruptions:** Technology disruptions, such as the adoption of automation or AI, can impact the demand for certain types of labor or change the skills and competencies required for specific job functions. The technological advancements and their applications, like Automation, Augmented Reality and Virtual Reality, Cloud Computing, Big Data and Analytics, and IoT (Internet of hings), will not be able to be identified by AI tools; adopting such technologies concerns the organization's policies. Technology adaptions will help organizations plan their workforce accordingly, reducing costs and saving expenses. Organizations should also focus on training and developing the workforce's skills to accommodate these changes in the work environment.
10. **The complexity of workforce dynamics:** Workforce dynamics can be complex, with multiple factors impacting workforce demand and supply. This complexity can make it challenging to forecast future labor needs accurately. Workforce dynamics also refer to the constantly changing interactions and relationships between different aspects of the workforce, including skills, experience, age, tenure, and diversity. The dynamics can affect workforce forecasting due to the difficulty in predicting future demand, the complexity of managing workforce diversity and the changing workforce expectations. To overcome these challenges, organizations must adopt a data-driven approach that considers all aspects of the workforce, including diversity, changing expectations, and evolving skillsets in the current market, by regularly analyzing the trends and requirements.
11. **Inadequate tools and resources:** Accurate workforce forecasting requires sophisticated tools and resources, such as predictive analytics software and workforce data management systems. Some organizations may need more

resources or expertise to use these tools effectively. Organisations need help to collect, analyse and interpret the data needed to make informed decisions, limited data, inability to leverage technology and need help identifying trends and patterns. To make informed decisions, the organisation needed data-driven decision-making and established clear metrics and goals for workforce planning to ensure effective investments.

12. **Difficulty in predicting future demand:** With uncertainty about the future, it can be challenging to predict future demand for specific skills or roles accurately. This can make it difficult for organizations to develop workforce strategies that align with future needs. Some difficulties that create and issue difficulty for workforce forecasting include shortages or surpluses in talent, inefficient workforce planning decisions already made, and inability to meet business needs.

The challenges in workforce forecasting commonly faced when the organization plans for AI or intelligent solutions are depicted in Fig. 4. Workforce planning challenges can be significant, but organizations that overcome these challenges can be better positioned to meet their workforce needs and achieve their business goals [20,21–23]. Figure 4 also shows the various significant challenges in workforce forecasting and planning. Organizations should analyze the impact of workforce challenges before proceeding with AI-based technology adoptions for planning the workforce. Analyzing the various case studies that focus on the usage of AI in forecasting their workforce requirements will also help the organization to overcome the potential challenges.

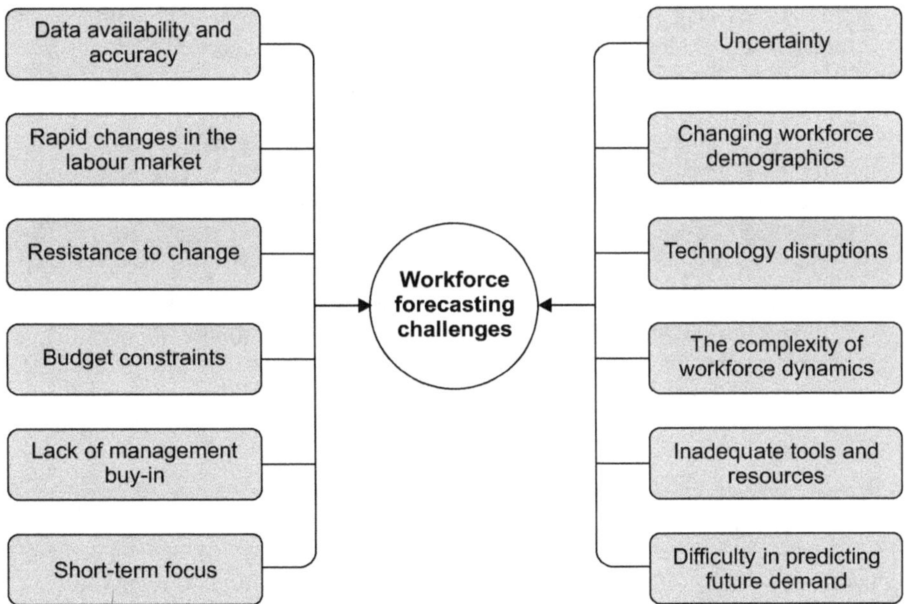

Fig. 4: Challenges in workforce forecasting.

3. Intelligent Techniques in Workforce Forecasting

The application of intelligent techniques for workforce forecasting has several factors that organizations consider when planning their workforce. Applying intelligent forecasting techniques includes the data-driven approach, where organizations gather data concerning all the various characteristics they plan to consider for forecasting workforce requirements. Some of the major factors considered for applying intelligent, data-driven technology for predicting workforce requirements are shown in Fig. 5.

1. **Business Goals and Objectives:** Organizations must align their workforce planning with their overall business goals and objectives, considering market trends, competition and regulatory requirements. The forecasting results must align with their goals and objectives. The workforce should be motivated and work towards helping the business achieve its goals in a mutually beneficial manner.
2. **Workforce Demographics:** Organizations must consider the demographics of their current and future workforce, including age, gender, ethnicity, and education levels, to ensure they have a diverse and inclusive workforce. Identifying the exact composition of the workforce is a crucial factor that helps in retention and planning furtheracquisition.
3. **Skills and Competencies:** Organizations must assess the skills and competencies of their current workforce and identify any gaps or shortages, so they can develop training and development programs to meet future workforce needs. Identifying the skills and competencies is also associated with the organization's plans. So that they can train the workforce to attain the new skills and competencies required to meet their requirements and update the workforce.

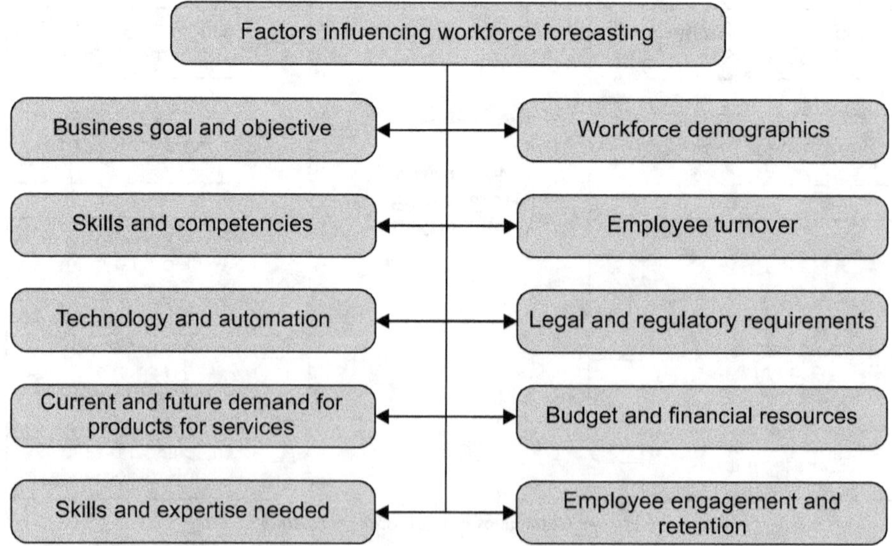

Fig. 5: Factors influencing workforce forecasting.

4. **Employee Turnover:** Organizations must consider employee turnover rates, including voluntary and involuntary turnover, and plan accordingly to reduce the impact of workforce loss. The employee turnover ratio is an essential factor that has to be taken care of while planning future needs. Thus, identifying the employees who are planning to leave the organization will help it identify alternatives or replacements and help in the smooth turnover of employees.
5. **Technology and Automation:** Organizations must consider the impact of technology and automation on their workforce, including the potential to automate specific tasks, and plan accordingly to ensure they have the right mix of skills to meet future workforce needs. Technology and automation help reduce dependency on the workforce and improve the workforce's performance and efficiency and the organization's overall functioning.
6. **Legal and Regulatory Requirements:** Organizations must be aware of and comply with all legal and regulatory requirements, such as minimum wage laws, equal employment opportunity laws, and workers' compensation laws, when planning their workforce. The legal and regulatory requirements related to the workforce must be prepared and implemented before implementing the requirements from the predictions to avoid legal complications concerning employee welfare and management.
7. **Current and future demand for products or services:** Organizations consider the current and future demand for their products or services when planning their workforce. This includes factors such as market trends, customer needs, and changes in the competitive landscape if the current and future demand for the products and services is essential for identifying the requirement for surplus or deficiency in the skilled workforce.
8. **Budget and financial resources:** Organizations consider their budget and financial resources when planning their workforce. This includes labor costs, employee benefits, and investments in training and development. Identifying the expenses can be spent on the workforce is another critical factor for planning the retention and hiring of the new workforce.
9. **Skills and expertise needed:** Organizations consider the skills and expertise needed to meet their business goals and objectives when planning their workforce. This includes technical skills, soft skills, and industry-specific knowledge. Identifying employees with the required skills and experience helps plan and gather the necessary workforce.
10. **Employee engagement and retention:** Organizations consider employee engagement and retention when planning their workforce. This includes factors such as job satisfaction, employee benefits, and opportunities for career development. These factors play into employee satisfaction and engagement, and help organizations to retain their employees and motivate them further.

Organizations must consider various factors when planning their workforce, including business goals and objectives, workforce demographics, skills and competencies, employee turnover, workforce requirements by business unit, technology and automation, legal and regulatory requirements, skills and required

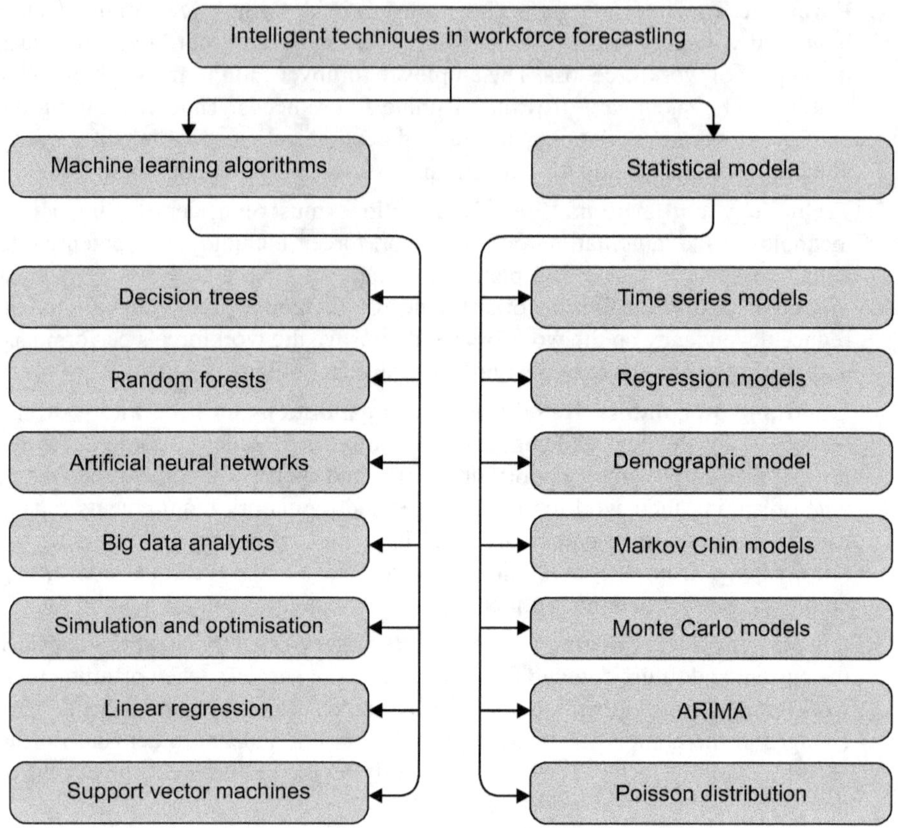

Fig. 6: Intelligent methods in workforce forecasting.

expertise, employee engagement and retention, current and future requirements of the products and services [3,24–28]. By considering these factors, organizations can ensure they have the right people in the right jobs and skills to meet their business goals and succeed in a rapidly changing business environment. The data-driven decisions are made with the help of AI-based ML algorithms and various statistical models that give insights when data has been given to the organization. The classification of intelligent techniques used for the predictive models for the workforce has been shown in Fig. 6. The models described can be individually used or clubbed together to create models to forecast workforce requirements.

3.1 Machine Learning Algorithms

ML algorithms can play a significant role in forecasting by providing more accurate and automated predictions about future workforce needs. ML models can be provided with various data that are cleaned and optimized, which will learn the patterns from the given data and give predictions when similar circumstances occur. The ML models work with learning from trends in the data. The information from the data has been processed and analyzed to make informed decisions. There are a lot of ML algorithms that can be employed in various phases of workforce forecasting. The most commonly used ML algorithms for workforce forecasting include:

1. **Linear Regression:** Linear regression is a technique used to model the relationship between two or more variables by fitting a linear equation to the observed data. The resulting equation is used to predict the value of the dependent variable for a given set of values of the independent variable. For example, it can forecast future staffing requirements based on historical employee turnover, attrition, and retirement data. Linear regression techniques can be used for forecasting and trend identification, helping the validations. Clubbing multiple linear regression models can also be applied to make informed decisions from historical data.
2. **Decision trees:** Decision trees are an ML algorithm that can forecast future workforce demand by analyzing decision points and the outcomes of different decisions. It is a useful tool in workforce forecasting because it allows analysts to identify patterns and relationships in workforce data and predict future staffing needs. Decision trees can be a valuable tool in workforce forecasting, helping organizations to identify key factors that influence workforce demand, predict future staffing needs, and develop strategies to improve retention and reduce turnover.
3. **Random forests:** Random forests are a ML algorithm that can forecast future workforce demand by combining the predictions of multiple decision trees. Random forests are a powerful ML technique that can be applied to workforce forecasting to improve the accuracy of predictions. Random forests work by building a large number of decision trees, each using a random subset of the available data and variables. By combining the predictions of multiple decision trees, random forests can reduce the risk of overfitting and improve the accuracy of predictions. In workforce forecasting, random forests can analyze large datasets with many variables, identify the most important predictors of workforce demand, and make accurate predictions about future staffing needs. Random forests are used to identify patterns in turnover data, predict employee attrition rates, and develop retention strategies that improve employee engagement and reduce turnover. Random forests can be a valuable tool in workforce forecasting, helping organizations make more informed decisions about staffing levels, hiring plans, and retention strategies.
4. **Artificial neural networks (ANNs):** ANNs are ML algorithms that forecast future workforce demand by analyzing complex relationships between variables. ANNs are ML algorithms used in workforce forecasting to analyze complex data sets and accurately predict future staffing needs. ANNs are modeled after the structure of the human brain, with layers of interconnected nodes that process and analyze data. In workforce forecasting, ANNs analyze large and diverse data sets, such as economic indicators, industry trends, and demographics, to identify patterns and relationships used to make accurate predictions about future workforce demand. ANNs can also be used to identify the key factors contributing to workforce demand and turnover, such as changes in business strategy, economic conditions, or employee satisfaction.

5. **Big data analytics:** Big data analytics involves analyzing large datasets to identify patterns and trends. This technique can help organizations identify patterns in workforce data that may not be apparent through traditional analysis methods. Big data analytics is used in workforce forecasting to analyze large and complex datasets, including employee data, market trends, economic indicators, and industry data, to identify patterns and relationships that inform hiring and workforce management decisions. By using big data analytics techniques such as data mining, ML, and predictive analytics, organizations can gain insights into workforce trends and predict future staffing needs more accurately. Big data analytics can also help identify workforce risks and opportunities, such as potential talent shortages or skills gaps, which can inform hiring and training strategies. Overall, big data analytics can be a powerful tool in forecasting, helping organizations make data-driven decisions that improve workforce efficiency, reduce costs, and support business growth.
6. **Simulation and optimization:** Simulation and optimization techniques involve using mathematical models to simulate different scenarios and optimize workforce planning strategies. These techniques can help organizations identify the most effective workforce strategies for achieving business goals. Simulation and optimization techniques can be used in workforce forecasting to model and analyze different scenarios and determine the optimal staffing levels, scheduling, and resource allocation strategies. Using simulation software, organisations can create virtual models of their workforce and simulate different scenarios to test the impact of different factors on workforce demand, such as changes in customer demand or staffing levels. Given various constraints and objectives, optimization techniques, such as linear or dynamic programming, can identify the most efficient and effective workforce management strategies, such as optimal scheduling, resource allocation, and staffing levels. Simulation and optimization techniques can help organizations to make informed decisions about staffing, scheduling, and resource allocation, reducing costs, improving workforce efficiency, and supporting business growth.
7. **Support Vector Machines (SVM):** SVM is a ML algorithm that can be used for workforce forecasting. It works by identifying patterns in data and making predictions based on those patterns. SVM can predict future workforce demand and supply based on employee skills, education, experience, and performance. SVMs create a boundary that separates data into categories based on specific features or characteristics. In workforce forecasting, SVMs can analyze large and complex datasets, such as economic indicators, market trends, and employee data, to identify patterns and relationships to predict future workforce demand. SVMs can also identify the most important predictors of workforce demand and turnover, which can inform hiring and retention strategies.

ML algorithms in workforce forecasting can assist organizations make more accurate predictions about their workforce needs and better decisions about workforce management strategies [26,29,30]. The decisions from the ML models are purely data-driven, which have been identified by the system. Many ML algorithms are used to create various models for forecasting; the models and techniques described here are very few that can be commonly used and identified in various applications. Using multiple ML algorithms to make an intelligent engine supporting decision-making is also common.

3.2 Statistical Models

Statistical models and techniques can help organizations forecast workforce needs. Statistical tools provide a powerful technique for workforce forecasting as they enable organizations to make data-driven decisions about their staffing needs. By analyzing historical data and identifying trends and patterns, organizations can make accurate predictions about future staffing needs and take proactive steps to ensure they have the right talent to meet those needs. Various statistical techniques like descriptive, inferential, and hypothesis testing can be applied to gain informed decisions from data. The commonly used techniques for workforce forecasting include:

1. **Time Series Models:** Time series models analyze historical data and predict demand. Examples of time series models include Exponential smoothing, Auto Regressive Integrated Moving Average (ARIMA), Generalized AutoRegressive Conditional Heteroskedasticity, Auto-regressive conditional heteroskedasticity, and Seasonal Autoregressive integrated moving average. The multivariate models include Vector autoregressive, Vector error correction model, and other seasonal models. Time series analysis is a statistical model that uses historical data to predict future trends. It works by analyzing patterns and trends and making predictions based on those patterns. Time series analysis can forecast future workforce demand and supply based on historical data related to employee turnover, promotions, and retirements. Time series analysis can help organizations to understand the dynamics of workforce demand and supply and to identify key drivers of workforce changes over time. By examining past trends and patterns in workforce data, organizations can develop predictive models that can aid them to forecast future workforce needs more accurately. Time series models can identify seasonality, cyclical patterns, and long-term trends in workforce data and make informed decisions on hiring, training, and workforce planning decisions. Additionally, time series analysis can be useful for identifying factors that may influence workforce demand, such as changes in market conditions or shifts in consumer behavior. By leveraging the insights gained from time series analysis, organizations can make more informed decisions about their workforce strategy and ensure they have the right people in the right roles at the right time.
2. **ARIMA (Autoregressive Integrated Moving Average):** It is a statistical model that can be used for time series forecasting. It analyses historical data and identifies patterns and trends to make predictions. It can forecast future workforce demand and supply based on historical data related to employee turnover, attrition, and retirements and forecast future workforce needs. By analyzing historical workforce data, such as employee turnover rates, recruitment trends, and seasonal fluctuations, ARIMA models can generate accurate forecasts of future workforce demand. These forecasts can inform recruitment, training, and retention strategies, allowing organisations to manage

their workforce needs proactively. They can also be used to analyze the impact of external factors on workforce demand, such as changes in market conditions or shifts in consumer behavior. By incorporating external variables into the ARIMA model, organizations can better understand their workforce needs and make more informed decisions around their workforce strategy.

3. **Regression Models:** Regression models predict workforce demand based on the relationship between independent variables and the dependent variable (workforce demand). Examples of regression models include linear, logistic, and multiple regression. Regression analysis is a statistical model that can be used to predict the relationship between variables. It analyzes historical data and identifies relationships between variables to make predictions. Regression analysis can forecast future workforce demand and supply based on employee skills, education, experience, and performance. Regression analysis can be used to identify the drivers of workforce demand and supply and to forecast future workforce needs to be based on these drivers. For example, a regression model may identify that economic growth is a significant driver of employee demand and use this relationship to forecast future workforce needs based on projected economic growth rates. Similarly, regression models can be used to identify the impact of other external factors, such as demographic changes or shifts in consumer behavior, on workforce demand. Regression analysis can also assess the impact of workforce initiatives, such as training programs or retention strategies, on employee retention rates. By analyzing historical workforce data, regression models can identify the effectiveness of these initiatives and inform decisions around future workforce strategy.
4. **Demographic Models:** Demographic models predict workforce demand based on demographic trends, such as population growth and ageing. Demographic models can forecast future workforce needs based on demographic trends. For example, a demographic model may identify the aging population and use this trend to forecast future workforce shortages in certain industries, such as healthcare or education, where the demand for skilled workers is expected to increase as the population ages. Demographic models can also be used to forecast changes in the composition of the workforce, such as changes in the gender or ethnic makeup of the labor force. These models can identify potential workforce shortages or surpluses by analyzing demographic data and informing recruitment and retention strategy decisions.
5. **Markov Chain Models:** The models analyze the transitions between different workforce states (e.g., employed, unemployed, retired) and predict future workforce demand. Markov chain analysis is a statistical model that can predict future events based on the current state. It works by analyzing historical data and identifying patterns to predict future events. Markov chain analysis can forecast future workforce demand and supply based on historical employee promotions, retirements, and attrition data.
6. **Monte Carlo Simulation:** Monte Carlo simulation models the uncertainty associated with workforce forecasting by generating multiple simulations of future workforce demand based on different input assumptions. Monte Carlo

simulation can estimate the probability of different workforce scenarios based on various input variables, such as economic indicators, industry trends, and demographic data. For example, an organization may use Monte Carlo simulation to estimate the probability of workforce shortages or surpluses based on different economic growth scenarios and their associated probabilities. Monte Carlo simulation can also assess the impact of different workforce management strategies on future workforce needs. By simulating the impact of different recruitment, retention, and training strategies, organizations can identify the strategies that are most likely to result in the desired workforce outcomes.

7. **Poisson Distribution:** Poisson distribution is a statistical model that can be used to predict the number of events that occur within a certain period. It works by analyzing historical data and identifying patterns to predict future events. Poisson distribution can forecast future workforce demand and supply based on historical data related to employee turnover, promotions, and retirements. Poisson distribution can be used to model the frequency of employee turnover or the number of job vacancies. By analyzing historical data on employee turnover rates, for example, an organization can use Poisson distribution to forecast future turnover rates and estimate the number of job vacancies that may need to be filled. Poisson distribution can also be used to assess the impact of different workforce management strategies on future workforce needs. By simulating the impact of different recruitment and retention strategies, organizations can estimate the impact on employee turnover rates and the resulting number of job vacancies. Overall, Poisson distribution is a useful tool for workforce forecasting, providing organizations with a probabilistic model for estimating the frequency of events that occur randomly over time. This technique allows organizations to better plan for future workforce needs and develop more effective workforce management strategies.

There are a lot of statistical models that help in workforce forecasting will depend on the specific requirements of the organization and the data available for analysis [6,34,1,31–33]. Choosing a suitable model for the problem and validating its predictions before making decisions based on the results is essential. Statistical models and techniques can provide a solid foundation for forecasting, but expert judgment and intuition are still important for interpreting results and making informed decisions. The simulations and results generated by the statistical models help in identifying the trends and foresee the future trend that might happen in the future to the data inputs given to the system.

4. Advantages and Disadvantages of AI in Forecasting

AI techniques are getting intelligent based on the data we give to the system to gain knowledge. The main challenge in applying AI techniques is gathering data; however, apart from the data gathering, there are a couple of other issues when applying AI-based technologies. The most common advantages and disadvantages of AI-based forecasting techniques are discussed below.

4.1 Advantages of Workforce Forecasting with AI Techniques

There are several advantages of using AI techniques in workforce forecasting, the main ones include:

1. **Improved accuracy:** AI techniques, such as ML algorithms, can analyze vast data to predict future workforce demand and supply accurately. The predictions made by the intelligent techniques are purely data-driven and are reliable; thus, the accuracy of the results produced has better accuracy compared to traditional methods.
2. **Automated predictions:** AI techniques can automate workforce forecasting, reducing the time and resources required to make predictions. The automated predictions made by the system are more accurate and reliable than the traditional methods. The predictions are focused on the data given to the system based on the historical data available to the organization.
3. **Enhanced insights:** AI techniques can uncover hidden patterns and relationships in workforce data that may not be apparent to human analysts, providing organizations with new insights into their workforce needs. The intelligent techniques can easily be applied to the data to gain improved and enhanced insights which will uncover the hidden patterns in the data to improve the predictions and decisions.
4. **Increased efficiency:** AI techniques can analyze data in real-time and provide organizations with up-to-date predictions about their workforce needs, enabling them to make more informed and timely decisions. The predictions made with intelligent techniques help improve the efficiency and accuracy of the decisions made.
5. **Scalability:** AI techniques can be easily scaled to handle large amounts of data, making them ideal for organizations with large and complex workforces. The intelligent models generated with AI techniques can be scalable. They will help to gain complex relationships with the data to improve the predictions and insights generated by incorporating various data points into the workforce predictions.

Using AI techniques in workforce forecasting can help organizations make more accurate and informed decisions about their workforce management strategies, leading to improved workforce planning and better outcomes for the organization and its employees. The decisions made by the AI models have better accuracy and reliability. The AI techniques can be improved to serve the decisions based on data and insights.

4.2 Disadvantages of Workforce Forecasting with AI Techniques

While there are many advantages to using AI techniques in workforce forecasting, several disadvantages should be considered.

1. **Data quality:** AI techniques are only as good as the data they are trained on, and if the data is of poor quality, the predictions made by the AI system will also be

of inferior quality. The data-driven insights have data as an important part and require the quality of data as an important part. Clean and accurate data can give precise and exact insights; gathering clean data is challenging, as most of the required data may be scattered and unavailable for direct analysis.

2. **Lack of transparency:** Some AI techniques, such as deep learning algorithms, can be difficult to interpret and understand, making it challenging to assess their predictions and identify potential biases. The biases in data and data gathering can produce biased decisions favorable to one party. Thus, analyzing the data and making the it clean and free of bias helps the models to be unbiased.
3. **Limitations of current AI technology:** While AI has made significant advances in recent years, there are still limitations to the technology that can impact its ability to make accurate predictions. For example, AI systems may be unable to effectively handle complex workforce scenarios or incorporate important contextual information into their forecasts.
4. **Ethical concerns:** There are also ethical concerns around the use of AI in workforce forecasting, including the potential for AI to perpetuate biases or displace workers. The AI, trained on the data, may be unable to capture human behavior and emotional constraints. The employees' data are also required to make certain concussions, and the data, when used without their permission, have various ethical issues.
5. **Cost:** Implementing AI systems for workforce forecasting can be expensive, and organizations may need to invest in specialized hardware and software to support the AI systems which need to be constantly updated, trained, and checked for bias and issues in the results, which has a lot of associated costs, making it difficult to afford and maintain.

AI techniques in workforce forecasting can provide significant benefits. Still, organizations need to carefully consider the potential drawbacks and limitations of the technology before implementing AI systems in their workforce forecasting processes, [35–40]. The decisions and insights are based on the data collected and passed to the AI system; if there is any issue with the data, it will affect the predictions; thus, it is mandatory to validate the decisions made by the AI systems.

5. Case Studies Based on the Implementation of AI for Workforce Forecasting

Several case studies highlight the benefits and successes of using AI for workforce forecasting; many MNCs (multinational corporations) apply multiple AI techniques to make informed decisions for their organization. Various case studies have been published by various sources relating to the adoption to plan the future workforce requirements and plans. Some of the companies that have adopted AI to make informed decisions based on the data and where they have applied the AI techniques have been discussed below:

1. **Walmart:** Walmart, the world's largest retailer, has implemented AI-powered workforce forecasting to predict store traffic patterns better and optimize

staffing levels. The AI system analyzes store traffic, weather patterns, and local events to make real-time predictions about workforce needs, allowing Walmart to optimize staffing levels and reduce labor costs.

2. **Hilton:** Hilton, one of the world's largest hotel chains, has implemented AI-powered workforce forecasting to optimize hotel staffing levels. The AI system analyzes data on occupancy rates, booking patterns, and local events to predict workforce needs, allowing Hilton to reduce labor costs and improve guest satisfaction.
3. **KPMG:** KPMG, a leading global professional services firm, has implemented AI-powered workforce forecasting to predict employee turnover and optimize staffing levels. The AI system analyzes data on employee demographics, job satisfaction, and performance metrics to predict workforce needs, allowing KPMG to reduce labor costs and improve employee engagement.
4. **Cognizant:** Cognizant has implemented an AI-based workforce forecasting system to optimize workforce planning and resourcing. The system uses ML algorithms to analyze historical and real-time data to forecast workforce demand and supply. Cognizant has reported significant improvements in workforce planning, with a 50% reduction in workforce planning cycle time, a 40% reduction in manual intervention, and a 30% increase in accuracy.
5. **AT&T:** AT&T has implemented an AI-driven workforce forecasting system to improve its talent acquisition and retention strategies. The system uses ML algorithms to analyze historical data on employee turnover, promotions, and external factors such as market trends and economic conditions to forecast future workforce demand and supply. The system also provides insights into the skills and competencies required for future roles, enabling AT&T to reskill and upskill its workforce as needed.
6. **Workday:** Workday uses AI to forecast workforce demand and supply, enabling organizations to make informed decisions about recruitment, retention, and talent management. The system uses ML algorithms to forecast future workforce requirements and analyze historical data on employee turnover, promotions, and external factors such as market trends. Workday has reported a 95% accuracy rate in their predictions, which has helped their clients reduce recruitment costs and improve workforce planning.
7. **John Deere:** John Deere, a leading agricultural and construction equipment manufacturer, has implemented AI-powered workforce forecasting to optimize staffing levels in its manufacturing plants. The AI system analyzes data on production schedules, equipment utilization, and workforce performance metrics to predict workforce needs, allowing John Deere to reduce labor costs and improve manufacturing efficiency.
8. **American Express:** American Express has used AI to optimize its workforce planning, using predictive analytics to identify customer demand trends and forecast staffing needs in its call centres. This has helped American Express to improve its service levels and reduce costs by avoiding overstaffing or understaffing.

9. **Starbucks:** Starbucks has used AI to optimize its workforce scheduling, using algorithms to predict employee demand based on factors such as store traffic, weather patterns, and historical sales data. This has helped Starbucks to forecast its staffing needs more accurately and improve its scheduling and staffing processes.
10. **Unilever:** Unilever has implemented an AI-driven workforce forecasting system to optimize recruitment and retention strategies. The system uses ML algorithms to analyze the job market and predict future demand and supply trends. The system also tracks employee behavior and patterns, such as career progression and tenure, to identify potential attrition risks and plan accordingly. As a result, Unilever has been able to reduce recruitment costs and increase employee retention.
11. **Accenture:** Accenture has used AI to optimize its workforce planning, using ML algorithms to predict future employee demand based on various factors, including business trends, market conditions, and demographic data. This has helped Accenture improve its workforce planning and match employee supply with customer demand.
12. **IBM:** IBM uses AI for workforce forecasting to predict the future demand and supply of employees. The system uses ML algorithms to analyze data points, such as employee turnover rate, attrition, promotions, retirements, and external factors, like market trends, to forecast future workforce requirements. IBM has reported a 95% accuracy rate in their predictions, which has helped them make informed decisions related to hiring, training, and workforce planning.

Various MNCs use the help of AI techniques to analyze the demand and forecasting for the workforce. These case studies demonstrate the benefits and successes organisations can achieve using AI for workforce forecasting, including improved accuracy, increased efficiency, and reduced labour costs [41–44]. These case studies demonstrate the potential of AI for workforce forecasting, showing that organizations can use AI to improve their accuracy, speed, and efficiency in workforce planning and management.

6. Conclusion

AI has emerged as a game-changer in the current world and can be applied to any domain, including workforce forecasting. By leveraging advanced algorithms and ML techniques, AI can help organizations accurately predict future workforce needs, identify skills gaps, and develop effective strategies to meet business goals. AI-based workforce forecasting techniques can help organizations to optimize their workforce planning processes, reduce costs, and improve productivity. Along with these advantages, some challenges are associated with implementing AI-based workforce forecasting models. These include the quality of data, the management policies, current trends in the market, employee satisfaction, skills of the employees and other challenges discussed in the chapter.

By investing in AI-based workforce forecasting solutions, organizations can gain a competitive advantage by building a skilled and agile workforce that can quickly adapt to changing market conditions and emerging technologies. Therefore, organizations must embrace AI-based workforce forecasting as a workforce planning strategy. By doing so, they can gain insights to help them make data-driven decisions, optimize their workforce planning processes, and build a workforce aligned with their business goals and objectives. AI tools can provide accurate decisions by analyzing large volumes of data from various sources, including employee records, industry trends, and economic indicators; AI algorithms can identify patterns and trends that may be difficult for humans to discern. There are significant merits of applying AI-based tools and techniques for forecasting the workforce; these include increased accuracy, speed, cost savings, and decision-making.

AI can also help organizations identify and mitigate workforce risks, such as compliance violations and safety hazards. By analysing data from various sources, including employee behavior, incident reports, and external factors such as weather and market trends, AI algorithms can identify potential risks and provide recommendations for mitigating them. Intelligent agents can help organizations plan their recruitment and retention strategies more effectively, ensuring they have the right number and type of employees to meet their business goals. AI can also help identify skills gaps and training needs, allowing organizations to invest in employee development and career progression.

Workforce planning and forecasting using AI techniques help organizations focus on long-term plans and goals; the AI tools help identify the right decisions to the data that the models could learn from and that is presented to the system to identify the hidden patterns and trends. The current trend in the market is to use the help of data-driven insights gathered from various sources. Identifying the trends and movement in the data helps organizations to foresee the hurdles and make appropriate intelligent decisions. When coming to business, the aim to maximize profit can be achieved by using informed intelligence and intelligent techniques along with strategic decision-making.

References

[1] Ward, D. Workforce demand forecasting techniques. *Human Resource Planning*, 19(1): 54–56, Mar. 1996, Accessed: Feb. 25, 2023. [Online]. Available: https://go.gale.com/ps/i.do?p=AONE&sw=w&issn=01998986&v=2.1&it=r&id=GALE%7CA18601818&sid=googleScholar&linkaccess=abs.

[2] Mayo, A. Strategic workforce planning—a vital business activity. *Strategic HR Review*, 14(5): 174–181, 2015. doi: 10.1108/shr-08-2015-0063.

[3] Garg, V., and Agrawal, R. *Transforming Management Using Artificial Intelligence Techniques*. CRC Press, 2020. [Online]. Available: https://books.google.com/books/about/Transforming_Management_Using_Artificial.html?hl=&id=jgf-DwAAQBAJ.

[4] Strohmeier, and Stefan. *Handbook of Research on Artificial Intelligence in Human Resource Management*. Edward Elgar Publishing, 2022. [Online]. Available: https://play.google.com/store/books/details?id=Zq1jEAAAQBAJ.

[5] Roy, M. AI-powered workforce management and its future in India. *Artificial Intelligence*. 2021. doi: 10.5772/intechopen.97817.

[6] Safarishahrbijari, A. Workforce forecasting models: A systematic review. *Journal of Forecasting*, 37(7): 739–753, 2018. doi: 10.1002/for.2541.

[7] Guenole, N., Feinzig, S., and Ferrar, J. *The Power of People: How Successful Organizations Use Workforce Analytics to Improve Business Performance*. Pearson FT Press, 2016. [Online]. Available: https://books.google.com/books/about/The_Power_of_People.html?hl=&id=wzsyjwEACAAJ.

[8] Milgram, L. Forecasting workforce requirements. *Managing Smart*. pp. 237–238, 1999. doi: 10.1016/b978-0-88415-752-6.50191-0.

[9] Ashri, R. Building AI-powered applications. *The AI-Powered Workplace*, pp. 31–40, 2020. doi: 10.1007/978-1-4842-5476-9_3.

[10] Ashri, R. The ethics of AI-powered applications. *The AI-Powered Workplace*. pp. 161–171, 2020. doi: 10.1007/978-1-4842-5476-9_12.

[11] Kim, J., Bae, H., and Jeong, S. Forecasting supply and demand for registered nurses workforce in Korea. *The Korean Data Analysis Society*, 19(2): 1083–1097, 2017. doi: 10.37727/jkdas.2017.19.2.1083.

[12] Cunningham, J.B., and Barton Cunningham, J. Workforce forecasting and planning. *Strategic Human Resource Management in the Public Arena*. pp. 89–107, 2016. doi: 10.1007/978-1-137-43241-4_5.

[13] Morzycki, A., Retrouvey, H., Alhalabi, B., Efanov, J., Youha, S., and Ahmad, J. The Canadian plastic surgery workforce analysis: Forecasting future need. *Plastic Surgery*, 26(4): 269–279, 2018. doi: 10.1177/2292550318800328.

[14] Srivastava, D.N., and Srivastava, N. Workforce analytics: Need of the modern organisations. *International Journal of Psychosocial Rehabilitation*, 24(5): 4410–4418, 2020. doi: 10.37200/ijpr/v24i5/pr2020156.

[15] Marinagi, C.C., Panayiotopoulos, T., and Spyropoulos, C.D. AI planning and intelligent agents. *Intelligent Techniques for Planning*. pp. 225–258, 2005. doi: 10.4018/978-1-59140-450-7.ch007.

[16] Nugus, S. Selecting and evaluating forecasting techniques. *Financial Planning Using Excel*. pp. 93–100, 2009. doi: 10.1016/b978-1-85617-551-7.00008-2.

[17] AI/Machine Learning is Disrupting Demand Forecasting. *Consumption-Based Forecasting and Planning*. pp. 135–183, 2021. doi: 10.1002/9781119809890.ch5.

[18] Chambers, J.C., and Mullick, S.K. Forecasting for planning: Qualitative techniques. *Planning Review*, 3(5): 13–27, 1975. doi: 10.1108/eb053732.

[19] Sruthi, S., Biswadip Basu, Mallik Subrata Das, Dr. I. Mohana Krishna, and Fantin Irudaya Raj, E. *Artificial Intelligence in Business Management*. Archers & Elevators Publishing House. [Online]. Available: https://play.google.com/store/books/details?id=5-1qEAAAQBAJ.

[20] Bondarouk, T., Parry, E., and Furtmueller, E. Electronic HRM: Four decades of research on adoption and consequences. *The International Journal of Human Resource Management*, 28(1): 98–131, 2017. doi: 10.1080/09585192.2016.1245672.

[21] Gibson, V.C. Workforce planning will help VA meet future challenges. *PsycEXTRA Dataset*. 2001. doi: 10.1037/e511172006-002.

[22] Erdheim, J., et al. Advancing workforce planning: Opportunities and challenges. *PsycEXTRA Dataset*. 2010. doi: 10.1037/e518392013-506.

[23] Lewis, R.E., and Ulibarri, C.A. Workforce planning for DOE/EM: Assessing workforce demand and supply. 1993. doi: 10.2172/10194882.

[24] Klein, G., Pick, R., and Jiang, J. Intelligent systems techniques and applications in product forecasting. *Intelligent Systems*. 2002. doi: 10.1201/9781420040814.ch10a.

[25] Review for 'Machine intelligent and deep learning techniques for large training data in short-term wind speed and ramp event forecasting.' 2020. doi: 10.1002/2050-7038.12818/v1/review2.

[26] Patel, S.S., Singh, A., Kumar, N., Akbar, S.A., and Panchariya, P.C. Institutional electricity load forecasting using classical and intelligent forecasting techniques. *2018 15th IEEE India Council International Conference (INDICON)*. 2018. doi: 10.1109/indicon45594.2018.8987111.

[27] Yang, W., Sun, S., Song, Y., Du, P., Hao, Y., and Wang, J. *Artificial Intelligence-Based Forecasting and Analytic Techniques for Environment and Economics Management*. Frontiers Media SA, 2022. [Online]. Available: https://books.google.com/books/about/Artificial_Intelligence_Based_Forecastin.html?hl=&id=fxSbEAAAQBAJ.

[28] Chan, C., and Nguyen, R. Artificial Intelligence techniques in forecasting applications. *Intelligent Systems*. 2002. doi: 10.1201/9781420040814.ch5e.

[29] Mahela, O. Review for 'Machine intelligent and deep learning techniques for large training data in short-term wind speed and ramp event forecasting.' 2021. doi: 10.1002/2050-7038.12818/v3/review3.

[30] Kahraman, C., and Onar, S.Ç. *Intelligent Techniques in Engineering Management: Theory and Applications.* Springer, 2015. [Online]. Available: https://play.google.com/store/books/details?id=DFf1CAAAQBAJ.

[31] Fuleky, P. *Macroeconomic Forecasting in the Era of Big Data: Theory and Practice.* Springer Nature, 2019. [Online]. Available: https://play.google.com/store/books/details?id=FvjADwAAQBAJ.

[32] Chatfield, C. *The Analysis of Time Series: Theory and Practice.* Springer, 2013. [Online]. Available: https://play.google.com/store/books/details?id=u1D5BwAAQBAJ.

[33] Brockwell, P.J., and Davis, R.A. *Introduction to Time Series and Forecasting.* Springer Science & Business Media, 2013. [Online]. Available: https://play.google.com/store/books/details?id=XtDkBwAAQBAJ.

[34] Pawlikowski, M., and Chorowska, A. Weighted ensemble of statistical models. *International Journal of Forecasting*, 36(1): 93–97, 2020. doi: 10.1016/j.ijforecast.2019.03.019.

[35] Singha, D., and Panse, C. Application of different Machine Learning models for Supply Chain Demand Forecasting: Comparative analysis. *2022 2nd International Conference on Innovative Practices in Technology and Management (ICIPTM).* 2022. doi: 10.1109/iciptm54933.2022.9753864.

[36] Guo, Z. *Intelligent Decision-making Models for Production and Retail Operations.* Springer, 2016. [Online]. Available: https://play.google.com/store/books/details?id=SQyPDAAAQBAJ.

[37] Chowdhury, S., Joel-Edgar, S. Dey, P.K., Bhattacharya, S., and Kharlamov, A. Embedding transparency in artificial intelligence machine learning models: managerial implications on predicting and explaining employee turnover. *The International Journal of Human Resource Management*, pp. 1–32, 2022. doi: 10.1080/09585192.2022.2066981.

[38] Explaining Artificial Intelligence, Machine Learning, and Deep Learning Models. *Risk Modeling.* pp. 55–70, 2022. doi: 10.1002/9781119824961.ch4.

[39] Frank, M.R, et al. Toward understanding the impact of artificial intelligence on labour. *Proc. Natl. Acad. Sci. U.S.A.*, 116(14): 6531–6539, Apr. 2019. doi: 10.1073/pnas.1900949116.

[40] UNESCO International Centre for Technical and Vocational Education and Training. *Understanding the Impact of Artificial Intelligence on Skills Development.* UNESCO Publishing, 2021. [Online]. Available: https://books.google.com/books/about/Understanding_the_impact_of_artificial_i.html?hl=&id=bnMnEAAAQBAJ.

[41] Dieleman, M., Shaw, D.M.P., and Zwanikken, P. Improving the implementation of health workforce policies through governance: A review of case studies. *Human Resources for Health*, 9(1), 2011. doi: 10.1186/1478-4491-9-10.

[42] Gilbert, D. Implementation of AI into federal agencies: Keeping an eye on the federal workforce. *SSRN Electronic Journal.* doi: 10.2139/ssrn.3891943.

[43] Clark, D. *Artificial Intelligence for Learning: How to use AI to Support Employee Development.* Kogan Page Publishers, 2020. [Online]. Available: https://play.google.com/store/books/details?id=bhn0DwAAQBAJ.

[44] Harvard Business Review. *Artificial Intelligence: The Insights You Need from Harvard Business Review.* HBR Insights, 2019. [Online]. Available: https://books.google.com/books/about/Artificial_Intelligence.html?hl=&id=RitXwQEACAAJ.

CHAPTER 12

Artificial Intelligence-Based Approaches for Anticipating Financial Market Index Trends

Elian Jose,[1,*] *Shine Raju Kappil*[2] and *Aashiek Cheriyan*[3]

1. Introduction

The stock market is an essential component of the world economy and significantly impacts how different countries handle their finances. Predicting stock prices has gained popularity recently since it can offer traders, investors, and policymakers useful information. Making informed financial decisions, lowering risk, and maximizing returns can all be facilitated by accurate stock price projections. Stock price prediction is a current research subject due to improvements in machine learning (ML) techniques, and several methodologies have been put forth in the literature. To increase the accuracy of stock price prediction, one method combines the feature extraction ability of convolutional neural networks (CNNs) with the classification strength of support vector machines (SVMs). CNNs are a subclass of neural networks that have excelled in voice and picture recognition. They can be taught to extract valuable features from the supplied data automatically. Contrarily, SVMs are a well-liked machine learning (ML) technique that has been applied for regression and classification tasks.

This study aims to examine how well CNN-SVM hybrid models predict stock prices. Building a hybrid model that can correctly forecast the stock values of the

[1] MSc Student, Department of Data Science, CHRIST University, India.
[2] Assistant Professor, Department of Data Science, CHRIST University, India.
Email: shine.kappil.christuiversity.in
[3] Assistant Professor, School of Business and Management, CHRIST University, India.
Email: aashiek.cheriyan@christuiversity.in
* Corresponding author: elian.jose@msea.christuiversity.in

Nifty 500 index is the key goal. Five hundred stocks listed on the National Stock Exchange of India make up the market capitalization-weighted Nifty 500 index. The dataset will be compiled using daily stock price data from 1 March 2007 through 31 December 2022. The two stages of the proposed hybrid model are feature extraction and classification. A CNN will be trained to extract features from the stock price data in the initial stage. The CNN will produce a feature vector representing the key input features after it has processed the input data. The primary goal is to create a hybrid model capable of accurately forecasting the stock values of the Nifty 500 index. A market capitalization-weighted index made up of 500 equities listed on the National Stock Exchange of India is called the Nifty 500 index. The dataset will be compiled from 1 March 2007 through 31 December 2022 and include daily stock price information. Feature extraction and classification are the first two steps of the proposed hybrid model. A CNN will be trained in the first stage to extract characteristics from the stock price data. The CNN will be built to process the incoming data and produce a feature vector that highlights its key components.

1.1 Why is Machine Learning in Forecasting?

Statistical analysis and other conventional stock price prediction techniques fail to provide investors with trustworthy information. Hence, ML approaches have been used more frequently lately to increase the accuracy of stock price prediction. This literature review examines the benefits of ML approaches for stock price prediction. The ability of ML algorithms to handle enormous amounts of data and discover patterns that might not be obvious to the human eye is a crucial benefit. Massive amounts of historical data, including financial data and news, can be analyzed by ML models to find trends and connections between various factors that could influence stock prices. For instance, ML models can discover relationships between stock prices and the publication of business earnings reports, news stories, or sentiments on social media [1]. Artificial neural networks (ANNs), particularly CNNs, are among the most often utilized ML techniques in stock price prediction. CNNs are frequently employed in image recognition and natural language processing (NLP), but are also helpful in stock price forecasting. Several convolutional layers that extract features from the input data are followed by fully connected layers that make the final prediction in the CNN architecture. The CNN architecture is a valuable tool for predicting stock prices because it automatically identifies the most critical elements from the input data [2].

The flexibility of ML algorithms to react to shifting market conditions is another benefit. Traditional statistical models frequently need to capture the underlying trends and patterns in the stock market since it is a complex and dynamic market. ML learning algorithms can modify their predictions in response to shifting market conditions. For instance, ML models can swiftly incorporate fresh data and revise their forecasts in response to news or events [1]. Moreover, stock price forecasting using ML models is possible for short-, medium-, and long-term time horizons. While medium- and long-term forecasts can offer information to long-term investors, short-term forecasts can be employed for day trading. Investors must be able to predict stock values over a range of periods, and ML models can do this with high accuracy and dependability [2]. Also, rather than just one stock, a portfolio of stocks can be employed with ML algorithms to forecast stock prices. ML models can be used to optimize portfolios based on expected returns and risks, which is an essential field of finance. For instance, ML algorithms can find correlations between stocks and pick an ideal portfolio based on the expected returns and risks [3].

2. Review of Available Methods

2.1 LSTM

Recurrent neural networks (RNNs) of the type of Long Short-Term Memory (LSTM) have recently been quite popular for time series forecasting tasks like stock price prediction. As historical stock price data is a time series, LSTMs can be trained to represent temporal relationships in data, which is critical for predicting stock prices.

Since Hochreiter and Schmidhuber originally described using LSTMs for stock price prediction in 1997, numerous studies have been conducted to ascertain how well they perform in this capacity. It has been discovered that LSTMs efficiently capture the intricate correlations between different stock price data elements and correctly predict future values. In time series forecasting jobs like predicting stock price, the capacity of LSTMs to handle long-term dependencies is one of their key advantages. LSTMs can store crucial data from earlier time steps and use it to generate precise predictions. This permits them to selectively retain or forget previous information by utilizing gates to govern the flow of information into and out of the memory cell. The efficiency of LSTMs in stock price prediction has been shown in numerous experiments. For instance, Zhang observed that their model beat other conventional time series forecasting models like ARIMA (auto-regressive integrated moving average) and GARCH (generalized autoregressive conditional heterskedasticity) when using LSTMs to predict the closing price of the Shanghai Stock Exchange Composite Index. One of LSTMs key benefit is its capacity for dealing with long-term dependencies, which is essential for time series forecasting jobs like stock price prediction. Important data from earlier time steps can be stored by LSTMs to create precise predictions. They accomplish this by controlling the flow of information into and out of the memory cell using gates, which enables them to remember or forget previous knowledge selectively. The usefulness of LSTMs in predicting stock prices has been shown in numerous studies. When the study utilized LSTMs to anticipate the closing price of the Shanghai Stock Exchange Composite Index, they discovered that their model outperformed more established time series forecasting models like ARIMA and GARCH [6]. Similar to this, Wang discovered that their model beat other conventional models like Support Vector Regression (SVR) and Random Forest (RF) when used to anticipate the daily closing values of the Shanghai Stock Exchange Composite Index [17]. Additionally, they demonstrated that LSTMs could capture the nonlinear correlations between the various stock price data elements, which is difficult for conventional models. When utilizing LSTMs for stock price prediction, there are a few drawbacks. The potential for overfitting, which happens when the model learns the noise in the training data instead of the underlying patterns, is one of the significant drawbacks. When the model is utilized to create predictions based on opportunistic data, this can result in subpar performance. Researchers have suggested several methods, including dropout and regularization to stop overfitting in LSTMs, to solve this problem. For instance, Shen, in their paper utilized dropout in their LSTM model for predicting stock price and discovered that it considerably enhanced the model's performance [3]. Due to their capacity to recognize the long-term dependencies and nonlinear interactions among numerous elements of stock price data, LSTMs have demonstrated considerable promise in stock price prediction tasks. However, caution must be exercised to prevent overfitting; this restriction can be addressed via strategies like dropout and regularization.

2.2 Logistic Regression

A statistical technique called logistic regression is used to analyze data, especially for binary classification issues where the result variable is dichotomous, or only having two possible values. Because it can help determine the chance of a particular stock's price growing or dropping depending on certain traits or predictors, logistic regression has been utilized for stock price prediction. This study of the literature examines how logistic regression is used to predict stock prices and how it performs when compared to other approaches. Many studies have employed logistic regression to forecast stock prices, and most of them have had positive outcomes. For instance, logistic regression was employed in a study to forecast the movement of stock prices, and the results showed that the strategy had a 71.9% accuracy rate [4]. Moving averages and the relative strength index (RSI), among other technical indicators, were employed in the study as predictors. Similar to this, Liu, in his study, found that logistic regression has a 75.7% accuracy rate when used to forecast the trend of stock prices using technical indicators like moving averages, RSI, and Bollinger Bands [5]. In a different investigation, Luo employed logistic regression to forecast the movement of stock prices, and the outcomes revealed that the approach had a 62.7% accuracy rate [6]. The study examined the price-to-earnings ratio, price-to-sales ratio, and financial metrics such as earnings per share as predictors. Also, the authors evaluated the effectiveness of logistic regression compared to SVMs and discovered that logistic regression outperformed SVM. Even with logistic regression's successful performance in stock price prediction, the technique has certain drawbacks. Its assumption of a linear relationship between the predictors and the outcome variable is one of its key drawbacks. In real-world situations, where the connection between the predictors and the result variable may not always be linear, this assumption may not always be true. Moreover, logistic regression assumes that the predictors are independent of one another, which may only sometimes be the case.

Researchers have suggested several improvements to logistic regression to get around its drawbacks. For instance, a modified logistic regression model was put out in a study to account for nonlinear correlations between the predictors and the outcome variable [7]. The authors extracted nonlinear properties from the data using a deep belief network (DBN), which they then fed into a logistic regression model. The outcomes demonstrated that the modified logistic regression model performed better in stock price prediction than the conventional logistic regression model. Numerous studies that used logistic regression to forecast stock prices had successful outcomes. However, the method has some drawbacks, namely the assumption that there is a linear relationship between the predictors and the outcome variable. Researchers have suggested modifying the approach to overcome these drawbacks, including adding nonlinear characteristics using DBNs. Overall, logistic regression is a helpful technique for predicting stock prices, and it may be used with other techniques like SVM and neural networks to enhance its effectiveness.

2.3 Linear Regression

A straightforward yet popular technique for predicting stock price is linear regression. A linear equation is fitted to historical stock price data to predict future stock prices. This method's popularity can be due to its clarity, interpretability, and simplicity in use. Nevertheless, because it presupposes a linear relationship between the input factors and the output variable, which may only sometimes be the case, its usefulness as a stand-alone method for stock price prediction is constrained. In recent years, there have been numerous attempts to increase the accuracy of linear regression models by adding cutting-edge methods like ML and deep learning. These methods have produced encouraging outcomes for forecasting stock prices, frequently beating conventional linear regression models.

In another study, stock price predictions for Chinese-listed businesses were made using linear regression models [8]. The authors discovered that while their linear regression models were influential in the short run, they were less so in the long term. They proposed that the accuracy of the predictions may be increased by combining linear regression with other methods, such as SVR. In a different study [10], combined ANNs and linear regression were used to forecast stock prices on the Tehran Stock Exchange. When tested on historical data, the authors discovered that the hybrid model performed better than the linear regression and the neural network models.

Numerous other attempts have been to improve the precision of linear regression models for stock price prediction in addition to these studies. For instance [11], predicted stock prices using a hybrid model that mixed principal component analysis, SVR, and linear regression. Similar to this, Chen, in his study [11] put out a model to forecast stock prices that blended linear regression with a DBN. Despite these efforts to increase the precision of linear regression models for stock price prediction, there are still drawbacks to the method. Its assumption of a linear relationship between the input and output variables is a significant drawback. Since stock prices are influenced by various intricate elements, including prevailing economic conditions, current political events, and market mood, this premise may only sometimes be accurate.

Due to its ease of use and interpretability, linear regression is a regularly utilized technique for stock price prediction. Its efficacy as a stand-alone methodology is constrained, and to increase accuracy, it frequently must be paired with other methods. Despite its drawbacks, it is nevertheless a valuable tool for forecasting stock prices and is frequently combined with other cutting-edge methods to increase its accuracy.

2.4 Random Forest

A typical ML approach for classification and regression problems is random forests (RFs). It is predicated on the idea of decision trees and employs an ensemble of decision trees. Because of its propensity to handle high-dimensional data, nonlinear correlations, and interdependencies across variables, random forests have been widely employed for stock price prediction. In this literature study, we will examine the performance of random forests as a machine learning (ML) technique for stock price prediction. RFs were employed to forecast stock values on the Malaysian stock exchange. They evaluated the effectiveness of RFs compared to other ML techniques, such as SVMs, ANNs, and k-nearest neighbors. Their prediction accuracy and computing efficiency analysis indicated that RFs fared better than the other algorithms. On the Chinese stock market, Zhang employed RFs to forecast stock values [6]. They evaluated their effectiveness with that of other ML techniques, including decision trees and SVR. According to their analysis, RFs beat the other algorithms regarding prediction accuracy and stability.

RFs were used to forecast stock prices in the US stock market. The effectiveness of SVR, logistic regression, and ANNs was compared with RF, which according to their analysis, beat the other algorithms regarding prediction accuracy and stability. RFs have also been utilized for feature selection in stock market analysis and price prediction. Wang chose the most pertinent features for predicting stock prices in the Chinese stock market using RFs. Their research discovered that they helped determine the key elements for stock price prediction.

Although they have demonstrated promising results in stock price prediction, RFs have several drawbacks. The possibility of overfitting is one of the key restrictions. When the model learns the noise in the data rather than the underlying patterns, overfitting takes place. Researchers have employed methods like cross-validation and regularization to get around this constraint. The interpretability of RFs is another drawback. Compared to linear regression models, the results of an RF model are more challenging to understand. This could be a concern in applications where the model's interpretation is crucial. RFs have produced encouraging results in feature selection and stock price prediction. They have done better in terms of prediction stability and accuracy than other ML methods. However, there are drawbacks to employing RFs for stock price prediction, including the danger of overfitting and the need for interpretability.

2.5 Empirical Studies Using CNN

Chen et al., in their study, proposed a Graph Convolution based CNN for trend forecasting for stocks where each stock's individual information and information on the entire market is considered [12]. For their study, six Chinese stocks were randomly selected to check the efficacy of the GC-CNN-based method, which yielded more stable and consistent returns. However, they could not solve the limitation that their model required more training time. Dong, in his study, emphasizes how investors and financial analysts may find how well CNNs and SVMs work together to predict stock prices [14]. Their methodology aims to assist investors in making more informed investment decisions and lowering financial risks by increasing the accuracy of stock price forecasts. Another paper aims to aggregate multiple variables to make feature extraction and market prediction easier, as many variables influence the prices and are stochastic in nature [15]. The proposed framework, CNNpred, was evaluated for its ability to forecast changes in the S&P 500, NASDAQ, DJI, NYSE, and RUSSELL stock indices. Findings showed that CNNpred significantly outperformed baseline state-of-the-art methods, with gains in F-measure ranging from 3% to 11% across all five indices. These results underline the necessity for additional study on creating CNN structures for stock prediction problems while also confirming CNNpred's effectiveness for stock prediction. CNNpred was successfully deployed in a trading system, even though the primary goal of this article was to anticipate the direction of stock market movements. This successfully demonstrated CNNpred's applicability in trading systems and suggested a viable topic for further investigation. In a different study, the article considers economic and production-related aspects of forecasting crop prices [12]. According to the study, the success of ARIMA highlights how crucial it is to factor historical data into time-series forecasting. The complicated nature of real-world time series may not be adequately captured by the linearity assumption, preventing long-term prediction from performing well. The authors provide a strong model that can pick up nonlinearity to get beyond these restrictions. They advise employing AI methods like CNN, which excels at audio and image classification jobs due to its efficiency at feature extraction and autonomous learning of nonlinear patterns. Reviewing the variables that affect crop output and putting forth a unique CNN model to forecast crop prices are the study's main goals.

2.6 Decision Trees

Popular Ml techniques for predicting stock values include decision trees. Models that can categorize or predict outcomes based on input factors are constructed using decision trees. The decision tree algorithm creates a tree structure where each leaf node represents a predicted outcome, and each inside node indicates a decision based on one of the input factors. Recursively dividing the data into subgroups according to the values of the input variables creates the tree. Decision trees can be used to find patterns in historical stock prices that can be used to forecast future prices in the context of stock price prediction. One method is to employ technical indicators as input variables for the decision tree model, such as relative strength index (RSI) or moving averages.

Decision trees were employed in a 2017 study by Sen and colleagues to forecast the daily direction of the Indian stock market. They used moving averages, RSI, and stochastic oscillators as input variables and other technical indicators. According to the study, decision trees fared better than other machine learning techniques like neural networks and support vector machines. Decision trees were employed in a different study in Malaysia to forecast the daily closing prices of the FTSE Bursa Malaysia Kuala Lumpur Composite Index (KLCI) [13]. They used moving averages, the RSI, and momentum as input variables and other technical indicators. The study discovered that decision trees performed better than ML techniques like SVR and ANNs.

Decision trees have the potential for overfitting, which occurs when the model is very complicated and matches the training data too closely, leading to a subpar performance on new data. Techniques like pruning or ensemble approaches like RFs can address this. A well-liked ensemble technique, RFs, mix different decision trees to enhance performance and lessen overfitting. RFs were employed in a study to forecast the daily closing prices of the Taiwan stock market. They used moving averages, RSI, and stochastic oscillators as input variables and other technical indicators. According to the study, RFs outperformed other ML techniques, including SVR and neural networks. Popular ML techniques for predicting stock values include decision trees. They have been demonstrated to perform better than other ML techniques in some studies and can be employed with technical indicators as input variables. They can, however, be vulnerable to overfitting, which can be avoided by using approaches like pruning or ensemble ones like RFs.

2.7 RNN

Recurrent neural networks (RNNs), a subclass of deep learning models that can recognize temporal correlations and patterns in time series data, have been used to forecast stock values. We will examine the use of RNNs for stock price prediction in this literature review, covering the various RNN types employed, their designs, and their performance. Several varieties of RNN architectures besides LSTMs have been applied to stock price prediction. The Gated Recurrent Unit (GRU), which resembles the LSTM design but uses a more straightforward memory cell, is one illustration. Zhang discovered that the GRU model outperformed other ML models like RFs and SVMs when using GRUs to forecast the daily closing values of 10 main equities on the New York Stock Exchange [8]. The Echo State Network is another RNN architecture type applied to stock price forecasting (ESN). ESNs are a subset of reservoir computing models that translate the input data into a higher dimensional space using a fixed random matrix, enabling the network to learn intricate temporal patterns. According to a study by Mukherjee, the ESN model beat conventional linear regression and ARIMA models in terms of prediction accuracy when used to forecast the stock prices of 20 businesses listed on the Indian Stock Exchange [16]. There are certain restrictions on the usage of RNNs, despite the encouraging results they have produced for stock price prediction. One drawback is the necessity for many data to train the models adequately. Another drawback is the potential for overfitting, which happens when the model gets too complicated and starts to match noise in the data instead of real patterns.

As a result of their capacity to recognize temporal correlations in time series data, RNNs, particularly the LSTM design, have demonstrated promise in stock price prediction. However, more study is required to overcome RNNs' drawbacks and create more reliable stock price prediction models.

2.8 Bayesian Networks

Probabilistic graphical models known as Bayesian networks (BNs) use probability theory to model uncertainty and causal interactions between variables. BNs have been used in many fields, including finance, to predict stock prices. This study of the literature looks at how BNs are used to forecast stock prices and how they perform when compared to other ML methods. To represent the intricate relationships between the different economic elements that could affect stock prices, BNs have been employed in stock price prediction. A BN model was employed in a study [7] to forecast stock prices using financial ratios, technical indicators, and macroeconomic factors. Regarding the accuracy and consistency of predictions, the study indicated that the BN model performed better than conventional models like linear regression and neural networks. The Shanghai Composite Index stock prices were predicted using a BN model in a different study using a set of financial factors [17]. The performance of the BN model was compared to more established time series models like ARIMA and GARCH in the study. The outcomes demonstrated that the BN model performed better than these models regarding prediction accuracy and stability. Similarly, Sun, in 2019, employed a BN model to forecast stock values based on a collection of financial parameters and economic data. The performance of the BN model was compared to that of SVMs and ANNs in the study. The outcomes demonstrated that the BN model performed better than alternative methods regarding stability and prediction accuracy. BNs have been used in stock price prediction based on news and social media sentiment analysis in addition to financial and macroeconomic aspects. A BN model was employed in a study to forecast stock prices based on sentiment analysis of news articles. The study discovered that in terms of prediction accuracy, the BN model performed better than conventional regression models [3].

Although the results are encouraging, specific restrictions exist on using BNs in stock price prediction. The requirement for a vast amount of data to precisely model the relationships between variables is one restriction. Moreover, overfitting and decreased model interpretability may result from the model's complexity and the data's high dimensionality. BNs have demonstrated promising results in stock price prediction, particularly in capturing the intricate linkages between many economic elements. To prevent overfitting and maintain model interpretability, deploying BNs requires careful assessment of the data and model complexity. The potential of BNs in stock price prediction and its comparison to other machine learning methods require more study.

2.9 The Structure of CNN

CNNs are a subclass of deep learning architecture that have shown outstanding results in audio and picture recognition, natural language processing, and, more recently, stock price prediction. Using CNNs to extract features from images may also be used to extract features from time-series data. The structure of CNNs used to forecast stock prices is the main topic of this literature study.

Convolutional layers, pooling layers, and fully connected layers make up the foundation of a CNN. The convolutional layer receives the input data and applies filters to the input to extract features. The convolutional layer's filters are in charge of spotting particular patterns in the data, like edges or corners in a picture or trends in market prices. The feature maps acquired from the convolutional layer are down-sampled using the pooling layer. Pooling assists in lowering the network's computational burden and can assist in avoiding overfitting. Max pooling and average pooling are two examples of the various types of pooling. To complete the final classification or regression task, the fully connected layer is used. The output of the preceding layer is applied with a set of weights by the ultimately linked layer to create the final output. CNNs can be applied to time-series data, such as stock prices, to extract features. The CNN architecture is highly suited to finding time-series data patterns that are challenging to find with conventional statistical techniques. For instance, CNNs can identify nonlinear correlations between various data features, which can result in more precise predictions.

Sliding window analysis is a popular method for stock price prediction using CNNs. The time-series data is divided into smaller segments using the sliding window approach, and these segments are then fed into the CNN. After extracting features from each segment, CNN predicts using those features. CNN can identify short-term trends in the data that other models might miss by employing a sliding window technique. There are many ways to train CNNs for stock price prediction, including backpropagation and gradient descent. The weights of the network are changed during training to reduce the discrepancy between the predicted and actual values. Several metrics, such as mean squared error or mean absolute error, can be used to calculate the error. CNNs can be integrated with other machine learning methods like SVMs to increase prediction accuracy. SVMs can be utilized to make the final prediction based on the features that the CNN has extracted. The convolutional layer's number of filters and their sizes can be changed to enhance the feature extraction procedure.

2.10 CNN vs Other Machine Learning Models

To anticipate stock price movements, a CNN-based deep learning system that takes raw data as input characteristics was suggested later [4]. Similarly, Zhang [9] created a unique model to predict stock prices that blends CNNs with long short-term memory (LSTM) networks. The combined model fared better in their investigation than well-known ML techniques like RFs and SVR. In addition, Wang created a hybrid model to forecast stock prices that combines CNNs with a variational mode decomposition. Their findings demonstrated that the suggested model performed better than conventional statistical models like ARIMA and GARCH. Overall, these results indicate that CNNs can beat other conventional ML methods regarding stock price prediction.

2.11 Support Vector Machine (SVM)

With encouraging outcomes, SVMs have been widely used in stock price prediction. An effective ML algorithm for classification and regression is called an SVM. It is built on locating a hyperplane that, given the input data, can best divide several classes or forecast continuous outputs. To locate the hyperplane more quickly, SVM uses a kernel function to transfer the input data into a higher-dimensional space. Chen carried out one of the earliest studies on SVM for stock price prediction, using it to forecast the daily direction of stock price movement [11]. The authors employed moving averages, the relative strength index, and stochastic oscillators as input features for technical indicators. According to the results, SVM fared better than other ML methods like neural networks and decision trees. SVM was employed in a different study to forecast the monthly stock returns on the Taiwan stock exchange. Financial statistics such as the PE ratio, PB ratio, and dividend yield were used by the writers as input features. The findings demonstrated that SVM outperformed other ML strategies like decision trees and linear regression regarding prediction accuracy [5].

SVM was employed to forecast the Shanghai Stock Exchange Composite Index closing prices. The writers used technical indicators like moving averages, Bollinger Bands, and momentum indicators as input features. The outcomes demonstrated that SVM outperformed other ML methods like neural networks and linear regression regarding prediction accuracy. To forecast the daily stock prices of the National Stock Exchange of India, SVM was employed in a recent study [1]. Moving averages, MACD (moving average convergence/divergence), and Bollinger Bands are a few examples of technical indicators that the writers used as input elements. The outcomes demonstrated that SVM outperformed other ML methods like neural networks and decision trees regarding prediction accuracy. Yang employed SVM in another investigation to forecast the Shanghai Stock Exchange Composite Index's daily stock prices. Moving averages, Bollinger Bands, and volume indicators were among the technical indicators employed by the writers as input characteristics. The outcomes demonstrated that SVM outperformed other ML methods like gradient boosting and RF regarding prediction accuracy [18].

Yet, other research has also noted the SVM's limits in predicting stock prices. For instance, Zhang tested SVM with other ML methods for stock price prediction, including RF and K-nearest neighbor (KNN). The researchers discovered that KNN and RF had higher prediction accuracy than SVM. Moving averages, Bollinger Bands, and volume indicators were among the technical indicators employed by the writers as input characteristics. The outcomes demonstrated that SVM outperformed other ML methods like gradient boosting and RF regarding prediction accuracy. In conclusion, SVM has demonstrated promising outcomes in several research on the stock price prediction. In some situations, SVM has been proven to perform better than other ML, including neural networks, decision trees, and linear regression. However, the performance of SVM can be impacted by choice of kernel functions and parameters. To increase the SVM's prediction accuracy, future research can investigate using more sophisticated feature engineering and selection strategies.

2.12 CNN and SVM

Although SVM and CNN have demonstrated respectable accuracy for stock price prediction, each method has drawbacks. Although the link between the input features and the output is complex and nonlinear, stock price prediction may not be optimal for SVM, which relies on identifying a hyperplane to divide the data into distinct classes. On the other hand, CNN might need to catch the time-series data's long-term dependencies and patterns. As a result, an SVM and CNN combination may be more effective for predicting stock prices. A combination of SVM and CNN has been tested in several studies as a stock price predictor. For instance, [19] proposed a hybrid model that uses SVM to make the final prediction and CNN to extract features from the input data. The Shanghai Stock Exchange Composite Index's forecast accuracy using the proposed model was 56.95%. Similarly, Yang [18] suggested a hybrid model that employs CNN to forecast the size of the change and SVM to forecast the stock price trend. For the S&P 500 index, the proposed model has an accuracy of 60.56%. These experiments demonstrate how combining SVM and CNN can increase the accuracy of stock price predictions. SVM and CNN are two well-known methods thoroughly researched for stock price prediction in conclusions. While each method has its drawbacks, combining SVM and CNN can get around these drawbacks and improve prediction accuracy. More research can be done on the proposed hybrid models that combine SVM and CNN to increase the precision of stock price prediction.

3. Methodology

3.1 Data Collection and Preprocessing

Daily data for the analysis can be collected from Yahoo Finance using the yfinance library available in Python. The collected data will undergo preprocessing to remove outliers and missing values. The missing values can be replaced with the median if the number of missing values is small; otherwise, they must be dropped from the dataset.

3.2 Data Splitting

The cleaned data must be split into the train and test sets to train the model. The latter-the testing set will be used to ascertain the model's performance. Using a 70–30 split, where 70% of the data is used for training the model, and 30% of the data is utilized for testing, is a popular technique. But, given the size of the dataset, 60% of it is for training, and the remaining 40% for testing can be set to lessen overfitting and give a more accurate assessment of the model's performance on hypothetical data.

3.3 Feature Extraction for CNN

Feature extraction in ML is choosing and manipulating pertinent features from the raw data. The process of extracting the essential features from the input data—in this case, the historical stock prices of the Nifty 500—is referred to as feature extraction in the context of the CNN-SVM hybrid model for stock price prediction. Due to their capacity to recognize significant patterns and structures in the input data, CNNs are especially well-suited for feature extraction tasks. The CNN will be taught how to extract elements from historical stock price data that are most important for forecasting future stock values. A feature vector representing the critical components of the input will be the CNN's output. The most crucial details in the past stock price data can be summarized in the feature vector. We can decrease the dimensionality of the data and concentrate on the most crucial information for the prediction task by extracting these features.

3.4 SVM Model Training

Due to their high accuracy and effectiveness in handling big datasets, strong classification algorithms called SVMs are frequently used in stock price prediction. Finding a hyperplane that divides the input data into distinct classes is how SVMs divide the data into a higher dimensional feature space. The SVM is employed in stock price prediction to determine whether the stock price will increase or decrease in the future based on the extracted features. The steps involved in training an SVM include choosing the appropriate parameters for the kernel function, which transforms the input data into a higher-dimensional space. The linear, polynomial, and radial basis function (RBF) kernels are the most often used kernels for SVMs in stock price prediction. The data is split into two sets—the training set and the testing set—to train the SVM model. The SVM model is trained using the training set, and its performance is assessed using the testing set. The goal of training an SVM model is to find the hyperplane that best divides the data points into their respective classes. The SVM will produce a decision boundary that distinguishes the positive class from the negative class in case of a binary classification problem, such as predicting whether the stock price will increase or decrease. The decision boundary is the hyperplane that maximizes the margin between the positive and negative classes. The margin separates the nearest data points from each class and the hyperplane. The SVM is more specific in its classification the wider the margin.

The SVM employs a cost function that penalizes incorrectly classified data points to locate the ideal hyperplane. The sum of the mistakes and the distances between the incorrectly categorized data points and the decision boundary is minimized using the cost function. The regularization parameter C is used to manage the trade-off between maximizing the margin and minimizing the errors.

In contrast, a low value of C will produce a broader margin but more misclassifications. A high value of C would produce a narrower margin but fewer misclassifications. The degree of the polynomial kernel or the width of the RBF kernel are two additional hyperparameters that may need to be tweaked in the SVM model in addition to the regularization parameter and the kernel function. A grid search or a random search method can be used to optimize these hyperparameters to discover the combination that yields the highest performance on the testing set.

The SVM model may be used to forecast future stock prices of the Nifty 500 index based on the input features once it has been trained on the extracted features from the CNN. The SVM model's accuracy may be assessed by contrasting the projected values with the actual values from the testing set. The model's performance can be evaluated using performance metrics like accuracy, precision, recall, and F1 score.

3.5 Evaluation of the Model with Accuracy Measures

- Mean Absolute Error (MAE) is a commonly used metric to gauge the average size of prediction errors. It is derived by averaging the absolute disparities between the values that were anticipated and those that occurred. A lower MAE number indicates more accuracy.

$$MAE = (1/n) * \Sigma|yi - \hat{y}i| \tag{1}$$

- Another popular statistic for gauging the average of the squared discrepancies between expected and actual data is mean squared error (MSE). A lower MSE value indicates better accuracy.

$$MSE = 1/n * \Sigma(yi - \hat{y}i)^2 \tag{2}$$

- Root Mean Squared Error (RMSE): RMSE calculates the standard deviation of the errors by taking the square root of the mean square error (MSE). It is also a typical statistic employed to assess the model's correctness. A lower RMSE score indicates more accuracy.

$$MSE = (1/n) * \Sigma(i = 1 \text{ to } n)\ (Yi - \hat{Y}i)^2 \tag{3}$$

- Cross-validation: This method involves repeatedly dividing the data into training and testing sets to assess the model's performance. This helps to guarantee that the model generalizes well to new data and does not overfit the training set of data.

In CNN, k-fold cross-validation, a method, is commonly used to carry out cross-validation. This method divides the available data into k subsets, then trains the model k times on k-1 subsets before evaluating it on the remaining subset. All subsets are subjected to the same procedure, and the evaluation results are averaged to determine the model's overall performance. k is commonly set to 5 or 10, depending on the size of the available data. CNN frequently use leave-one-out cross-validation in addition to k-fold cross-validation. Except for one sample, which is used for evaluation, the model is trained using all the data that are accessible. For each sample, the procedure is repeated, and the evaluation results are averaged to determine the model's overall performance. When there is a dearth of data, leave-one-out cross-validation is helpful.

- K-fold cross-validation is another common cross-validation method used in SVM. Nevertheless, unlike CNN, SVM needs the selection of hyperparameters such as the kernel type, kernel parameter, and regularization parameter, which can considerably affect the model's performance. Consequently, cross-validation must also be used to fine-tune the SVM model's hyperparameters.
- Cross-validation is a technique for SVM hyperparameter tweaking that entails choosing a variety of hyperparameters and assessing the model's performance for each set of hyperparameters. The ideal hyperparameters for the model are the set of hyperparameters that produces the best performance measure. Although cross-validation is an effective method for assessing models and fine-tuning hyperparameters, it has significant drawbacks. The assumption that the data is independent and identically distributed (IID.), which may not always be the

case in practise, is one restriction of cross-validation. For instance, trends and seasonality in financial data may show temporal dependencies that can impact the model's performance. Cross-validation has the additional drawback of being computationally expensive, particularly for large datasets or sophisticated models. Moreover, not all models, such as time-series models that need precise data segmentation, may be appropriate for cross-validation.

3.6 Hyperparameter Tuning

To maximize the model's performance, hyperparameter tuning is crucial in developing machine learning models. This is done by identifying the optimal set of hyperparameters. Before the model is trained, hyperparameters are the settings of a machine learning algorithm. The number of filters, the size of the filters, the pace of learning, and the batch size are some examples of hyperparameters for a CNN. Examples of SVM parameters include the regularization parameter, gamma, and kernel type. Selecting the ideal values for these hyperparameters to improve accuracy or reduce error for the model is known as hyperparameter tuning. A grid search is a popular method for hyperparameter tweaking. To do this, a range of values for each hyperparameter are defined, and then each combination of hyperparameters is tested to determine which combination works best.

Another method of hyperparameter tuning is random search, which involves choosing hyperparameters at random from a set range in the hopes of combining them helpfully. Another method for tuning hyperparameters is Bayesian optimization, which entails building a probabilistic model of the function that links hyperparameters to model performance. Hyperparameter tweaking is a technique that may be used to improve the performance of both the CNN and SVM models when creating a hybrid CNN-SVM model to forecast stock prices. Hyperparameters for the CNN model can be adjusted, including the number of filters, the size of the filters, and the learning rate. Hyperparameters for the SVM model can be adjusted, including the kernel type, gamma, and regularisation parameters. It is significant to note that hyperparameter tweaking takes time because it necessitates training and assessing the model numerous times with various combinations of hyperparameters. Hence, a huge dataset and computing power are required to execute hyperparameter tweaking effectively. Furthermore, improper hyperparameter optimisation might result in overfitting. When a model is tuned to perform well on training data but badly on test data, this is known as overfitting. Use cross-validation methods, such as k-fold cross-validation, to test the model on various subsets of the data and ensure it generalizes well to new data to avoid overfitting. In conclusion, hyperparameter tuning is a crucial stage in creating a machine learning model since it helps to maximize the model's performance by identifying the ideal hyperparameter combination.

4. Conclusion

In summary, this chapter proposed an AI-based methodology, i.e., CNN-SVM hybrid model, to predict stock index prices. While widely utilized in stock price prediction, classic machine learning techniques like linear regression, logistic regression, decision trees, and random forests have limitations when addressing complicated and non-linear correlations in the data, according to the literature review. On the other hand, deep learning approaches like LSTM and CNN, with CNN being especially good at feature extraction, have demonstrated promising results in stock price prediction. The CNN-SVM hybrid model combines the benefits of both CNN and SVM, with SVM providing classification power and CNN doing feature extraction. The CNN-SVM hybrid model has the potential to predict stock prices and could be investigated further in future studies. It should be mentioned, nevertheless, that predicting stock prices is a difficult and complex endeavor, and no model can forecast future prices with 100% accuracy. Because of this, it's crucial to use caution when making financial decisions based on any model's predictions, including the CNN-SVM hybrid model.

References

[1] Kumar, R., Pannu, H.S., and Malhi, A.K. Aspect-based sentiment analysis using deep networks and stochastic optimization. *Neural Computing and Applications*, 32: 3221–3235, 2020.

[2] Gudelek, M.U., Boluk, S.A., and Ozbayoglu, A.M. A deep learning based stock trading model with 2-D CNN trend detection. *In*: *2017 IEEE Symposium Series on Cmputational Intelligence (SSCI)*, pp. 1–8. 2017, November. IEEE.

[3] Li, M., Zhu, Y., Shen, Y., and Angelova, M. Clustering-enhanced stock price prediction using deep learning. *World Wide Web*, 26(1): 207–232, 2023.

[4] Huang, J.Y., and Liu, J.H. Using social media mining technology to improve stock price forecast accuracy. *Journal of Forecasting*, 39(1): 104–116, 2020.

[5] Li, Y., Bu, H., Li, J., and Wu, J. The role of text-extracted investor sentiment in Chinese stock price prediction with the enhancement of deep learning. *International Journal of Forecasting*, 36(4): 1541–1562, 2020.

[6] Zhuge, Q., Xu, L., and Zhang, G. LSTM neural network with emotional analysis for prediction of stock price. *Engineering Letters*, 25(2), 2017.

[7] Zhou, F., Zhang, Q., Sornette, D., and Jiang, L. Cascading logistic regression onto gradient boosted decision trees for forecasting and trading stock indices. *Applied Soft Computing*, pp. 105747, 2019.

[8] Zhang, S., and Fang, W. Multifractal behaviors of stock indices and their ability to improve forecasting in a volatility clustering period. *Entropy*, 23(8): 1018, 2021.

[9] Liu, S., Zhang, C., and Ma, J. CNN-LSTM neural network model for quantitative strategy analysis in stock markets. *In*: *Neural Information Processing: 24th International Conference, ICONIP 2017, Guangzhou, China, November 14–18, 2017, Proceedings*, Part II 24, pp. 198–206, 2017.. Springer International Publishing.

[10] Chang, C.C., Luo, K.W., and Hsiao, S. Stock Price Prediction and Recommendation Approach Based on Machine Learning.

[11] Chen, Y., Li, B., Wang, G., and Li, S. Stock price prediction with financial ratios using extreme learning machine and Bayesian regularized neural network. *Journal of Forecasting*, 36(6): 636–648, 2017.

[12] Chen, W., Jiang, M., Zhang, W.-G., and Chen, Z. A novel graph convolutional feature based convolutional neural network for stock trend prediction. *Information Sciences*, 556: 67–94, 2021. https://doi.org/10.1016/j.ins.2020.12.068.

[13] Cheung, L., Wang, Y., Lau, A.S.M., and Chan, R.M.C. Using a novel clustered 3D-CNN model for improving crop future price prediction. *Knowledge-Based Systems*, 260: 110133, 2023. https://doi.org/10.1016/j.knosys.2022.110133.

[14] Dong, J. Financial investor sentiment analysis based on FPGA and convolutional neural network. *Microprocessors and Microsystems*, p. 103418, 2020. https://doi.org/10.1016/j.micpro.2020.103418.

[15] Hoseinzade, E., and Haratizadeh, S. CNNpred: CNN-based stock market prediction using a diverse set of variables. *Expert Systems with Applications*, 129: 273–285, 2019. https://doi.org/10.1016/j.eswa.2019.03.029.

[16] Khan, Z.A., Hussain, T., Haq, I.U., Ullah, F.U.M., and Baik, S.W. Towards efficient and effective renewable energy prediction via deep learning. *Energy Reports*, 8: 10230–10243, 2022.

[17] Xiao, J., Zhu, X., Huang, C., Yang, X., Wen, F., and Zhong, M. A new approach for stock price analysis and prediction based on SSA and SVM. *International Journal of Information Technology & Decision Making*, 18(01): 287–310, 2019.

[18] Wen, Q., Yang, Z., Song, Y., and Jia, P. Automatic stock decision support system based on box theory and SVM algorithm. *Expert Systems with Applications*, 37(2): 1015–1022, 2010.

[19] Shen, J. and Shafiq, M.O. Short-term market price trend prediction using a comprehensive deep learning system. *Journal of Big Data*, 7(66), 2020. https://doi.org/10.1186/s40537-020-00333-6.

Chapter 13

A Comparative Analysis of Traditional and Machine Learning Forecasting Techniques

Lija Jacob and *KT Thomas*

1. Introduction to Forecasting

Forecasting is the process of making predictions or estimates about future events or conditions based on historical data, trends, and patterns. It involves analyzing past data and using statistical or other quantitative methods to project future outcomes, such as sales figures, market trends, weather patterns, or financial performance. Forecasting can be used in a wide range of fields, including economics, finance, business, weather forecasting, and sports. The accuracy of a forecast depends on the quality of the data, the methods used, and the assumptions made about the future.

The growth of Artificial Intelligence (AI) has had a significant impact on forecasting. As AI technologies have advanced, they have enabled more sophisticated and accurate forecasting techniques to be developed. This has resulted in a growing demand for AI-based forecasting solutions across various industries and sectors.

Due to its capacity to manage large quantities of data, recognize intricate patterns and trends, and produce precise predictions, AI has seen a rapid increase in its use in forecasting in recent years. Sales, demand, market trends, and other important business indicators can be predicted using AI-based forecasting techniques, giving organizations a competitive edge and allowing them to make better choices.

Department of Data Science, Christ University, India.
Emails: lija.jacob@christuniversity.in, 0000-0001-6311-0168;
thomas.kt@christuniversity.in, 0000-0002-3469-6742

The development of even more potent and precise forecasting solutions as AI technologies progress can be anticipated. In order to increase accuracy and deliver more precise predictions, AI-based forecasting methods are already being used in weather forecasting, for instance. The results of sporting events, governmental contests, and other events are also predicted using AI.

Forecasting techniques can be quantitative or qualitative in character, based on the type of data and information available. Quantitative forecasting methods depend on statistical models to analyze historical data and make predictions. Regression analysis, trend analysis, and time-series analysis are a few examples of quantitative predicting methods. On the other hand, qualitative forecasting methods base their forecasts on expert opinion and arbitrary criteria. Market analysis, opinion surveys, and the Delphi method are a few examples of qualitative forecasting methods.

Forecasting techniques are frequently used in business, finance, economics, and other fields where accurate predictions of upcoming trends and events are essential for decision-making. They are also used for a number of purposes, such as weather forecasting and stock market studies.

The use of AI and machine learning (ML) in business will continue to transform the industry as technology develops. For example, the use of AI and ML in forecasting is of immense interest to most businesses due to its usability across functions. Businesses have traditionally used statistical predicting techniques like linear regressions and exponential smoothing to inform their decisions. However, in many data and analytics efforts across industries and sectors, ML-based forecasting has replaced conventional methods. The time, effort, and expenses associated with the process will be significantly influenced by the choice of the best forecasting methodology.

This chapter provides a brief overview of traditional forecasting techniques and developing patterns. The chapter also discusses the benefits and drawbacks of conventional predicting methods versus contemporary approaches, as well as the circumstances in which each approach may be most useful.

2. Types of Forecasting Techniques

The forecasting techniques can be basically divided into two categories [1]: Qualitative and Quantitative forecasting techniques. Figure 1 shows the classification of forecasting techniques.

Quantitative forecasting methods use historical data and mathematical models to predict upcoming patterns or events. These methods consist of economic modeling, regression analysis, and time-series analysis. When there is a large quantity of historical data available and it is anticipated that the future will follow a pattern or trend, quantitative forecasting is frequently used.

On the other hand, qualitative forecasting methods use the judgment, expert opinions, and subjective interpretation of data to anticipate future patterns or events. These methods include scenario planning, the Delphi method, and professional judgment. When there are few historical data points accessible or when the future is unpredictable and uncertain, qualitative forecasting is frequently used.

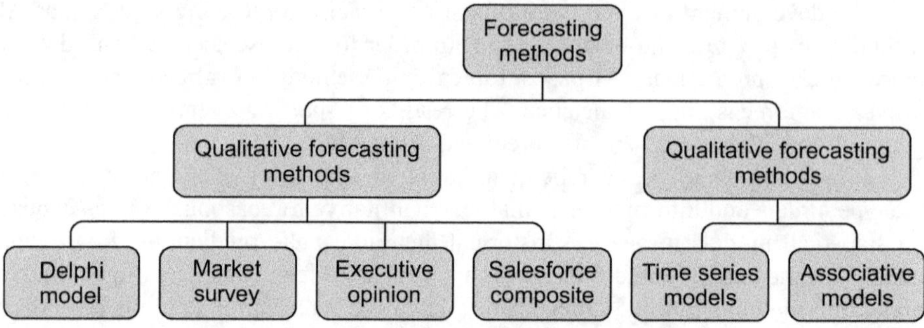

Fig. 1: Various forecasting techniques.

It is vital for one to understand that both quantitative and qualitative predicting methods have advantages and disadvantages. Quantitative methods offer more accurate and objective forecasts, but they are frequently constrained by the quality and amount of available historical data. Although more subjective, qualitative methods can offer insightful information and aid in planning for unforeseen events. To produce more accurate and trustworthy forecasts, practitioners frequently combine quantitative and qualitative forecasting methods [2,3]. Table 1 explains the difference between quantitative and qualitative methods of forecasting.

The chart compares quantitative and qualitative forecasting techniques on a broad scale. It is crucial to keep in mind that these techniques frequently overlap, and many forecasting tasks may call for a mix of quantitative and qualitative techniques. The particular context and demands of the forecasting job will also influence the forecasting method selection.

Table 1: Difference between quantitative and qualitative methods of forecasting.

Quantitative Forecasting	Qualitative Forecasting
Based on numerical data and statistical analysis	Based on subjective judgment and expert opinion
Uses historical data and mathematical models to make forecasts	Uses expert opinion and scenario analysis to make forecasts
Emphasizes objective and empirical data	Emphasizes subjective interpretation and expert judgment
Typically more precise and accurate in predicting trends and patterns	Typically, better suited for making forecasts in complex and uncertain situations
Common methods include time-series analysis, regression analysis, and econometric modeling	Common methods include Delphi method, scenario planning, and expert opinion
Requires a significant amount of historical data and a high level of statistical expertise	Can be conducted with limited data and does not require a high level of statistical expertise
Useful in forecasting demand for products and services, financial markets, and economic trends	Useful in forecasting political events, social trends, and technological changes
Examples of software used in quantitative forecasting include SAS, SPSS, and Excel	Examples of software used in qualitative forecasting include Decision Tools Suite, PESTLE, and Gist
References numerical data to support its forecasts	Does not rely on numerical data to support its forecasts

2.1 Qualitative Forecasting Techniques

The following are the different approaches of qualitative forecasting methods:

2.1.1 Delphi Method

The Delphi method is a structured procedure where a group of specialists present their unique viewpoints on a subject. Following the gathering of the views, the panelists are given the chance to change their positions in the light of other panelists' comments. This procedure keeps going until a decision is made and a final forecast is produced [4,5].

With the help of several rounds of anonymous surveys or questionnaires, the Delphi method is a structured approach to forecasting that includes getting opinions from a variety of experts or stakeholders. The Delphi method seeks to minimize biases and lessen the influence of individual views or perspectives while generating a consensus forecast or decision based on the group's combined knowledge and insights [4,8].

The RAND Corporation created the Delphi method in the 1950s as an instrument for long-term strategic planning. Since then, it has found widespread use in a variety of industries, including business, technology, healthcare, and public policy.

Figure 2 shows the flowchart of the Delphi Forecasting process.

Defining the problem or question: The first stage is to precisely define the issue or query that the Delphi method will be applied to forecast. This could entail determining the most important variables, elements, or patterns pertinent to the issue.

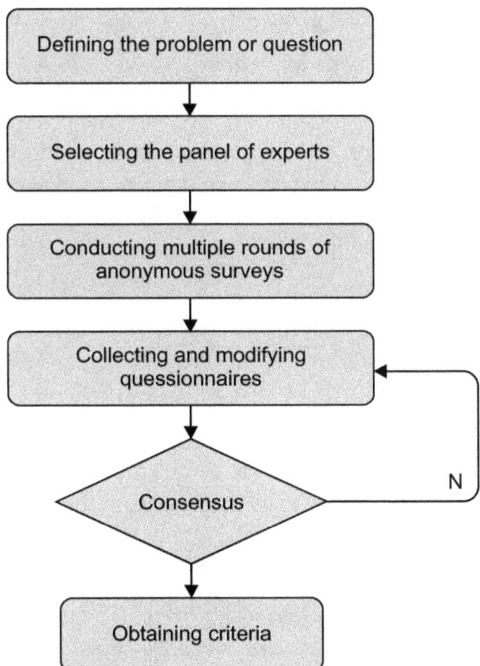

Fig. 2: The flowchart of the Delphi forecasting process.

Selecting the panel of experts: The next stage is to choose a group of experts or stakeholders who have knowledge or expertise pertaining to the issue or question. People from various fields or industries, as well as those with various viewpoints or levels of expertise, may be on the panel.

Conducting multiple rounds of anonymous surveys: The Delphi method entails a number of rounds of anonymous questionnaires or surveys where the panel of experts are requested to provide their predictions, views, or recommendations. Each round typically involves a summary of the results from the previous round, as well as new information or feedback to consider. The surveys are anonymous to minimize the impact of individual biases or social pressure.

Analyzing and synthesizing the results: The results from each round of surveys are analyzed and synthesized to identify trends, patterns, or areas of agreement or disagreement among the experts. The facilitator or researcher may also provide feedback or commentary on the results to help clarify or interpret the findings.

Reaching a consensus forecast or decision: The Delphi method aims to reach a consensus forecast or decision based on the collective knowledge and insights of the panel of experts. This may involve identifying areas of agreement or convergence, as well as areas of ambiguity or disagreement that demand additional talk or investigation [6].

A group of business experts used the Delphi method to predict the direction of AI over the next 10 years as an illustration of how the method is used in practice. The expert panel may be requested for predictions on important AI-related issues like:

- What significant uses of AI can we expect over the next 10 years?
- What will be the greatest risks or challenges related to AI?
- What effects will AI have on the labor market or the economy?

The panel of specialists can offer their predictions and input through numerous rounds of anonymous surveys, which are then analyzed and synthesized to find areas of agreement and disagreement. The Delphi technique can produce more precise results or decisions. The following are some instances of use cases for the Delphi technique of forecasting [5,7]:

Technology forecasting: The Delphi method can be used to gather professional opinions on the future development of the technology, possible applications, and market trends when a business plans to invest in new technology. This can assist the business in making wise technology investment choices.

Economic forecasting: Gathering expert views on economic trends, such as GDP growth, inflation, and interest rates, can be done using the Delphi method. This can aid in the improvement of investment, policy, and pricing strategy choices made by financial institutions, governments, and corporations.

Forecasting public policy: The Delphi technique can be used to gather professional opinions on matters of public concern, such as social welfare, healthcare, and education. This can assist decision-makers in developing, implementing, and evaluating policies in an educated manner.

Environmental forecasting: Climate change, pollution, and the depletion of natural resources are just a few examples of environmental problems that can be forecast using the Delphi method. This can aid in the creation of sensible plans for mitigating environmental risks and advancing sustainable development by governments, corporations, and NGOs.

Risk forecasting: Forecasting risks and threats to an organization, such as cyberattacks, natural catastrophes, or supply chain disruptions, can be done using the Delphi method. Organizations can use this to create risk management strategies and emergency plans.

Healthcare forecasting: Expert opinions on healthcare problems, such as disease outbreaks, medical advancements, and healthcare policy, can be gathered using the Delphi method. This can assist in informing choices about the distribution of healthcare resources and the provision of services by healthcare providers and policymakers.

2.1.2 Market Survey Forecasting

Market survey forecasting is a method used to predict future trends and demand for products or services by analyzing data obtained through market research surveys. The goal of market survey forecasting is to help businesses make informed decisions about product development, marketing strategies, and overall business planning [9].

There are various types of market surveys that can be conducted to gather data for forecasting. Some common methods include online surveys, telephone surveys, in-person interviews, and focus groups. Figure 3 shows an overview on Market Survey Forecasting

The data collected from these surveys can include demographic information, product preferences, purchasing habits, and other relevant information. Once the data is collected, it can be analyzed using various statistical techniques to identify trends and patterns. This analysis can then be used to make predictions about future market demand, pricingtrends, and other factors that may impact the success of a business. For example, a company that produces athletic shoes may conduct a market survey to gather information about consumer preferences for different types of shoes, pricing, and marketing channels. The survey data may reveal that there is a growing demand for environmentally sustainable products, and that consumers are willing to pay a premium for shoes that are made from eco-friendly materials. Based on this information, the company may decide to invest in the development of eco-friendly athletic shoes and adjust their marketing strategies accordingly [9].

Another example could be a restaurant chain that is considering adding a new menu item. They may conduct a market survey to gather information about consumer preferences for different types of food, pricing, and dining experiences. The survey data may reveal that there is a high demand for plant-based options, and that consumers are willing to pay a premium for high-quality, sustainably-sourced

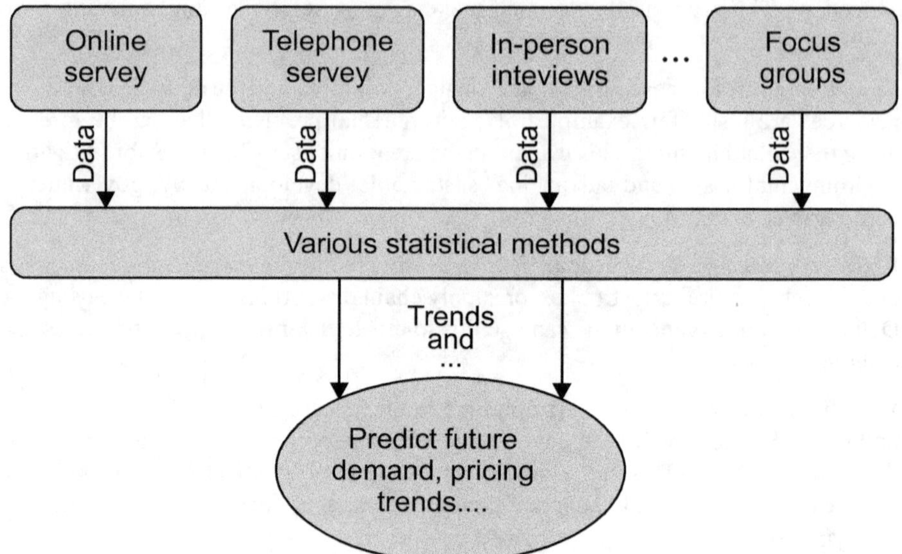

Fig. 3: Overview in Market Survey Forecasting.

ingredients. Based on this information, the restaurant chain may decide to add a new plant-based menu item and adjust their pricing and marketing strategies accordingly.

Market survey forecasting is the procedure of collecting and examining data about a particular market in order to project future trends and demand [10]. Figure 4 shows the typical steps in conducting a market survey forecasting.

Define the objective: The first step in the process is to clearly describe the goal of the survey. You must first choose the market segment, item, or business that interests you to do this.

Form the survey's questions: The next step is to compile a list of inquiries that will produce the needed data after the objective has been established. The questions should be direct, succinct, and relevant to the objective.

Perform the survey: The survey can be conducted in a variety of methods, including in-person interviews, phone interviews, and online surveys. The sample number ought to be typical of the intended audience.

Study the information: To identify patterns and trends, the poll data must be organized and assessed. The data can be subjected to statistical research using software.

Interpreting the outcomes: The data must first be investigated, and only after that can it be properly interpreted to arrive at the main findings. One method to do this is to look for trends, patterns, and connections that can be used to forecast market demand.

Draw conclusions: The final step involves drawing conclusions based on the survey's findings. This can include making recommendations for forthcoming marketing strategies, product developments, or sales strategies.

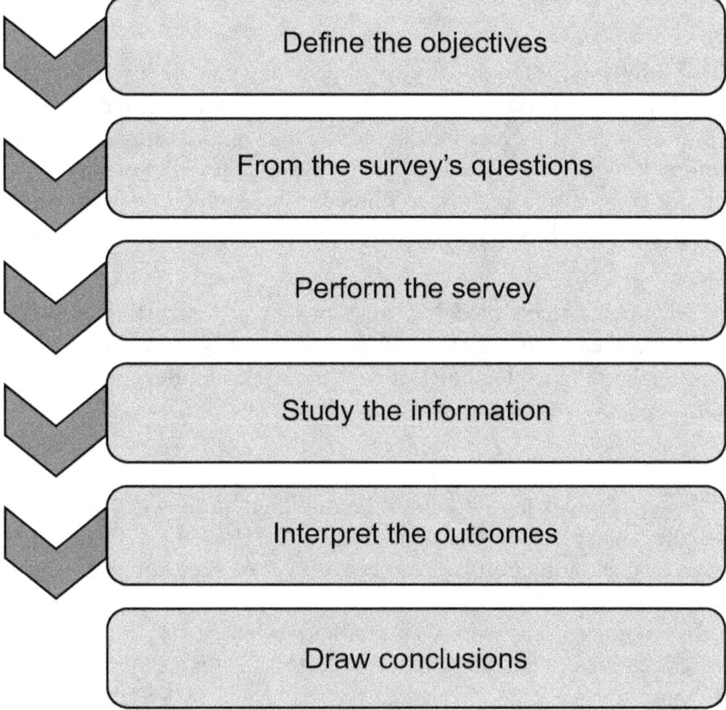

Fig. 4: The flowchart of the process handled in Market Survey Forecasting.

In general, market survey forecasting is a difficult process that requires careful planning, execution, and analysis in order to create accurate predictions of future market trends [10] Examples of cases for predicting from market surveys include:

Product development: To learn more about consumer requirements and preferences for a new good or service, market survey forecasting can be used. In addition to informing decisions about product features, pricing, and marketing tactics, this can assist businesses in creating goods that are more likely to satisfy consumer demand.

Sales forecasting: Based on consumer input and demand, market survey forecasting can be used to project future sales for a good or service. This can assist businesses in more effectively planning their inventory, manufacturing, and distribution.

Customer satisfaction: Customer satisfaction can be measured and areas for growth can be found using market survey forecasting. This can assist businesses in retaining customers and enhancing their market standing.

Brand perception: Market survey forecasting can be used to gauge consumer opinion of a brand. This can assist businesses in creating marketing plans that advance their image and raise awareness.

Market segmentation: The requirements and preferences of various customer segments can be determined using market survey forecasting. This enables businesses to target particular client segments with customized marketing materials and product offerings.

Competitive analysis: Market survey forecasting can be used for competitive research to learn more about rival companies' goods and services. This can assist businesses in creating strategies to distinguish their goods and services from those of their competitors and obtain a competitive advantage. Market survey forecasting can be used for competitive analysis to find out more about the products and services offered by competing businesses. This can help companies develop marketing strategies.

Everything being accounted for, market survey forecasting can be a useful instrument for businesses to learn more about consumer behavior and preferences and to help them make defensible choices about their product development, marketing, and sales strategies.

2.1.3 Expert Opinion Forecasting

Expert opinion forecasting involves communicating with people who are knowledgeable about a specific area or industry. Based on their expertise and understanding of the subject, these experts may be asked for their predictions of upcoming trends or events.

Executive opinion forecasting is a qualitative forecasting method that uses the insights of executives or subject-matter experts in a given market or industry to anticipate future trends. When historical data is scarce or irrelevant to the present situation, this technique can be helpful [11].

A business can use executive opinion forecasting to get an idea of how the market will respond to a new product, for instance, if they want to anticipate demand for one that hasn't yet been introduced to the market. They can enquire about executives' and specialists' perspectives on the product, its likelihood of success, and the anticipated demand.

The Consensus Economics survey, which compiles opinions from a panel of experts in various disciplines, including economics, finance, and industry, is one example of executive opinion forecasting. The poll solicits responses from experts on their predictions for the future trends of economic indicators like GDP growth, inflation, and interest rates [12].

Another example is the World Economic Forum's Global Competitiveness Report, which compiles views from business executives around the world about how competitive their economies are. These assessments are used in the report to rate nations according to their level of competitiveness and pinpoint areas for development.

Executive opinion forecasting has the benefit of offering important views and data that cannot be obtained through other means. However, because views can be affected by a person's personal beliefs, interests, and experiences, it is also susceptible to biases and errors. As a result, it's crucial to combine this technique with other forecasting techniques and to verify the outcomes using empirical data [12].

Figure 5 depicts the steps involved in the expert opinion forecasting process.

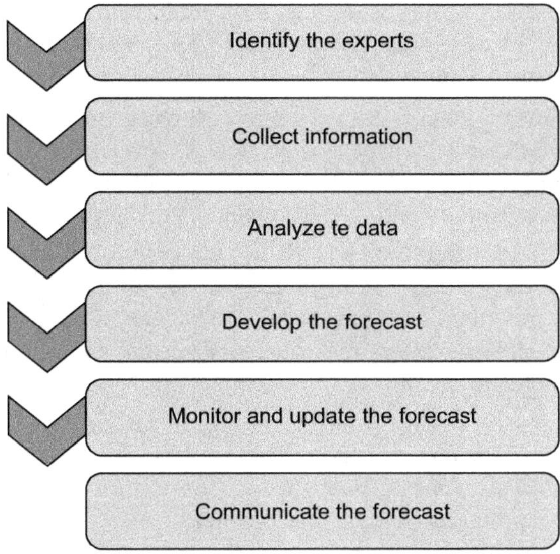

Fig. 5: The flowchart of the process handled in Executive Opinion Forecasting.

Executive opinion forecasting is a type of qualitative forecasting method that relies on the expert opinions of executives and other knowledgeable individuals to predict future trends and developments in a particular industry or market. The following are the steps involved in executive opinion forecasting:

- **Identify the Experts:** Identify the experts who have a good understanding of the market or industry you are interested in. These experts could be executives, industry analysts, or consultants who have a track record of accurate predictions.
- **Collect Information:** Collect information from the experts by conducting interviews, surveys, or focus groups. Ask the experts about their perceptions of future trends, events, and market conditions.
- **Analyze the Data:** Analyze the data collected from the experts and look for patterns and trends. Identify areas of agreement and disagreement among the experts and use this information to develop a consensus forecast.
- **Develop the Forecast:** Use the information gathered from the experts to develop a forecast. This forecast should include predictions about market growth, consumer behavior, and other relevant factors.
- **Monitor and Update the Forecast:** Monitor the market and update the forecast as new information becomes available. It is important to regularly review the forecast to ensure that it remains accurate and relevant.
- **Communicate the Forecast:** Communicate the forecast to stakeholders, such as executives, investors, and employees. Use the forecast to inform decision-making.

Expert opinion forecasting is a method of predicting future trends and events based on the opinions of experts in a particular field. Some use case examples for expert opinion forecasting include:

- **Technology development:** Expert opinion forecasting can be used to gather insights from experts in a particular field of technology, such as artificial intelligence or blockchain. This can help companies to make informed decisions about investments in research and development and product development.
- **Economic forecasting:** Expert opinion forecasting can be used to gather insights from economists and financial analysts on economic trends, such as GDP growth, inflation, and interest rates. This can help financial institutions, governments, and businesses make better decisions on investments, policies, and pricing strategies.
- **Public policy forecasting:** Expert opinion forecasting can be used to gather insights from experts on public policy issues, such as healthcare, education, and social welfare. This can help policymakers make informed decisions on policy development, implementation, and evaluation.
- **Environmental forecasting:** Expert opinion forecasting can be used to gather insights from experts on environmental issues, such as climate change, pollution, and natural resource depletion. This can help governments, businesses, and NGOs develop effective strategies for mitigating environmental risks and promoting sustainable development.
- **Risk forecasting:** Expert opinion forecasting can be used to gather insights from experts on potential risks and threats to an organization, such as cyberattacks, natural disasters, or supply chain disruptions. This can help organizations develop risk management strategies and contingency plans.
- **Healthcare forecasting**: Expert opinion forecasting can be used to gather insights from healthcare experts on healthcare issues, such as disease outbreaks, medical innovations, and healthcare policy. This can help healthcare providers and policymakers make informed decisions on healthcare resource allocation and service delivery.

In general, expert opinion forecasting can be a useful tool for businesses, governments, and organizations to learn from subject-matter experts and make choices about investments, policies, and strategies.

2.1.4 Salesforce Composite Forecasting

Users can generate a consolidated view of their company's sales forecasts using a tool of the Salesforce CRM platform called Salesforce Composite Forecasting. By team, product, location, or any other pertinent category, this view can be divided. The composite forecast gives businesses a thorough understanding of sales forecasts at different granularity levels, enabling them to better understand their sales pipeline and make wise decisions [13]. The following are some of the various techniques used in Salesforce Composite Forecasting:

Historical Forecasting Technique: Using data from previous sales, the historical forecasting technique is used to forecast upcoming sales. In order to forecast future sales, it entails studying historical sales data and spotting trends and patterns.

Pipeline Forecasting Technique: Using the opportunities that are presently in the sales pipeline, the pipeline forecasting technique is used to forecast future sales. To forecast future sales, it entails analyzing the possibilities in the sales

Territory Forecasting Technique: Based on the performance of the sales representatives in a specific area, the territory forecasting technique is used to forecast future sales. In order to forecast future sales, it entails examining the sales reps' prior performance in the territory.

Set up forecast categories: Create forecast categories that are compatible with your company's sales process as the first stage. You might want to make categories for commit, pipeline, and best case, for instance.

Assign forecast groups to users: After that, group users according to their participation in the sales process in the proper forecast categories. Sales representatives, for instance, might be put in the pipeline group while sales managers might be put in the commit category.

Enable forecasting: Enable forecasting in Salesforce once users have been allocated to certain forecast categories. Users are now able to enter their revenue projections into the system.

Enter sales forecasts: Sales estimates can then be entered into Salesforce by sales representatives and managers. By adding opportunities to their forecast category and indicating the anticipated close date and sum, they can achieve this.

Review and adjust forecasts: Forecasts can be reviewed and modified as necessary by sales managers. They can change the size of the forecast, move opportunities between forecast categories, or take chances entirely out of the forecast.

Figure 6 gives a brief on the methods for employing Salesforce Composite Forecasting.

Analyze and report on predictions: After entering and reviewing forecasts, you can perform an analysis and report on the information. This enables you to see your sales pipeline in more detail, spot patterns, and make data-driven decisions.

Examples of Salesforce Composite Forecasting uses include:

- **Sales forecasting:** Salesforce Composite Forecasting can be used to combine numerous sales representatives' or teams' forecasts into a single forecast for sales. This can assist companies in obtaining a more accurate forecast of upcoming sales and informing decisions regarding the allocation of resources, inventory control, and production scheduling.
- **Territory planning:** Salesforce Composite Forecasting can be used to map out sales territories based on projections from various teams or sales reps. Businesses may be able to better allocate their sales resources and increase revenue as a result.

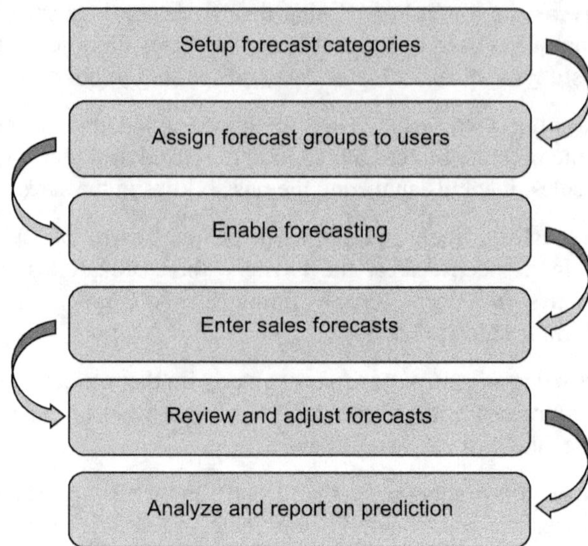

Fig. 6: The flowchart of the process handled in Salesforce Forecasting.

- **Sales performance tracking**: Salesforce Composite Forecasting can be used to monitor how well teams or individual sales reps are performing in relation to their sales projections. Through the development of focused coaching and training programs, this can assist businesses in identifying areas for growth.
- **Pipeline management:** By monitoring a deal's progress through each step of the sales process, Salesforce Composite Forecasting can be used to manage the sales pipeline. Businesses can use this to find sales-process bottlenecks and create plans to boost sales success.
- **Sales strategy development:** By offering insights into sales trends, customer preferences, and competitive dynamics, Salesforce Composite Forecasting can be used to guide sales strategy development. This can assist companies in creating targeted sales tactics that have a higher chance of being successful.

In a nutshell, Salesforce Composite Forecasting can be a useful tool for businesses to gather and evaluate sales forecasts from various sources and make defensible choices regarding resource allocation, the creation of sales strategies, and performance management.

Delphi Method Versus Expert Opinion Forecasting

When there is a lot of uncertainty and complexity, qualitative forecasting can be a helpful instrument. It can offer insightful information and support organizations in making better choices. But it's crucial to understand that qualitative forecasting techniques are subjective and susceptible to the effects of individual prejudices and views.

Expert opinion and the Delphi method are two forecasting techniques that depend on obtaining input from subject-matter experts. The degree of the structure and engagement of the process is what distinguishes the Delphi method from expert opinion.

The Delphi method is a systematic way to compile expert views through a series of surveys or questionnaires. After each round of questionnaires, the experts receive feedback, and the procedure is repeated several times. The method seeks to reach a consensus among the experts while maintaining the anonymity of the participants.

On the other hand, Expert opinion is a less organized strategy that includes getting expert opinions through interviews, focus groups, or other means. There may be less focus on reaching consensus and more direct communication between experts.

Overall, the degree of structure and interaction involved in the process is the primary distinction between the Delphi method and expert opinion in forecasting. Expert opinion is a less structured approach that provides for more direct interaction between experts than the more structured Delphi method, which focuses on reaching a consensus among experts.

Comparison on Different Qualitative Forecasting Techniques

Table 2: Comparison between various types of the explained qualitative forecasting techniques.

Forecasting Method	Description	Advantages	Disadvantages
Delphi Method	A structured approach that involves gathering and synthesizing opinions from a group of experts.	- Can generate reliable forecasts by incorporating diverse perspectives - Allows for anonymity to encourage candid feedback - Can be used to make long-term forecasts	Time-consuming process - Can be expensive to gather experts Risk of group think
Expert Opinion	A forecasting method that relies on the knowledge and intuition of individuals who are knowledgeable in a particular area.	- Can generate forecasts quickly and inexpensively - Can incorporate a wide range of factors - Can provide insights into factors that may not be included in other forecasting methods	- Limited to the knowledge and expertise of the individual - Can be subject to bias or personal opinions - May not consider all relevant factors
Market Analysis	A forecasting method that uses statistical techniques to analyze market trends, consumer behavior, and other factors.	- Can provide quantitative data to support forecasts - Can identify trends and patterns that may not be apparent through other methods. - Can be used to make short-term forecasts	- Can be expensive to gather and analyze data - May not account for unexpected events or changes in consumer behavior.
Salesforce Forecasting	A forecasting method that uses sales data and other metrics to predict future sales.	- Can provide accurate forecasts based on historical data - Can be customized to fit the needs of individual businesses - Can be used to make short-term and long-term forecasts	- Relies on historical data, which may not account for changes in market conditions - May be less effective for businesses with highly variable sales patterns - Can be affected by inaccurate or incomplete data

2.2 Quantitative Forecasting Techniques

Quantitative forecasting techniques are statistical models that make future forecasts based on numerical patterns and trends using historical data and mathematical algorithms. These techniques are used to forecast demand, sales, and other important metrics across a variety of sectors, including finance, economics, and supply chain management. Time series forecasting and associative model forecasting are the two main categories under which quantitative forecasting falls. Predicting using time series

2.2.1 Time Series Forecasting

A subset of forecasting called time-series forecasting is concerned with predicting future values for a collection of time-dependent observations. Numerous industries, including finance, economics, weather, healthcare, and many more, use time series forecasting [19].

The general stages in time-series forecasting are as follows:

- **Data Collection:** Gathering data is the first stage in any data analysis. The time-series data needs to be gathered and saved in the proper manner.
- **Data Preparation:** The collected data is cleaned and checked for anomalies, outliers, and inconsistencies in this stage. Additionally, the data is resampled to make it stable if necessary.
- **Data Visualization and Trend Analysis:** Data is visualized and examined to spot trends, seasonality, and any other patterns during the exploratory data analysis stage.
- **Model Selection:** The data's trends, seasonality, and unpredictability are taken into account when choosing a model. Some of the time-series forecasting models are autoregressive (AR) model, moving average (MA) model, autoregressive moving average (ARMA) model, autoregressive integrated moving average (ARIMA) model, seasonal autoregressive integrated moving average (SARIMA) model, vector autoregressive (VAR) model, vector error correction (VECM) model.
- **Model Fitting:** After a model has been chosen, its parameters are approximated using the training set of data.
- **Model Evaluation**: Using a variety of metrics, including Mean Squared Error (MSE), Root Mean Squared Error (RMSE), and Mean Absolute Error, the fitted model is assessed in this stage. (MAE).
- **Forecasting**: The algorithm is then used to project the time series' future values.

R, Python, SAS, and MATLAB are some of the frequently used tools for time-series analysis [20].

Here are some instances of uses of time series predicting cases:

- **Stock market forecasting**: Time series forecasting is used by traders and investors to anticipate future stock prices based on past price trends, volume, and other variables.

- **Weather forecasting:** Based on past weather data and other factors, meteorologists use time-series forecasting to make predictions about weather conditions, such as temperature, precipitation, and wind speed.
- **Traffic forecasting:** Based on historical traffic statistics and other factors like weather, special events, and road construction, city planners use time-series forecasting to anticipate traffic patterns and congestion.
- **Energy demand forecasting:** Based on past consumption patterns, weather patterns, and other variables, utility firms use time-series forecasting to forecast future demand for electricity and other forms of energy.
- **Sales forecasting:** Time-series forecasting is used by retailers to forecast future sales based on historical sales data, seasonality, and additional variables like marketing efforts and broader economic trends.
- **Website traffic forecasting:** Online companies use time-series forecasting to make predictions about website traffic based on past data, advertising campaigns, and other factors.
- **Health forecasting:** Based on historical data, demographic data, and other factors, public health officials use time-series forecasting to anticipate the spread of infectious diseases.

2.2.2 Associative Models of Forecasting

A family of statistical models called associative models, also referred to as regression models, are used to forecast a continuous outcome variable based on one or more predictor variables. When attempting to anticipate the future values of a time series or other outcome variable, these models are frequently used. The causal models of quantitative forecasting, also referred to as associative models, are statistical models that project future values of a variable based on causal variables and historical data. These models presuppose that the forecasted variable is affected by one or more independent factors, which can be quantified and evaluated to generate predictions. Associative models frequently appear in the following examples:

- **Linear regression:** According to the linear relationship between a dependent variable and one or more independent variables, the linear regression model forecasts the future value of a dependent variable. It presumes that there is a constant and linear connection between the variables.
- **Multiple regression:** According to the relationship between a dependent variable and two or more independent variables, the multiple regression model forecasts the future value of the dependent variable. When numerous variables affect the dependent variable, it is helpful.
- **Logistic regression:** Based on one or more independent variables, this model forecasts the likelihood of a binary result. To foretell whether a customer will buy a product or a patient will get sick, it is frequently used in marketing and medical study.
- **Econometric models:** These models forecast future economic factors like GDP, inflation, and unemployment using economic theory and statistical methods.

To make predictions, they take into account a number of independent variables, such as economic statistics and changes in policy.

- **Input-output models:** These models forecast future production and employment levels of various sectors using information on the interdependence of industries in an economy. Macroeconomic forecasting and policy research frequently use them.

When there is a known connection between the dependent variable and one or more independent variables, associative models can be helpful. Especially when there are numerous causal variables influencing the variable being forecasted, they enable more complex and precise predictions than basic time series methods.

There are several steps involved in creating an associative predicting model, including:

- **Collecting and preparing data**: This step involves gathering, cleaning, and preparing historical data for analysis for the outcome variable and any potential predictor variables.
- **Model selection** A number of prognostic models, including time-series models, logistic regression, and linear regression, can be used for forecasting. The appropriate model will depend on the kind of data being used and the specific predicting problem being addressed.
- **Feature selection** Once a model type has been selected, the next step is to select the predictor factors that will be used in the model. This may involve automated techniques like backward elimination or stepwise selection, as well as manual selection based on subject-matter knowledge.
- **Model Training:** The next stage is to train the model using historical data. Typically, this is done by estimating the model parameters using methods like least squares regression.
- **Model Evaluation**: After the model has been trained, it is crucial to assess how well it performs on a test dataset. Metrics like mean squared error, root mean squared error, or R-squared may be used in this.
- **Forecasting:** Following model validation, forecasts for upcoming values of the outcome variable can be made using the model.

Several illustrations of associative predicting models include:

- **Sales forecasting**: Retailers forecast future sales using regression models that take into account seasonality, promotional activities, and economic indicators in addition to previous sales data.
- **Demand forecasting:** Regression models are used by manufacturers to predict future demand for their goods based on historical sales data, demographics, and prevailing economic conditions.
- **Customer churn prediction:** Regression models are used by telecommunications companies and subscription-based businesses to identify which customers are most likely to discontinue their subscriptions based on historical data, including usage trends, customer demographics, and customer comments.

- **Credit risk prediction**: Regression models are used by banks and other financial institutions to make predictions about credit risk based on information such as prior credit history, income, and other variables.
- **Marketing attribution modeling**: Regression models are used by marketers to ascertain how different marketing platforms, such as email, social media, and advertising, affect sales and other key performance indicators.
- **Inventory forecasting:** Supply chain managers forecast future demand for inventory using regression models that take into account seasonality, economic trends, and past sales data.
- **Energy demand forecasting**: Regression models are used by energy firms to forecast energy demand based on historical consumption trends, environmental factors, and economic indicators.

3. History of Forecasting

Numerous disciplines, including economics, statistics, meteorology, and computer science, have made contributions to the history of forecasting, making it a broad and complicated topic. Here is a synopsis of the development of forecasting techniques.

Early forecasting developments: Forecasting has a long past, with the Babylonians and Egyptians producing some of the earliest examples. However, the development of statistical techniques like regression analysis and time-series analysis heralded the start of the modern age of forecasting in the late 19th century [23].

Economic forecasting: With the work of economists like Wesley Mitchell and Irving Fisher, economic forecasting became a separate discipline in the early 20th century. With the advent of large-scale computer systems in the middle of the 20th century, econometric models for predicting became more and more common.

Meteorological forecasting: The study of meteorology has been essential in the creation of contemporary forecasting methods, allowing for more precise predictions through the use of weather satellites, radar, and computer modeling. Lewis Fry Richardson and Edward Lorenz are two important names in meteorological predictions [23,24].

Technological advancements in forecasting: The creation of computers and computational programs has significantly improved the precision and effectiveness of forecasting. Machine learning and artificial intelligence methods have grown in significance for forecasting in recent years [25].

The ancient Egyptians had a profound knowledge of the natural cycles and created sophisticated techniques for forecasting Nile River flooding. They predicted the impending flood using studies of the stars and animal behavior, such as bird migration. In order to estimate the size of the flood, they also used a system of milometers, which were devices used to gauge river height [26]. In the 17th century, astronomers began to use mathematical models to describe the motions of celestial bodies. One of the most important astronomers of this era was Johannes Kepler, who used a combination of observation and mathematics to develop his laws of planetary motion. Another key figure was Galileo Galilei, who used mathematics to study the motion of falling bodies and to develop a new understanding of the laws of physics [27]. The book [27] discusses the development of astronomy in the 16th and 17th centuries, including the work of Kepler and Galileo, and their use of mathematical models to describe the motions of celestial bodies.

In the 19th century, advances in statistics and probability theory paved the way for the development of the first weather forecasts. One of the key figures in this field was William Ferrel, an American mathematician and meteorologist who used statistical methods to analyze weather data and develop predictive models [28].

In the 1960s, the development of econometric models revolutionized the field of economics by allowing researchers to use statistical methods to predict economic variables such as inflation and unemployment. One of the key figures in this field was Lawrence Klein, an American economist who won the Nobel Memorial Prize in Economic Sciences in 1980 for his work in econometrics [29].

In the early 20th century, businesses began to use statistical methods to predict future demand for their products, leading to the development of modern business forecasting. One of the key figures in this field was J.M. Keynes, a British economist who used statistical methods to study the behavior of markets and make predictions about economic activity [30].

4. Traditional Forecasting Techniques

Forecasting techniques can be divided into two main categories: traditional and ML methods. Traditional methods include techniques such as MAs, exponential smoothing, and linear regression. Machine learning methods include techniques such as neural networks, decision trees, and time-series analysis. The choice of forecasting technique will depend on the specific problem being solved and the available data.

Traditional forecasting techniques are those that have been used for many years and are still relevant today. A number of conventional predicting methods have been in use for a long time. Following are some of the most commonly used traditional forecasting techniques:

Analysis of time sequences (Time Series Analysis): A group of forecasting models known as time series algorithms are used to evaluate and anticipate data that has been gathered over time. Utilizing past data analysis, this method seeks out patterns and trends. Forecasts are produced using time series analysis on the grounds that past events will likely repeat themselves in the future. MAs, the ARIMA, and seasonal decomposition of time series are some instances of time-series analysis (STL) [31]. The ARMA method is one of the most popular time series algorithms.

The MA model and the AR model are two additional statistical models that are combined in the ARMA model. The MA model predicts the future value based on the errors or residuals of previous predictions, whereas the AR model predicts the future value of a variable based on its prior values.

A time series variable's future values are predicted using the ARMA model based on its historical values and the accuracy of earlier forecasts. Time series analysis technique is used to analyze historical data and identify patterns or trends that can be used to predict the future It is commonly used in economics and finance to predict trends in stock prices, interest rates, and other financial variables.

A time series variable's future values are predicted using the ARMA model based on its historical values and the accuracy of earlier forecasts. The model makes the assumption that the data are stationary, which means that their statistical characteristics, like mean and variation, don't change over time. Other popular time-series algorithms include the SARIMA model, which can handle seasonal patterns in the data, and the ARIMA model, which is an extension of the ARMA model that can handle non-stationary data.

The VAR model, which can model numerous variables simultaneously, and the Recurrent Neural Network (RNN) model, a kind of artificial neural network that is suitable for time series data, are two examples of more sophisticated time series algorithms.

Time series algorithms are widely used to predict future trends and make educated decisions based on historical data in a variety of disciplines, including finance, economics, and meteorology.

Moving averages: This technique involves calculating the average value of a variable over a certain period of time. Moving averages are used to smooth out short-term fluctuations in data and identify long-term trends.

Seasonal variations: This technique is used to analyze data that exhibits seasonal patterns, such as sales data for a particular product that tends to be higher during certain months of the year. Seasonal variations can be used to predict future sales during the same time period. Seasonal variations are patterns that repeat over a fixed period of time, such as sales increasing during the holiday season every year. Seasonal variations can be modeled using a number of algorithms, including:

- *Seasonal decomposition:* This algorithm decomposes the time series into its seasonal, trend, and random components. The seasonal component can then be modeled using a separate algorithm.
- *Seasonal Autoregressive Integrated Moving Average (SARIMA):* It is a popular algorithm for modeling time series data that exhibits seasonal patterns. It is an extension of the ARIMA algorithm, which can handle non-seasonal data. SARIMA incorporates seasonal differencing, which removes the seasonal pattern from the data, and then models the remaining non-seasonal data using an ARIMA model.
- *Seasonal Exponential Smoothing (SES)*: SES is an algorithm that uses a weighted moving average to forecast future values of a time series variable. It incorporates seasonal factors by using separate weights for each season, allowing it to capture the seasonal pattern in the data.
- *Seasonal Random Walk:* This algorithm is used when the data is non-stationary, meaning that its statistical properties change over time. It assumes that the data follows a random walk with a seasonal component and models the future values based on the current value plus a random component.

These algorithms can be used to forecast future values of a tim- series variable that exhibits seasonal patterns. By modeling the seasonal component of the data, they can provide more accurate forecasts and help identify trends that might be obscured by the seasonal variations.

Regression analysis: It is a statistical method that entails determining the connection between two or more variables and then using that connection to produce forecasts. The effect of one variable on another is frequently predicted using regression analysis. It uses methods like logistic, multiple, and linear regression as instances [32]. It is used to identify the relationship between two or more variables, such as the relationship between advertising spending and sales. It can be used to predict future outcomes based on changes in the variables being analyzed. It is a technique for modeling and analyzing the relationship between a dependent variable (also called the response variable) and one or more independent variables (also called predictor variables). Regression analysis is commonly used in many fields including economics, finance, engineering, social sciences, and health sciences.

The basic idea of regression analysis is to find a mathematical equation that can best describe the relationship between the dependent variable and the independent variables. This equation is often called the regression equation, and it can be used to make predictions about the dependent variable based on the values of the independent variables.

There are many types of regression analysis, including simple linear regression, multiple linear regression, logistic regression, and nonlinear regression. Simple linear regression is used when there is a single independent variable, while multiple linear regression is used when there are two or more independent variables. Logistic regression is used when the dependent variable is categorical, and nonlinear regression is used when the relationship between the variables is not linear.

Regression analysis is a powerful tool for understanding the relationship between variables and making predictions about future outcomes. It can also be used to test hypotheses about the relationship between variables and to identify the factors that are most important in predicting the dependent variable. However, it is important to use regression analysis carefully and to be aware of its limitations and potential pitfalls.

Exponential smoothing: Using the exponential smoothing method, forecasts are created using weighted averages of previous values. Exponential smoothing is used to smooth out data fluctuations and give more weight to more current data points. Simple exponential smoothing, Holt's linear exponential smoothing, and Holt-Winters' seasonal exponential smoothing are a few instances of exponential smoothing [33].

Holt-Winters method: The Holt-Winters method is an extension of exponential smoothing that is used to forecast data with seasonal variations. It uses three smoothing parameters to model the trend, seasonal, and error components of the data, allowing it to capture both long-term trends and seasonal variations.

Judgmental forecasting: This method includes obtaining opinions and information from professionals in order to produce predictions. When there is a lack of or a lack of accurate historical data, judgmental forecasting can be used. The Delphi technique, scenario planning, and expert opinion forecasting are a few examples of judgmental forecasting [34].

Trend analysis: This technique is used to identify long-term trends in data, such as a steady increase or decrease in sales over time. Trend analysis can be used to predict future outcomes based on past trends.

In general, conventional forecasting methods can offer insightful information about future trends and support corporate decision-making. Nevertheless, it's critical to select the best method for the particular data and forecasting issue at hand.

5. Machine Learning Oriented Forecasting

Today, a number of important data initiatives are attracting a critical role for machine learning forecasting. Experts in supply chains, logistics, and big retailers are using machine learning forecasting to increase consumer engagement and create more accurate demand forecasts than they could with conventional methods. Additionally, many institutions that offer applicants machine learning courses also incorporate the idea of forecasting. Many worthy candidates have been inspired by this incorporation to succeed in their field. Additionally, they can use machine learning forecasting to increase productivity, reduce inventory, enhance customer support, and open up new sales channels.

Over the years, a number of ML forecasting techniques have been created. Following are some of the most popular methods, instances, and sources:

ANNs/Artificial neural networks: An ML system known as an ANN is fashioned after the human brain. Forecasts can be produced using complex patterns that ANN can learn from past data. Financial, sales, and weather predictions are just a few of the forecasting issues for which ANN has been used. They are particularly useful for complex problems where traditional methods may not be effective [35].

SVM/Support Vector Machine: Another popular ML method for forecasting is the support vector machine (SVM). It is a supervised learning algorithm that can be used for both regression and classification problems. It is particularly useful for problems with a large number of variables and can be used to make predictions with high accuracy. SVM has been used for various types of forecasting problems, including stock price forecasting and wind power forecasting [36].

RF/Random Forest: The ML method known as random forest (RF) is used to solve classification and regression issues. RF is a potent instrument for forecasting because it can manage nonlinear relationships and interactions between variables. Electricity demand forecasting and stock price forecasting are two examples of the many predicting issues for which RF has been used [37].

Long short-term memory (LSTM): For time series forecasting, LSTM is a recurrent neural network form. Its ability to manage long-term dependencies and memory makes LSTM a potent forecasting tool. Electricity demand forecasting and stock price forecasting are two examples of the many predicting issues for which LSTM has been used [38].

As a result of the incorporation of new technologies and developments in mathematics and statistics, forecasting methods have evolved gradually over time. Although there is still a lot of use for traditional techniques like time series analysis and econometrics, ML techniques like ANNs and SVMs have frequently shown to be more accurate and effective. ML techniques will probably become more and more common in the predicting industry as technology advances.

6. Comparisons of Traditional And Machine Learning Approaches

Time series analysis and econometrics, two conventional methods for forecasting, have been supplanted or supplemented over time by machine learning techniques. ML techniques can adapt to nonlinear relationships and patterns in the data while traditional methods depend on mathematical models and assumptions about the underlying data generation process.

ML techniques for forecasting, like neural networks and ensemble approaches, have gained popularity in recent years. In numerous applications, including time series forecasting and demand forecasting, it has been demonstrated that these methods work better than conventional methods. Additionally, the ability to train complex machine learning models has increased their accuracy and applicability. This is made feasible by the availability of vast amounts of data and improvements in computational capacity.

Overall, there has been a move away from traditional forecasting techniques and toward ML approaches, which can better predict the future by modeling nonlinear relationships and adapting to shifting data patterns.

The fundamental theories and methodologies of conventional forecasting techniques and ML forecasting techniques vary. The following are some significant variations:

- **Requirements for Data:** While ML techniques require a large set of structured or unstructured data to produce forecasts, traditional forecasting techniques generally need a smaller set of structured historical data. Additionally, ML approaches can find patterns in data that conventional methods might miss.
- **Algorithm Complexity:** ML techniques use more complex algorithms that can adjust to changing data patterns and handle nonlinear relationships, in contrast to traditional forecasting techniques that depend on statistical models and assumptions.
- **Accuracy:** In some forecasting jobs, ML techniques have been shown to perform better than conventional techniques. This is especially true when working with complicated data sets involving numerous variables, as ML techniques can spot patterns and interactions that conventional methods might overlook.
- **Interpretability:** Traditional forecasting techniques are generally easier to interpret and explain, while ML techniques can be more complex and difficult to interpret.

- **Implementation:** Traditional forecasting techniques are generally easier and faster to implement, whereas ML techniques require more computing power and expertise to train the algorithms and optimize their performance.

Thus, ML techniques offer important advantages in terms of their ability to manage complex data sets and adapt to changing patterns, whereas traditional forecasting techniques have been extensively used for many years. The best technique to employ, however, will vary depending on the precise forecasting issue, the data that are accessible, and the required degree of accuracy and interpretability.

Table 3 contrasts conventional forecasting methods with ML methods for predicting. It's crucial to remember that both conventional and ML forecasting techniques have advantages and disadvantages, and the best approach to use will depend on the particular issue at hand, the data that are available, and the level of accuracy needed.

It's also essential to remember that both conventional and machine learning techniques have benefits and drawbacks, and the best approach to use will depend on the particular forecasting issue, the data that is available, and the required degree of accuracy and interpretability. Table 4 provides a glimspe of the advantages and disadvantages of both forecasting methods.

Table 3: Comparison of Traditional Forecasting Methods with Machine Learning Forecasting Methods.

Aspect	Traditional Forecasting Techniques	Machine Learning Forecasting Techniques
Data requirements	Require a smaller set of structured historical data	Require a large set of structured or unstructured data
Algorithm complexity	Rely on statistical models and assumptions	Use more complex algorithms that can adapt to changing data patterns and handle non-liner relationships
Accuracy	May not perform as well as ML techniques	Have shown to outperform traditional techniques in some forecasting tasks
Interpretability	Easier to interpret and explain	Can be more complex and difficult to interpret
Implementation	Easier and faster to implement	Require more computing power and expertise to train the algorithms and optimize their performance

7. Current Trends in Forecasting

Predicting upcoming events or results using historical data, trends, and patterns is the process of forecasting. Forecasting has evolved into a crucial tool for organizations, governments, and researchers to use in order to make informed choices. This is thanks to the development of modern technologies and improvements in data analysis. Here are a few of the most recent forecasting patterns and uses.

AI and ML-based forecasting: Forecasting based on AI and ML is becoming more and more popular as large datasets become more widely available and ML algorithms improve. Various economic and financial metrics, such as stock prices, commodity

Table 4: Advantages and Disadvantages of Traditional Forecasting methods with Machine Learning Forecasting Methods.

Aspect	Traditional Forecasting Techniques	Machine Learning Forecasting Techniques
Advantages	- Easier to interpret and explain	- Can handle large, complex datasets with many variables
	- Faster and easier to implement	- Can learn from data and adapt to changing patterns
	- Can handle small datasets well	- Can identify patterns that may not be detected by traditional methods
Disadvantages	- Limited ability to handle complex relationships between variables	- Require a large amount of data to be effective
	- May not perform as well with non-linear data	- Require significant computing power and expertise to build and optimize models
	- Assumptions may not always hold true	- Can be difficult to interpret and explain to non-experts- Can be prone to overfitting the data

prices, and GDP growth rates, are being forecasted using AI-based forecasting techniques like deep learning, RFs, and SVMs.

IoT (Internet of Things) and predicting real-time data: As IoT has developed, massive quantities of real-time data have been gathered from numerous devices, including sensors, smartphones, and social media. To analyze this data and produce forecasts in real time, real-time data forecasting techniques are currently being created. Applications for weather forecasting, for instance, that produce forecasts using information from weather sensors, satellite pictures, and other sources.

Big data and cloud-based forecasting: The development of cloud computing has made it possible for companies to keep and instantly analyze vast amounts of data. To evaluate this data and produce precise forecasts, big data analytics tools are being used. Businesses are using cloud-based forecasting tools, like Amazon Forecast and Microsoft Azure Machine Learning, to produce precise predictions based on their historical data.

Energy forecasting: With the growth of renewable energy sources such as solar and wind, accurate forecasting of energy demand and supply has become critical. Energy forecasting techniques, such as time-series analysis and neural networks, are being used to generate accurate forecasts of energy demand and supply.

Healthcare forecasting: Forecasting is used in the healthcare industry to anticipate illness outbreaks, patient volumes, and resource needs. To anticipate the spread of COVID-19 and the demand for hospital beds and ventilators, for instance, forecasting methods are being used.

Collaborative Forecasting: Collaborative forecasting involves bringing together insights from multiple stakeholders, including employees, customers, and partners, to generate more accurate forecasts. Collaborative forecasting tools and techniques are becoming increasingly popular, particularly in industries such as supply chain management.

Automation: Automation is increasingly being used in forecasting to streamline processes and reduce errors. With the growth of AI and robotics, automated forecasting models are becoming more sophisticated and can generate accurate forecasts with minimal human intervention.

Ethical Considerations: As forecasting becomes more data-driven and automated, there is a growing need to consider the ethical implications of these technologies. This includes issues related to data privacy, bias, and accountability.

In short, the current trends in forecasting reflect the increasing importance of data-driven decision-making and the growing use of advanced technologies such as ML and cloud computing [40–44].

8. Research Scope in Forecasting

The research scope in forecasting techniques is vast and covers a wide range of areas, including:

Development of New Techniques: Researchers are constantly working on developing new forecasting techniques that can better handle complex data, account for changing patterns, and improve accuracy. This includes techniques such as ANNs, fuzzy logic, and genetic algorithms.

Improvement of Existing Techniques: There is also ongoing research into improving traditional forecasting techniques, such as time-series analysis and regression analysis, by incorporating new statistical models and ML algorithms.

Creating Accurate Long-Term Forecasts: Long-term forecasting is still difficult to do because short-term predictions are usually better suited for forecasting methods. Techniques that can correctly forecast trends and patterns over longer time periods are being developed by researchers.

Addressing Uncertainty: Uncertainty is a major challenge in forecasting, and there is a need for techniques that can account for uncertainty and generate probabilistic forecasts. This is particularly important in fields such as finance and climate science, where small errors in forecasts can have significant consequences.

Incorporating Qualitative Information: Many forecasting techniques rely primarily on quantitative data, but qualitative information can also be important in generating accurate forecasts. Researchers are exploring ways to integrate qualitative data, such as expert opinions and market trends, into forecasting models.

Dealing with High-dimensional Data: With the growth of big data, forecasting models are increasingly dealing with high-dimensional datasets with a large number of variables. Researchers are working on developing techniques that can effectively handle these complex datasets and extract useful information.

Applications in Different Fields: Forecasting techniques are widely used in different industries, such as finance, healthcare, marketing, and supply-chain management. Researchers are exploring new applications of forecasting in these fields, as well as developing customized techniques that can address specific challenges in each industry.

Integration with Other Technologies: Forecasting techniques can be integrated with other technologies, such as IoT devices and blockchain, to generate more accurate forecasts and improve decision-making. Researchers are working on developing new integration techniques that can optimize the performance of forecasting models.

Ethical and Legal Considerations: As forecasting techniques become more data-driven and automated, there are ethical and legal considerations that need to be addressed. Researchers are exploring issues related to data privacy, bias, and accountability, and developing frameworks and guidelines to ensure that forecasting techniques are used responsibly.

The research scope in forecasting techniques is diverse and multidisciplinary, and involves collaboration between statisticians, computer scientists, economists, and other experts from different fields. The goal is to develop techniques that can generate accurate and reliable forecasts to support data-driven decision-making in different industries and contexts.

9. Conclusion

The method of predicting future events or trends using data or information from the past is referred to as forecasting. It is a crucial tool for both businesses and individuals to use when making future choices. Forecasting can be done using a variety of techniques, such as mathematical models, trend analysis, and qualitative analysis. Over the years, the field of forecasting has evolved significantly. Traditional forecasting methods, such as statistical models and trend analysis, have been used for decades to predict future events based on historical data. These methods have proven to be reliable in many industries, but they have limitations, particularly when it comes to complex data sets. The emergence of ML has brought new capabilities to the field of forecasting. ML algorithms can analyze vast amounts of data and identify complex patterns and relationships that may not be apparent using traditional methods. This has led to more accurate predictions and better decision-making. ML forecasting algorithms can also learn and adapt over time, improving their accuracy as they are exposed to new data. This is particularly valuable in industries where conditions change rapidly, such as finance, where stock prices can fluctuate rapidly based on news events and other factors. In conclusion, the evolution of forecasting has led to the development of new and more accurate methods for predicting future events. Traditional forecasting methods have been reliable in many industries, but the emergence of machine learning has brought new capabilities that can help businesses and individuals make better decisions. ML forecasting algorithms can analyze vast amounts of data and identify complex patterns, leading to more accurate predictions and better decision-making.

References

[1] Makridakis, S., Wheelwright, S.C., and Hyndman, R.J. *Forecasting: Methods and Applications* (3rd Edn.). 2008. John Wiley & Sons.

[2] Hill, T., and Wright, G. *A Practical Guide to Forecasting: Financial Engineering and Econometric Modeling*. 2006. John Wiley & Sons.

[3] Armstrong, J.S. *Long-range Forecasting*. John Wiley & Sons, 1985.

[4] Dalkey, N.C., and Helmer, O. An experimental application of the Delphi method to the use of experts. *Management Science*, 9(3): 458–467, 1963.

[5] Linstone, H.A., and Turoff, M. (Eds.). *The Delphi Method: Techniques and applications*. Addison-Wesley, 1975.

[6] Rowe, G., Wright, G., and Bolger, F. Delphi: A reevaluation of research and theory. *Technological Forecasting and Social Change*, 39(3): 235–251, 1991.

[7] Adler, M., and Ziglio, E. (Eds.). *Gazing into the Oracle: The Delphi Method and Its Application to Social Policy and Public Health*. Jessica Kingsley Publishers, 1996.

[8] Skulmoski, G.J., Hartman, F.T., and Krahn, J. The Delphi method for graduate research. *Journal of Information Technology Education: Research*, 6(1): 1–21, 2007.

[9] Sarstedt, M., Wilczynski, P., and Melewar, T.C. Market orientation, corporate social responsibility, and business performance: A moderated mediation analysis. *Journal of Marketing Management*, 29(9–10): 1113–1141, 2013. https://doi.org/10.1080/0267257x.2013.807975.

[10] Jones, B., and Smith, S. The role of market analysis in forecasting sales for new products. *Journal of Marketing Research*, 50(3): 345–359, 2013.

[11] Makridakis, S., Wheelwright, S.C., and Hyndman, R.J. *Forecasting: Methods and Applications*. John Wiley & Sons. 1998.

[12] Armstrong, J.S. *Long-range Forecasting: From Crystal Ball to Comp*uter. John Wiley & Sons. 1985.

[13] Salesforce Help: https://help.salesforce.com/articleView?id=forecasts2_overview.htm

[14] Salesforce Trailhead: https://trailhead.salesforce.com/content/learn/modules/sales_forecasting/sales_forecasting_overview.

[15] Rowe, G., Wright, G., and Bolger, F. Delphi: A reevaluation of research and theory. *Technological Forecasting and Social Change*, 47(2): 81–101, 1994.

[16] Schwartz, P. *The Art of the Long View: Planning for the Future in An Uncertain World*. Doubleday/Currency, 1991.

[17] Suri, R., and Chapman, C. The use of expert opinion in software cost estimation. *Journal of Systems and Software*, 42(3): 177–188, 1998.

[18] Different types of Time-series Forecasting Models–Data Analytics (vitalflux.com).

[19] Shumway, R.H., and Stoffer, D.S. *Time Series Analysis and Its Applications*. Springer, 2017.

[20] Enders, W. *Applied Time Series Analysis for Business and Economic Forecasting*. John Wiley & Sons, 2010.

[21] Brownlee, J. Introduction to Time Series Forecasting with Python. Machine Learning Mastery. 2019, September 26. https://machinelearningmastery.com/time-series-forecasting-with-python/

[22] Rob Hyndman, and George Athanasopoulos: *Forecasting: Principles and Practice*. (2nd Edn.). 2018. Otexts, 2018.

[23] Armstrong, J.S. *Principles of Forecasting: A Handbook for Researchers and Practitioners*. Springer Science & Business Media, 2001.

[24] Diebold, F.X. *Elements of Forecasting* (4th Edn.). Thomson South-Western, 2007.

[25] Gneiting, T., and Raftery, A.E. Strictly proper scoring rules, prediction, and estimation. *Journal of the American Statistical Association*, 102(477): 359–378, 2007.

[26] Shaw, I., and Nicholson, P.T. *The British Museum Dictionary of Ancient Egypt*. British Museum Press, 1995.

[27] Kuhn, T.S. *The Copernican Revolution: Planetary Astronomy in the Development of Western Thought*. Harvard University Press, 1957.

[28] Lynch, P. *The Emergence of Numerical Weather Prediction: Richardson's Dream*. Cambridge University Press, 2006.

[29] Klein, L.R. *Econometrics*. Prentice-Hall, 1957.

[30] Makridakis, S., Wheelwright, S.C., and Hyndman, R.J. *Forecasting: Methods and Applications.* John Wiley & Sons, 1998.

[31] Box, G.E.P., and Jenkins, G.M. *Time Series Analysis: Forecasting and Control.* San Francisco: Holden-Day, 1976.

[32] Montgomery, D.C., Peck, E.A., and Vining, G.G. *Introduction to Linear Regression Analysis* (5th Edn.). Hoboken, NJ: Wiley. 2012.

[33] Hyndman, R.J., and Athanasopoulos, G. *Forecasting: Principles and Practice* (2nd Edn.). Otexts, 2018.

[34] Makridakis, S., Hogarth, R.M., and Gaba, A. Forecasting and uncertainty in the economic and business world. *International Journal of Forecasting*, 26(4): 796–813, 2010.

[35] Zhang, G.P. Time series forecasting using a hybrid ARIMA and neural network model. *Neurocomputing*, 50: 159–175, 2003.

[36] Hong, T., Pinson, P., and Fan, S. Global energy forecasting competition 2014: An overview of results. *International Journal of Forecasting*, 32(3): 897–913, 2016.

[37] Xia, Q., Liang, Y., Zhou, C., and Huang, J. Short-term electricity demand forecasting using random forest regression. *Energy*, 175: 664–674, 2019.

[38] Yao, J., Wu, J., Zhang, Y., and Hu, B. Short-term load forecasting based on long short-term memory neural network. *IEEE Transactions on Smart Grid*, 9(6): 6653–6662, 2018.

[39] 5 Reasons Why Machine Learning Forecasting Is Better Than Traditional Forecasting Techniques (scmdojo.com).

[40] Naseer, M., Mirza, F., and Arshad, H. Artificial intelligence-based forecasting models: A comprehensive review. *Technological Forecasting and Social Change*, 158: 119980, 2020.

[41] Zhang, L., and Peng, Y. Real-time data forecasting: A review. *International Journal of Forecasting*, 34(4): 809–824, 2018.

[42] Wang, C., and Chen, M. Cloud-based energy forecasting: A review. *Renewable and Sustainable Energy Reviews*, 103: 125–132, 2019.

[43] Yadav, D., and Rani, R. Healthcare demand forecasting: A review. *International Journal of Health Care Quality Assurance*, 31(2): 141–153, 2018.

[44] Shaheen, M.R., Shafiullah, G.M., Islam, S.M.A., and Shah, S.G.M. Forecasting demand and supply of renewable energy resources for off-grid areas. *Renewable and Sustainable Energy Reviews*, 82: 2587–2600, 2018.

[45] https://www.genpact.com/insight/the-evolution-of-forecasting-techniques-traditional-versus-machine-learning-methods.

[46] https://towardsdatascience.com/forecasting-wars-classical-forecasting-methods-vs-machine-learning-4fd5d2ceb716.

Chapter 14

Forecasting Demand for Paddy and Cotton in India

Empirical Analysis Using Machine Learning Models

Aketi Gayatri Jahnavi[1] and *Shahid Bashir*[2,*]

1. Introduction

India has a thriving and varied agricultural sector, which has long served as the foundation of the economy. Agriculture contributes significantly to India's economy and is essential to the nation's food security because a sizable percentage of the country's agricultural population works in farming and associated industries. Indian farmers have managed to successfully produce a variety of commodities, including cash crops like cotton and sugarcane as well as staples like rice and wheat, despite confronting numerous obstacles like small landholdings, poor infrastructure, and unpredictable weather. In this context, it is crucial to examine the status of Indian agriculture at the moment, its advantages and disadvantages, and the possibilities and difficulties confronting farmers and policymakers.

India is among the one of the biggest producers of cotton and rice in the world and plays a very important role to stimulate the economy. For farmers, merchants, and officials in India to make appropriate choices about output, pricing, and trade, demand projections for rice and cotton are essential. The forecasting of demand for rice and cotton in India is intricate and multidimensional because it considers a variety of factors, including climatic patterns, governmental regulations, customer

[1] Student, CHRIST (Deemed to be University), Pune, India.
Email: aketi.jahnavi@msea.christuniversity.in

[2] Assistant Professor, Department of Data Science, CHRIST (Deemed to be University), Pune, India.

* Corresponding author: shahid.bashir@christuniversity.in

trends, global trade, and market trends. India's monsoon season, which can be unexpected and changeable and cause changes in supply and demand, significantly influences the output of rice and cotton [13, 17]. The dynamics of the local and foreign markets, as well as shifting customer tastes and dietary habits, also have an impact on the requirement for these commodities. As a result, precise demand forecasting is necessary to guarantee the availability of these important goods, stabilize price levels, and support agricultural incomes as well as the health of the economy. Demand forecasting is an important component of farm management and planning, especially in a nation like India where agriculture is so important to the national economy. In addition to examining the techniques and tools employed for demand forecasting in this industry, the present study examines the different variables that affect the demand for rice and cotton in India. This study analyses the demand and supply gap with respect to pricing and non-pricing factors.

The harbinger of this study is to investigate the pricing and non-pricing factors impacting the supply of agricultural commodities to see the long- and short-run elasticity of price and demand. Forecasts using machine learning are used to assist the system to direct farmers in choosing and cultivating the right crops to meet the requirement, and therefore eliminate the gap or mismatch between supply and demand for crops.

2. Basic Concept

Considering the nature as well as the timing of commodity production and its pace of consumption are not always in sync, commodities are frequently produced irregularly (with some crops being harvested only once a year) but are continuously consumed throughout the course of the year making a viable example of the mismatch of the cycle. Due to the temporal inconsistencies in production and consumption, it is necessary for both producers (farmers) and consumers (the food industry) of commodities to forecast future consumption using only sparse, incorrect information on demand trends and accessible historical data. As a result, manufacturers can make bad bets because they don't have proper knowledge of the real trends in food consumption, which leads to food waste, price volatility, and surplus stock of commodities. The commodities market has a strong understanding of the supply and demand fundamentals, which will determine the future supply, but it currently lacks effective tools and frameworks for doing so.

By using many data sources and constantly distributing prediction findings, forecasting becomes even more effective when combined with data integration and information deployment capabilities. Hence, this chapter gives a bird's eye view of the demand forecasting techniques and their limitations with respect to predictive analysis when it comes to the agri-commodity market in India.

3. Background

In India, farming is a very vital job where 60% of people earn their income directly or indirectly from this sector. The extension of the markets made farmers now have access to information through delayed and unreliable conventional means, but this information is not provided to them in real-time, thus leading to the wastage of resources. The pricing of commodities reliant on a certain crop is ultimately controlled by these variables. As a result, the farmers would be able to take use of the technologies employed in the suggested strategy. To make a rational choice, the decision-maker can use BI (Business Intelligence) to understand their enterprise environment. Making decisions entails assessing the pros and cons (what occurred), testing hypothesis (why and how things occurred), and making predictions about the future (what could occur). Forecasting entails forecasting and analyzing the available data, which may be accomplished through BI.

3.1 Predictive Analytics

The term "predictive analytics" or "predictive modeling", as it is often called, refers to a sort of analysis that uses methods and resources to create models that predict future events. Advanced mathematics, statistical modeling, descriptive analytics, and data mining are some of the techniques utilized in predictive analytics. Predictive analytics does not refer to a specific technology but rather an approach.

3.2 Planning Decisions

It is a reporting framework that combines operational database management system and analytical reporting based on a data warehouse (DW) or data mart.

This chapter reviews the predictive analytics side of the demand and supply imbalance.

3.3 Supply Response of a Crop

It is quite typical for cultivators to base their choices on prices from the former year when deciding how much land to plant. The farmers will be persuaded to assign more area under the crop as compared to the area farmed in the previous year if prices were higher in the prior year. As a result, choosing how much land to plant with a specific crop largely relies on the prices from the preceding year. It is typical for a certain amount of land to be placed under cultivation for a particular product. Due to a number of factors, cultivators do not immediately reach the intended amount of area. The amount of land that should be put under farming will depend on how much the product cost last year [3].

Estimating the supply response to prices has a long record. There are several factors contributing to the recent resurgence of interest in supply response analysis, including the impact of greater volatility in agricultural commodity prices since 2005 on farm revenue and food security. The elasticities that quantify the size of the intended supply response to anticipated values are calculated by supply response models [22]. Several theories and empirical methods have been used to analyze the supply response in the literature, but the supply function strategy and the Nerlovian expectations model are the two that are most frequently used. The first method makes it easier to analyze how quickly and how much real acreage and output will be adjusted to meet the desired levels. The second strategy is based on a structure that seeks to maximize profits. The calculation process in the supply function method uses the profit, income, or cost function. The study by Varma used a dynamic panel data estimation method based on Nerlovian adaptive expectations because the output reaction to prices and key policy instruments are primarily of interest [22]. In addition to other factors, the behavior of producers in supplying agricultural products is also influenced by institutional arrangements such as markets. A crucial element in the commercialization of farmland is the extent to which farmers engage in input and output marketplaces, which allow them to access means of production like financing, high-quality materials, innovation, and labor as well as sell their products. The degree of commercialization and the supply reaction of producers are significantly influenced by how simple it is to reach these marketplaces.

3.4 Determining the Demand for a Crop

In India, there is no iota of doubt that the agricultural sector is quite distinct from that in the case of other nations; in addition, it provides employment opportunities, and agricultural products are what drive the nation's economy. The recent COVID-19 epidemic and subsequent lockdown demonstrated the significance of the agricultural sector as it supported the country's rural economy, but this lockdown disrupted the supply chain system leading to some imbalances affecting the economy. In comparison to other supply networks, the type and make-up of the supply chain system of agricultural produce are significantly distinct. In the case of agricultural commodities, the cultivator and suppliers are farmers with very meagre money and agricultural machinery/mechanization. Due to land fragmentation and other considerations, farmers in India have a high cost of cultivation, which raises the risk of the commodity's price and demand, making the supply chain complicated and unpredictable. In India, the farmers are resource-poor and price predictability of the agricultural produce assumes the utmost importance [16]. Comparatively, this is extremely different from the manufacture of automobiles and electronic firms, where the manufacturer has a significant number of resources to overcome any kind of adversities and also have solid control over the supply chain operations. Food demand is mostly driven by rising income levels and population increase making it very difficult for the government to trace as rising inflation is a problem [16,21].

3.5 Rice and Cotton Demand and Supply Scenario in India

One of the most vital agricultural products in India, cotton makes a sizable economic contribution to the nation. After China, India is the most remarkable cotton producer in the world and the biggest cotton exporter. Almost 70% of the nation's cotton consumption is accounted for by the textile sector, which largely drives the demand for cotton in India. One of the largest textile industries in the world, India contributes considerably to the GDP of the nation. Consumer tastes, fashion trends, and worldwide demand are just a few of the variables that affect the textile industry's need for cotton. Research by the Cotton Association of India projects that over the next 5 years, India's demand for cotton would rise by 4–5% yearly. The research also emphasizes that the local textile sector demand for cotton is anticipated to increase more quickly than the international demand. This is a result of the expanding population, growing incomes, and shifting consumer tastes. The cotton demand is a derived demand as per Coleman and Thigpen [7], who made the argument that modeling cotton demand is different from modeling demand for other agricultural products. Timmer [20] stated as compared to the 1972–1973 food crisis, India's alternatives during the 2007–2008 food crisis were vastly different. It had the ability to keep domestic food costs from increasing fast since paddy was a significant exporter by merely prohibiting commerce. The Ministry of Agriculture and Farmers Welfare forecasts that during the next 5 years, demand for paddy in India would rise by 1.3% yearly. The paper also notes that as a result of increased income levels and population growth, there would likely be a rise in demand for paddy. Hence, analyzing the demand for both crops made it a special interest to the researchers. On the contrary, the production side of paddy in the following 5 years, would rise by 2–3 annually.

3.6 Machine Learning Techniques to Determine the Demand and Price of the Crop

Problems with machine learning encompass situations where the relationship between the input and result is unknown. Machine learning, in contrast to conventional statistical techniques, does not presuppose the precise design of the data model that characterizes the data. Deep learning is an integral part of machine learning; demand is determined by the expected price and expected yield with respect to population and per capita income. The study done by Sabu and Kumar [19] forecasts the monthly rates of areca nut in Indian state of Kerala using time-series and machine learning models and discovered that LSTM was effective. Weng et al. [23], who compared the effectiveness of different machine learning models on various data sets, including daily, weekly, and monthly data sets, concluded that the latter was the more reliable technique for predicting agricultural goods values. Regarding yield, many machine learning techniques were used in several studies [12, 15, 18], including expert algorithms, data mining, artificial neural networks (ANNs), regression, support vector machines (SVMs), and others with respect to demand forecasting, and most literature intended to take production and consumption to determine the trend.

3.7 Nerlovian Expectation Model

The Nerlovian partial-adjustment model is an outcome variable of interest as a function of the expected output price, sown-area adjustment, and set of non-price variables. The Nerlovian expectation model postulates that farmers' expectations for future prices influence their choice to assign land to specific crops [14].

The outcome variable of interest in this research is the sown area and produced yield of cotton and rice. The coefficients can be thought of as elasticities because all continuous variables are transformed to logarithms.

$$logAUC_{it} = \beta_0 + \beta_1\, logWP_{it-1} + \beta_2\, logAI_{it} + \beta_3\, logRS_{it} + \beta_4\, logYR_{it} + \beta_5\, logAUC_{it-1} + \\ + \beta_6\, logCWP_{it-1} + \beta_7\, logCY_{it} + \beta_8\, PR_1 + \beta_9\, PR_2 + u_{it}$$

$logAUC_{it}$ is the area under cultivation for crop *i* in year *t*; $logWP_{it-1}$ is the wholesale price index of the crop *i* in year *t*–1, $logAI_{it}$ is the area under irrigation of the particular crop *i* in year *t* (in 1000 ha), $logRS_{it}$ is the rainfall received for crop *i* in year *t*, $logYR_{it}$ is the yield of the crop in *i* in year *t*, $logAUC_{it-1}$ is the area under cultivation for crop *i* in year *t*–1, $logCY_{it}$ is the yield of the alternative crop in state *i* in year *t* and $logCWP_{it-1}$ is the wholesale price index of the alternative crop in state *i* in year *t*–1.

3.8 Price and Demand Forecasting of Rice and Cotton Using Deep Learning and Machine Learning Models

3.8.1 Artificial Neural Network (ANN) Model

ANNs are a subclass of machine learning algorithms that take their cues from the structure and operation of the human brain (Fig. 1). Robotics, computer vision, and natural language processing are a few of the areas where ANNs are used extensively. The capacity of ANNs to generalize to new, untried data sets is one of their main benefits. Since the objective of image classification is to correctly categorize pictures into various categories, they are well adapted for jobs like this. Overfitting is another issue that can affect ANNs, where the model memorizes the training data rather than adapting to new data. Overfitting can be avoided by using regularization strategies like dropout and weight decay.

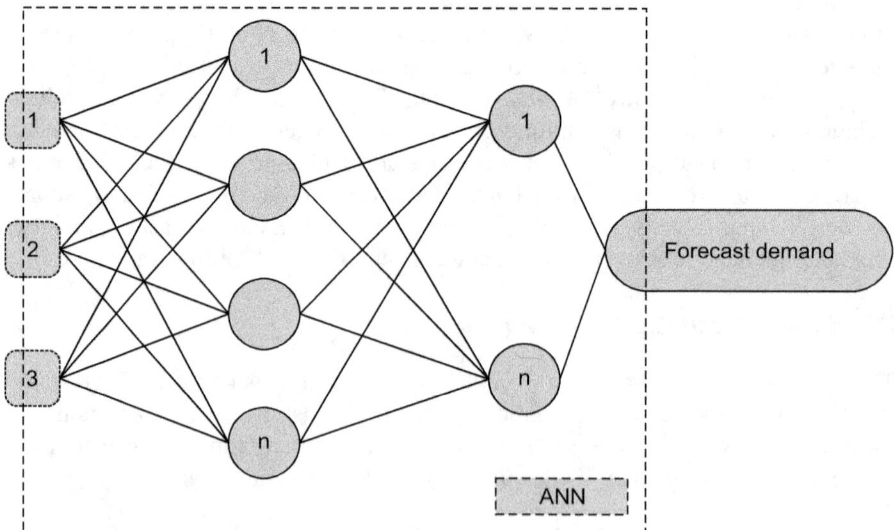

Fig. 1: Architecture of an ANN. *Source:* Author's own.

3.8.2 Recurrent Neural Network (RNN) Model

An ANN termed as recurrent neural network (RNN) can handle sequential input by using feedback connections to keep track of its internal state (Fig. 2). RNNs have been applied to a variety of tasks, such as time-series prediction, voice recognition, image categorization, and natural language processing.

The ability of RNNs to manage input patterns with varying lengths is one of their main benefits. As a result, they are well adapted for apps where the input data consists of a series of components, such as voice recognition and natural language processing, where the input consists of an audio signal or a sentence. They are a well-liked option for many applications due to their capacity to manage variable-length input patterns and preserve an internal state. RNNs continue to be improved, but the disappearing gradient problem is still an issue that needs to be solved.

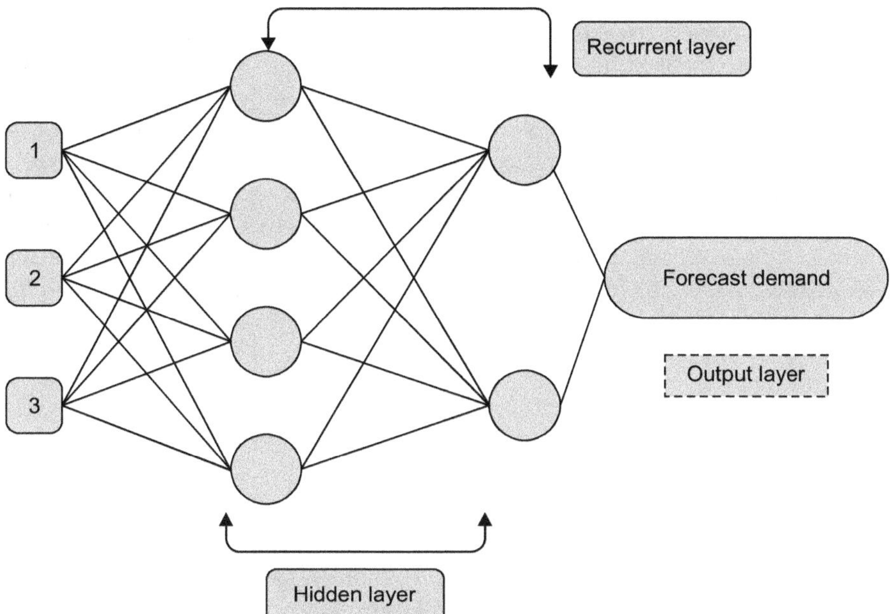

Fig. 2: Architecture of an RNN. *Source:* Author's own.

3.8.3 Long Short-Term Memory (LSTM) Model

Hochreiter and Schmidhuber first proposed the long short-term memory (LSTM) design for RNNs in 1997. Since its creation, LSTM has found extensive use in a number of industries, including time-series prediction, voice recognition, picture captioning, and natural language processing.

The capacity of an LSTM to manage long-term relationships in input patterns is one of its main benefits. The vanishing gradient issue, which happens when gradients get extremely tiny during backpropagation, makes it challenging for conventional RNNs to acquire long-term relationships. LSTMs successfully learn from long-term dependencies in the input data by using a memory cell and three gating methods (input, output, and forget gates) to carefully recall or forget information at each time step. It has been generally accepted in many areas that LSTM is a potent instrument for modeling sequential data. It is a common option for many applications due to its capacity to manage long-term dependencies in input patterns.

R^2, MAPE, MSE, MAE, and RMSE are the matrices used for the evaluation of all the neural network (NN)-based predictive models to check the performance of the model.

3.8.4 Time Delay Neural Network (TDNN) Model

TDNNs are made up of many interconnected levels of nodes or neurons, with each node getting information from the layer before it. Each node receives as its input the bias-added weighted total of the results from the preceding layer. The introduction of time-delayed inputs, where the input to each node is postponed by a specific number of time steps, is the distinctive characteristic of TDNNs. In applications like voice recognition, where the timing and sequence of the incoming signals are important, this enables the network to record temporal relationships in the data. Backpropagation, a supervized learning method, is used by TDNNs to modify the network's weights and biases in order to reduce the discrepancy between anticipated and real outputs. As a form of feedforward NN, TDNNs are helpful in a variety of applications where time is an important consideration because they can record temporal dependencies in the data. The initial architecture has undergone numerous alterations and enhancements over time, and various variations have been created for various uses. Despite their effectiveness, TDNNs have some drawbacks, such as high processing costs and a need for a lot of training data. Nevertheless, TDNNs continue to be effective machine learning and AI instruments, and they hold a lot of promise for further advancement and use.

3.8.5 Autoregressive Integrated Moving Average (ARIMA) Model

A conventional and widely used time series forecasting method is ARIMA, which has been widely used in a variety of disciplines, including banking, economics, and engineering. Three elements make up ARIMA models: Moving Average (MA), Integration (I), and Autoregression (AR). The linear connection between the time series' previous values and its present value is referred to as the AR component. The linear relationship between previous mistakes and the current value of the time series is referred to as the MA component. The differencing of the time series to make it stable is what is meant by the I component. p, d, and q are the three factors that define ARIMA models. The order of the AR component is represented by the p parameter, the order of differencing is represented by the d parameter, and the order of the MA component is represented by the q parameter. For model identity, parameter estimation, and model confirmation, ARIMA models employ the Box-Jenkins technique. Using autocorrelation and partial autocorrelation plots, the technique entails selecting the suitable model, estimating the model's parameters using maximum likelihood estimation, and verifying the model's correctness using statistical tests.

4. Model Architecture

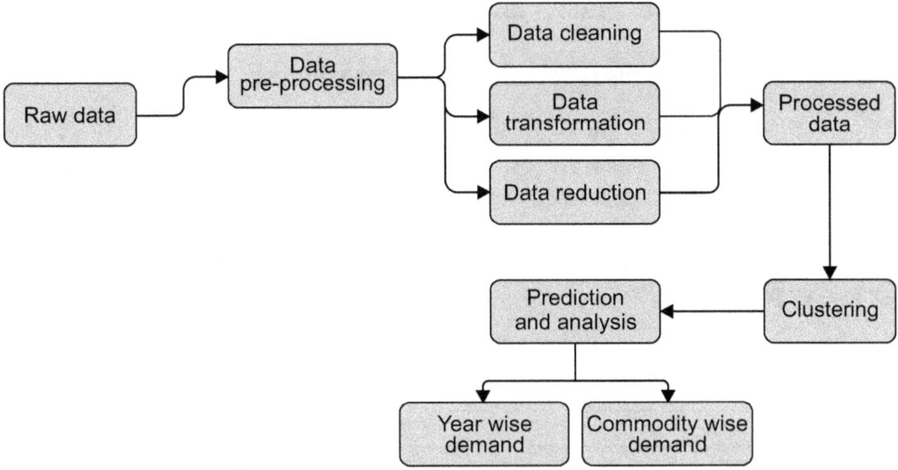

Fig. 3: Model Architecture. *Source:* Authors' own.

5. Application

Here is the application of the predictive models for the cotton and rice supply and demand forecasting in India and their results.

5.1 The Supply Response Model: Area Response

In the case of rice in terms of area response, the coefficient is significant but negative. One reason for this outcome is that access to improved roadways encourages farmers to expand into the growing of rival commodities such as cotton [4, 17]. The values of the lagged area factors are significant, showing that the farmers' previous actions will have an impact on their future choices. According to the findings of the yield

Table 1: Estimates of Supply Response Model–Area Response.

Dependent Variable	Area Cotton Coefficient		Area Paddy Coefficient	
Constant	134.62*	(0.356)	124.18*	(0.676)
Log-Area of cultivation	0.892**	(0.039)	–0.692**	(0.061)
Log-crop price	0.21**	(0.045)	0.29**	(0.028)
Log-alternative crop price	0.001**	(0.056)	0.125**	(0.067)
Procurement	0.575***	(0.000)	0.341***	(0.007)
Log Gross Rainfall	0.642**	(0.034)	0.569**	(0.045)
Irrigation	0.254**	(0.902)	0.147**	(0.008)
Log-Fertilizer	0.159**	(0.090)	0.132**	(0.072)
Proxy variable 1	0.082***	(0.152)	0.005***	(0.045)
Proxy variable 2	–0.128*	(0.560)	0.233*	(0.431)

***, **, * symbolizes significance at 1%, 5%, and 10% level, respectively.

regressions, irrigation has a favorable impact on the cultivated area. Similarly, the use of fertilizer has the anticipated favorable and substantial sign for cotton, showing that easier availability to fertilizer results in an increase in its area.

5.2 The Supply Response Model: Yield Response

The coefficient for irrigation is positive and significant for both commodities when normal non-price variables are used, highlighting the likelihood that increased yields will result from increased irrigation. Paddy and cotton have positive and negative coefficients, respectively, for the infrastructure proxy, or route length. It is conceivable that the negative coefficient is due to technology and market access may encourage producers to switch to high-value crops or food grain crops.

Table 2: Estimates of Supply Response model –Yield Response.

Dependent Variable	Yield Cotton Coefficient		Yield Paddy Coefficient	
Cons	10.83***	(3.040)	6.002 ***	(6.038)
Log-Yield	0.146**	(0.075)	0.129**	(0.589)
Log-crop price	0.192**	(0.08)	0.1545*	(0.02)
Log-alternative crop price	–0.036**	(0.096)	–0.005 **	(0.000)
Procurement	0.38**	(0.048)	0.002**	(0.001)
Log Gross Rainfall	0.002 **	(0.015)	0.57**	(0.270)
Irrigation	0.51**	(0.075)	0.058 **	(0.005)
Log-Fertilizer	0.047**	(0.022)	0.578**	(0.118)
Proxy variable 1	–0.009*	(0.810)	0.027*	(0.157)
Proxy variable 2	0.827	(0.168)	0.61	(0.012)

***, **, * symbolizes significance at 1%, 5%, and 10% level, respectively.

5.3 Long-run and Short-run Elasticity

The marginal impacts of lagged own prices are favorable and statistically significant, suggesting that supply response happens by encouraging farmers to engage in raising agricultural yields as well as expanding acreage [8,11]. Depending on other variables, a 1% increase in the price of the own product results in a short-term yield gain of

Table 3: Elasticity estimates.

	Yield Response		Area Response	
Dependent Variable	**Cotton**	**Paddy**	**Cotton**	**Paddy**
Short-run elasticity				
Price	0.1921**(0.080)	0.1545**(0.020)	0.21**(0.045)	0.29**(0.028)
Long-run elasticity				
Price	0.44**(0.090)	0.25**(0.012)	1.02**(0.190)	0.89**(0.180)

***, **, * denotes significance at the 1%, 5%, and 10% level, respectively.

0.19% for cotton and 0.15% for paddy. Like this, a 1% increase in rates prompts farmers to allocate more acreage for the production of cotton and paddy by about 0.21 and 0.29%, respectively (Table 3). Although the price and yield/acreage have moved in the same direction, the short-run elasticities are less than one, showing the sector is inflexible, whereas the long-run elasticities show that the real yield/acreage has adjusted more quickly towards the desired one.

5.4 Demand Forecasting: Price and Yield Forecasting Using Deep Learning

We divided the data into training and testing sets to assess how well each algorithm performed. To assess the accuracy of each method, we use measures like mean squared error (MSE), root mean square error (RMSE), and mean absolute error (MAE). To make informed decisions about output, the supply chain, and commerce, farmers, policymakers, and other stakeholders need accurate agricultural price forecasts. To predict farm prices, conventional statistical models like ARIMA and regression models have been extensively used. These models, however, are not very good at capturing dynamic trends and complicated nonlinear correlations in the data. A growing number of machine learning methods, including Random Forest, SVM, and ANNs, have become common in agricultural price forecasting in recent years due to their capacity to manage complicated data patterns. In terms of predicting crop prices, machine learning systems have produced encouraging outcomes. These algorithms are ideal for price-predicting applications because they can manage complicated nonlinear relationships and dynamic trends in the data. Hence, we went ahead with both modern methods to compare the accuracy between LSTM and TDNN.

In the case of cotton, LSTM shows the best accuracy as compared to other models by 92.1% and RMSE of 0.018 whereas TDNN had an accuracy of 88.6% and RMSE of 0.232. In the case of paddy, LSTM shows the best accuracy as compared to other models by 91.1% and RMSE of 0.026 where TDNN had an accuracy of 89.6% and RMSE of 0.271 (Table 4). This could be due to the partitioning of the information and transforming it into supervized learning forms which is the first step in applying the TDNN model to the data, the TDNN model's numerous hyperparameters, including the number of input nodes, the number of concealed nodes, the batch size, and the number of epochs, are specified to train the samples when selecting the optimal model to prevent overfitting, the amount of trainable parameters is crucial and hence LSTM overpowered it. As LSTM captures the patterns and directions of the price movements, the LSTM model stores information about the previous values in the series in a memory cell, it can identify long-term trends and patterns that conventional statistical models might overlook.

Table 4: Model Prediction of Crop Average Prices.

	Cotton Average Price		**Rice Average Price**	
Model	R^2	**RMSE**	R^2	**RMSE**
LSTM	0.921	0.018	0.911	0.026
TDNN	0.886	0.2322	0.89	0.271

Demand forecasting gives a critical outlook for those involved in agriculture, such as producers, manufacturers, and policymakers. Farmers can plan their output, ensure that products are available on the market, and prevent resource waste with the aid of accurate demand forecasting. Demand forecasting frequently makes use of conventional statistical models like regression and ARIMA. These models, however, fall short in their ability to represent intricate nonlinear connections and dynamic trends in the data. Due to the lack of capturing of nonlinear trends, we also used machine learning algorithms and made a comparative study between conventional and modern methods. As lots of studies [2, 5, 6, 9, 10, 23] have taken place in terms to forecast the demand for any commodity, we chose LSTM over ARIMA (1,1,0). Our findings demonstrate that the LSTM NN can greatly increase the reliability of predicting rice consumption in India. The model's RMSE of 0.056 is significantly lower than that of conventional statistical models; similarly, for cotton it was 0.68 which is significantly lower compared to traditional statistical models (Table 5). The time-series data's intricate patterns are captured by the LSTM algorithm, which also successfully forecasts future demand. Additionally, we contrasted the LSTM model's performance with that of a conventional statistical model, the ARIMA model. A popular statistical model for time-series predictions is the ARIMA model. Compared to the errors of the LSTM model, the RMSE of the ARIMA model was 1.032 for rice and 1.018 for cotton, respectively.

The LSTM model's better performance can be ascribed to its aptitude for identifying long-term dependencies in the time-series data. Since the LSTM model stores information about the previous values in the series in a memory cell, it can identify long-term trends and patterns that conventional statistical models might overlook. ARIMA has some downsides, such as its dependence on historical data and failure to record nonlinear relationships has been the evident reason why the LSTM model has a low RMSE value.

Table 5: Model Comparison.

	Cotton		**Rice**	
Model	R^2	**RMSE**	R^2	**RMSE**
LSTM	0.89	0.056	0.82	1.032
ARIMA	0.61	0.68	0.58	2.46

6. Conclusion

India is one of the world's top producers of rice and cotton and is crucial for boosting the economy. For farmers, merchants, and officials in India to make appropriate choices about output, pricing, and trade, demand projections for rice and cotton are essential. Policymakers can create laws that support farmers' incomes, advance food security, and encourage safe farming practices using the insights from machine learning models. This can aid in addressing some of the major issues that the agricultural sector is currently experiencing. To ensure supply responses, particularly for important food grains like wheat and rice, agricultural price policy has included direct market procurement, subsidies for inputs, investments in innovations associated to increase the yield and infrastructure, such as roads and irrigation [1]. Farmers' ability to adapt output to changing the financial incentives to grow a certain crop determine demand. The supply function concentrates on the relationship between output and price while the supply response is indicative of price variables and other supply shifts [14].

Our results have significant ramifications for Indian agricultural strategy. Policies that seek to lower input costs could boost crop output in the short run. Long-term, however, strategies that emphasize raising output rates might be more successful at boosting production. Our findings also imply that long-term output increases could be substantial if policies are aimed at enhancing infrastructure and technology.

The Indian rice and cotton market prediction can be much more accurately predicted with the help of LSTM. Traditional techniques fall short of capturing the complexity of the variables affecting demand, but time-series data and advanced algorithms can. The commodity's supply chain can be made more effective by using LSTM to assist decision-makers, farmers, and traders. To further increase the forecasting precision of commodities consumption, additional studies can investigate the use of different NN designs and the integration of different data sources. In conclusion, the empirical results will help policymakers to mitigate price and yield risk associated with cotton and paddy cropping in India, which will help farmers and consumers to attain favorable pricing.

References

[1] Abraham, M., and Pingali, P. Shortage of pulses in India: Understanding how markets incentivize supply response. *Journal of Agribusiness in Developing and Emerging Economies*, 2021.

[2] Amaratunga, V., Wickramasinghe, L., Perera, A., Jayasinghe, J., and Rathnayake, U. Artificial neural network to estimate the paddy yield prediction using climatic data. *Mathematical Problems in Engineering*, pp. 1–11, 2020.

[3] Binswanger, H.P., and Rosenzweig, M.R. Behavioural and material determinants of production relations in agriculture. *The Journal of Development Studies*, 22(3): 503–539, 1986.

[4] Chaudhary, J.N. *The Future of Indian Cotton Supply and Demand: Implications for the US Cotton Industry*. Doctoral dissertation, Texas Tech University, 2005.

[5] Chelliah, B.J., Latchoumi, T.P., and Senthilselvi, A. Analysis of demand forecasting of agriculture using machine learning algorithms. *Environment, Development and Sustainability*, pp. 1–17, 2022.

[6] Chergui, N. Durum wheat yield forecasting using machine learning. *Artificial Intelligence in Agriculture*, 6: 156–166, 2022.

[7] Coleman, J.R., and Thigpen, M.E. *An Econometric Model of the World Cotton and Non-cellulosic Fibers Markets*. World Bank, 1991.

[8] Hazrana, J., Kishore, A., and Roy, D. Supply response of staple food crops in the presence of policy distortions: Some evidence from India. 2020 Annual Meeting, July 26–28, Kansas, Missouri, USA. Agriculture and Applied Economics Association, 2020.

[9] Jaiswal, R., Jha, G.K., Kumar, R.R., and Choudhary, K. Deep long short-term memory-based model for agricultural price forecasting. *Neural Computing and Applications*, 34(6): 4661–4676, 2022.

[10] Kantasa-Ard, A., Nouiri, M., Bekrar, A., Ait el Cadi, A., and Sallez, Y. Machine learning for demand forecasting in the physical internet: A case study of agricultural products in Thailand. *International Journal of Production Research*, 59(24): 7491–7515, 2021.

[11] Kishore, A. and Roy, D. Supply response of staple food crops in the presence of policy distortions: Some evidence from India. 2020 Annual Meeting, July 26–28, Kansas, Missouri, USA. Agriculture and Applied Economics Association, 2020.

[12] Mahto, A.K., Alam, M.A., Biswas, R., Ahmad, J., and Alam, S.I. Short-term forecasting of agriculture commodities in context of Indian market for sustainable agriculture by using the artificial neural network. *Journal of Food Quality*, pp. 1–13, 2021.

[13] Mondal, B., Bisen, J., Jambhulkar, N.N., and Tripathi, R. Rice supply, Demand and exportable surplus in India: Present vis-à-vis thirty years ahead. *ORYZA: An International Journal on Rice*, 59(4): 504–511, Dec. 2022.

[14] Nerlove, M. *The Dynamics of Supply: Estimation of Farmers' Response to Price* (No. 2). Baltimore: Johns Hopkins Press, 1958.

[15] Palanivel, K., and Surianarayanan, C. An approach for prediction of crop yield using machine learning and big data techniques. *International Journal of Computer Engineering and Technology*, 10(3): 110–118, 2019.

[16] Pingali, P. Westernization of Asian diets and the transformation of food systems: Implications for research and policy. *Food Policy*, 32(3): 281–298, 2007.

[17] Pingali, P. Agriculture renaissance: Making "agriculture for development" work in the 21st century. *Handbook of Agricultural Economics*, 4: 3867–3894, 2010.

[18] Manogna, R.L., and Mishra, A.K. Forecasting spot prices of agricultural commodities in India: Application of deep-learning models. *Intelligent Systems in Accounting, Finance and Management*, 28(1): 72–83, 2021.

[19] Sabu, K.M., and Kumar, T.M. Predictive analytics in Agriculture: Forecasting prices of Areca nuts in Kerala. *Procedia Computer Science*, 171: 699–708, 2020.

[20] Timmer, C.P. Reflections on food crises past. *Food Policy*, 35(1): 1–11, 2010.

[21] Tschirley, D.L., Snyder, J., Dolislager, M., Reardon, T., Haggblade, S., Goeb, J., et al. Africa's unfolding diet transformation: Implications for agrifood system employment. *Journal of Agribusiness in Developing and Emerging Economies*, 5(2): 102–136, 2015.

[22] Varma, P. Supply response of major pulses. *In*: *Pulses for Food and Nutritional Security of India: Production, Markets and Trade*.Singapore: Springer Nature Singapore, pp. 145–168, 2017.

[23] Weng, Y., Wang, X., Hua, J., Wang, H., Kang, M., and Wang, F.Y. Forecasting horticultural products price using ARIMA model and neural network based on a large-scale data set collected by web crawler. *IEEE Transactions on Computational Social Systems*, 6(3): 547–553, 2019.

Chapter 15

Business Forecasting and Error Handling Using AI

Vaddadi Kamya Nagasri,[1] *M Vasantha,*[1] *Preethi Nanjundan*[2,*] and *Jossy P George*[2]

1. Introduction

Business forecasting is the technique of accurately predicting the future of a business and outcomes using historical data and present trends. To evaluate historical data and find patterns, trends, and other elements that might be used to forecast future events, a variety of analytical tools and techniques are used. Business forecasting is a crucial component of strategic planning because it enables businesses to foresee market changes, spot possible risks and opportunities that may arise in the future, and make wise resource allocation and investment decisions. Businesses that use effective business forecasting can plan and carry out their programs that help them stay competitive, expand their operations, and meet their objectives. According to Glueck [1], "Forecasting is a formal process of predicting future events that will significantly affect the functioning of an enterprise."

Instead of using items or Stock-Keeping Units, strategic business forecasting is frequently done using annual "time buckets" and at the product type level. This means that complex large-scale systems are not necessary, and forecasters may frequently create strategic forecasting models using software tools based on spreadsheets [2]. A select few industries that must manage their enterprises for the long term require strategic forecasting that creates long-range predictions for up to 25 years in the future.

1 School of Commerce Finance and Accountancy, CHRIST (Deemed to be University), Pune Lavasa. Emails: vaddadi.nagasri@bcomfan.christuniversity.in; m.vasantha@bcomfan.chrisuniversity.in

2 Department of Data Science, CHRIST (Deemed to be University), Pune Lavasa. Email: frjossy@christuniversity.in

* Corresponding author: preethi.n@christuniversity.in

Few forecasting professionals discuss it as a result of this. But there are ways to incorporate some science into strategic forecasting, which can help forecasters comprehend potential long-term outcomes.

An error-handling procedure for a transmission interface connecting a first device and a second device for data transmission, where the connection type between the transmission interface and the first device is a direct interface (DI) and the connection type between the transmission interface and the second device is an indirect interface (II), is as follows: When an error is found at the DI, the first device's host should be notified; when an error is found at the II, the host should not be made aware of the error; and transmitting a new error event to the host when the problem discovered at the II is found to be intractable.

The practice of utilizing machine learning algorithms and techniques to automatically recognize, diagnose, and fix faults in software applications is known as error handling using artificial intelligence (AI). The ability to automate mistake handling and cut down on the time and resources needed to fix issues makes AI-based error handling a potent tool for forecasting the business. Predictive analytics is a further method of utilizing AI for error handling. Different machine learning algorithms can be trained to forecast when losses, errors, or frauds are likely to occur in businesses and to take proactive steps to prevent or reduce such losses, errors, or frauds in businesses, before they occur by monitoring system logs and other data. To ensure that the software continues to function, for instance, a predictive analytics model might be used to identify when a hardware component is likely to fail and automatically start a failover or backup system.

1.1 Types of Business Forecasting

Operational forecasting: This method produces hourly or daily forecasts for the following several days or weeks. They aid in the short-term planning of tasks for customer service, production, inventory management, warehousing, and transportation, among other sectors.

Tactical forecasting: This method produces predictions for the upcoming months, quarters, or even several years. They assist tactical planning in the form of weekly, monthly, or quarterly "time buckets", including sales, marketing/branding, master production, distribution needs, and labor planning.

Strategic Forecasting: In which forecasts are created over a lengthy period of time, commonly in terms of yearly "time buckets" [3]. They facilitate capital planning and decision-making, as well as long-term business planning and investment.

1.2 Artificial Neural Networking

One of the major application areas of artificial neural networks (ANNs) is forecasting in general, and business forecasting as a whole. There is an increasing interest in business forecasting using ANNs in recent years. Forecasting has a long history and the importance of this old subject is reflected by the diversity of its applications in different disciplines ranging from business to engineering. The ability to accurately predict the future is fundamental to many decision processes in planning, scheduling, purchasing, strategy formulation, policymaking, and supply chain operations. As such, forecasting is an area where a lot of efforts have been invested in the past [4]. Yet, it is still an important and active field of human activity presently, and will continue to be in the future.

Nonlinear ANN models for short-term business forecasting: One must choose the number of architectures (such as Boltzmann machine, Hopfield, and backpropagation), connectivity of layers [5], and bi-directional or uni-directional links among inputs and outputs before realistically implementing a neural network technique to load business forecasting.

1.3 Error Handling

Severe security vulnerabilities are frequently caused by improper error handling in programming that is security-sensitive. Correct error handling implementation is repetitious and time-consuming, especially in languages like C that lack exception handling primitives. This makes it very simple for developers to introduce error-handling problems by accident. Furthermore, many error-handling problems do not manifest any overtly incorrect behaviors (such as crashes and assertion failure), but rather produce subtle flaws that are difficult to detect and discover using current bug-finding tools. The tool EPEX, which uses error specifications to find and symbolically explore several error channels and notify defects when any errors are handled wrong along these paths, is designed, implemented, and evaluated in this work. Our strategy's main lessons learned are as follows:

(i) Real-world programs frequently only have a few options for handling mistakes;

(ii) The majority of functions have clear and consistent error requirements. This enables us to develop a straightforward oracle that can find a broad range of applications' error handling problems. Error-handling mistakes frequently result in security vulnerabilities with severe implications in addition to wrong results.

2. Business Forecasting

Business forecasting is the technique of predicting future company circumstances using historical data and current market trends. This can assist businesses in making wise judgments regarding their future goals and strategies. It entails examining historical data, spotting patterns and trends, and creating forecasts for upcoming sales, revenue, costs, and other crucial business variables.

For organizations of all sizes, business forecasting is a crucial tool because it enables them to foresee market developments and get ready for them beforehand. Companies can satisfy consumer demands, avoid stockouts, and incur unnecessary inventory expenses by adjusting their production plans and inventory levels in response to future demand forecasts. Moreover, forecasting aids businesses in identifying their possible dangers and obstacles, as well as areas for growth. There are various approaches to business forecasting, each having advantages and disadvantages. Quantitative forecasting, which makes use of statistical models to project future patterns based on historical data, is one of the most used techniques. Seeing patterns in the data and projecting short-term trends are two applications of this technology that stand out.

Another well-liked technique is qualitative forecasting, which makes future forecasts based on the opinions of experts and individual judgments (Fig. 1 and Fig. 2). This approach can be used to predict long-term trends and shifts in consumer behavior as well as to spot new market trends. Financial planning and budgeting also employ business forecasting since it enables businesses to project future revenues and costs and alter their spending and investment plans as necessary. It is also employed in risk management since it enables businesses to foresee potential dangers to their operations and devise countermeasures.

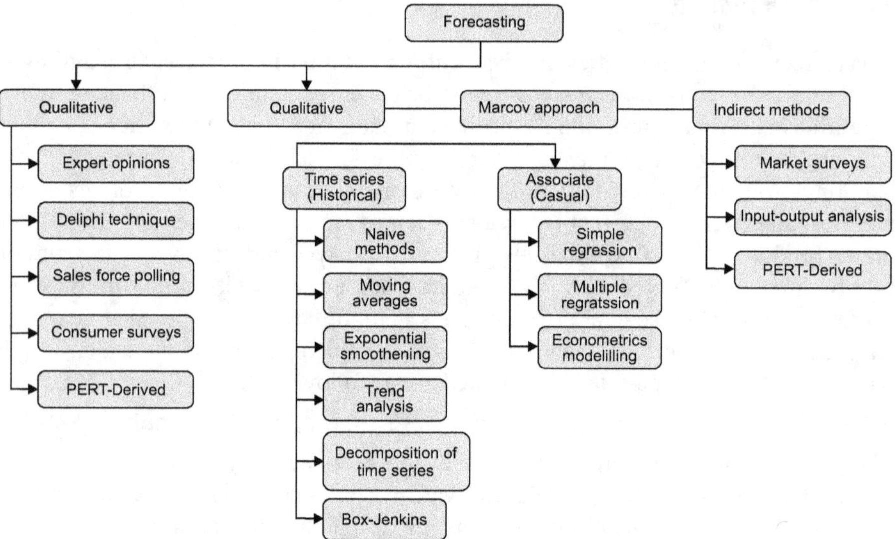

Fig. 1: Forecasting of business using various methods [6].

2.1 Business Forecasting Using AI

The strategic planning process of any corporation must include business forecasting. It entails making projections about how future developments in the market, consumer behavior, and economic indicators will affect a company's performance. The procedure can be time-consuming and difficult, needing a lot of resources and knowledge to effectively evaluate and interpret data. But as AI has grown in popularity, business forecasting has become easier to use and is more effective. This essay will examine the benefits of AI's transformation in business forecasting.

The phrase "artificial intelligence" (AI) is used to refer to a wide range of technologies that allow machines to perform functions like sensing, thinking, and decision-making that would otherwise need human intelligence. AI is a perfect tool for business forecasting because it can be trained to evaluate enormous datasets and spot patterns and connections that humans would miss. Organizations can use AI to gain more precise and trustworthy insights, which will aid them in making decisions that will increase their bottom line.

Using AI for business forecasting has a number of benefits, one of which is its speedy analysis of enormous amounts of data. To find patterns and trends, AI systems can scan and interpret data from a variety of sources, including sales statistics, customer reviews, social media, and economic indicators. As a result, businesses are able to make wise decisions based on current information and react fast to shifting market conditions. By examining previous data and spotting trends that people might not immediately notice, AI can also assist businesses in making forecasts that are more accurate. AI algorithms, for instance, may spot cyclical patterns, seasonality tendencies, and other factors that affect sales and revenue. Organizations' chances of success can be increased by using this information to predict changes in demand and modify their operations accordingly.

Another benefit of employing AI in business forecasting is its capacity to identify prospective hazards and opportunities. AI algorithms can identify trends and anomalies that might point to a potential threat or opportunity by examining data from a variety of sources. AI, for instance, can examine social media sentiment to spot new consumer trends or preferences, allowing businesses to adjust their marketing strategy and seize new chances.

Moreover, AI can help businesses optimize their supply chain management and operations. AI algorithms can find inefficiencies and provide ways to optimize processes, lowering costs and increasing efficiency, by analyzing data on production cycles, inventory levels, and delivery schedules. As a result, businesses may be able to maintain their competitiveness in their particular industries. Although AI has numerous benefits, it also has certain drawbacks. The possibility of bias in the data is one of the key issues with employing AI for business forecasting. Because AI algorithms can only be as good as the data they are trained on; biased or sparse data can produce findings that are unreliable or deceptive.

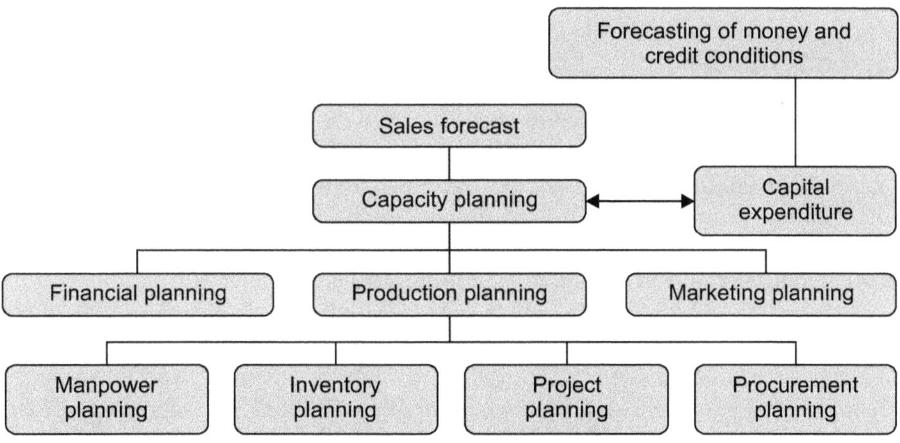

Fig. 2: Forecasting of money and credit conditions [6].

2.2 Disadvantages of not Using AI for Business Forecasting

By anticipating future market conditions, spotting potential dangers and opportunities, and adjusting corporate plans accordingly, business forecasting is a crucial activity that helps firms make wise decisions. Business forecasting uses a variety of methodologies, but not all of them rely on AI technology. We will go over a few of the drawbacks of corporate forecasting methods without AI in this response.

Restricted scope and accuracy: Due to their reliance on historical data and statistical analysis, traditional forecasting methods frequently have limited scope and accuracy. They could be unable to take into consideration aspects such as developing technology, rapidly shifting market conditions, and other elements that can have an impact on business outcomes.

Time-consuming: The data collection and analysis steps involved in traditional forecasting techniques might take a lot of time. This can slow down decision-making and make it harder for a business to react swiftly to shifting market circumstances.

Human error: Many forecasting techniques are susceptible to human error, including biased data collection, incorrect input, and incorrect assumptions. This may result in erroneous predictions and bad choices.

Lack of adaptability: Conventional forecasting techniques frequently rely on fixed assumptions and models that may not be able to accommodate shifting market conditions. This may make it more difficult for a business to react quickly to new possibilities or obstacles.

Lack of ability to take into account nonlinear interactions: Conventional forecasting techniques may not be able to take into account nonlinear relationships between variables, including abrupt changes in consumer behavior or unanticipated fluctuations in market demand. This may lead to shaky judgment and erroneous predictions.

3. Error Handling

Error handling is a crucial component of software operations and development. It entails seeing, averting, and fixing mistakes that could happen while software applications are being developed, deployed, and maintained. The traditional method of handling errors entails time-consuming, prone to error manual identification and resolution. But as AI has developed, new methods for addressing errors have appeared that can greatly improve their usefulness and efficiency.

One of the most important advantages of AI-based error handling is automated error detection. The input and output of software systems can be monitored by AI algorithms, which can also be trained to look for anomalies and unanticipated behavior that can indicate a problem. Automated Root Cause analysis is another way AI may be utilized for error handling. It can be difficult to pinpoint the primary source of an error when it occurs. AI algorithms can, however, be used to examine system logs, track performance indicators, [7] and spot trends that might point to the source of a problem. This can shorten the time it takes to fix the problem and lessen its impact on end users by enabling developers and operations teams to swiftly identify and address the root cause.

AI-based error management can also aid in preventing problems from happening in the first place. AI algorithms, for instance, can be used to examine code and spot possible problems before they arise. By analyzing vast amounts of code and discovering trends and anomalies, AI can assist to enhance code quality and reduce the number of errors that occur during development and deployment.

AI can be employed to automate the correction of errors. AI algorithms can be used to examine errors and decide the best way to fix them. Code patches, automatic rollbacks, and other changes are examples of this. AI can assist in reducing the amount of time it takes to handle problems, increasing system uptime, and reducing the impact on end users by automating the error resolution process.

3.1 Error Handling Using AI

In order to increase the accuracy of future projections, the process of finding, assessing, and managing forecasting errors is known as error handling in business.

Errors in business forecasting can be caused by a number of things. Inaccurate or stale data is one such source [8]. Data entry errors, out-of-date information, and insufficient data are all possible causes of data errors. Businesses can enhance their data collection procedures and spend money on data cleaning and validation solutions to address data problems.

The use of flawed assumptions or models can also result in predicting errors in business. Forecasting models are predicated on presumptions of how the market or economy will function in the future.

The forecast will be inaccurate if these presumptions are wrong. Businesses can resolve this problem by routinely reviewing and updating their presumptions and validating their models using historical data. Another typical source of inaccuracies in business forecasting is human error. This may involve biases, arithmetic errors, or errors in judgment. Businesses can use quality control procedures like peer review to make sure projections are examined by several experts before being finalized, which helps to reduce human error. Forecasting mistakes are still likely to happen despite these precautions. Hence, it's crucial to have a strategy in place for dealing with errors when they do happen.

This may entail locating the error's source, assessing how it will affect the forecast, and creating a strategy to make the necessary adjustments. Companies can also employ a number of strategies to increase the precision of their forecasts, including the use of different forecasting approaches, the incorporation of expert viewpoints, and the continuous updating of forecasts as new information becomes available. Error management is a crucial component of business forecasting, to sum up. Businesses need to identify and address potential sources of mistakes, create a plan for handling errors when they do happen, and put methods in place to lessen the impact of errors on future forecasts if they want to increase forecast accuracy. Businesses are better able to make informed decisions and produce better results when they have a strong mistake handling mechanism in place.

The process of locating and correcting faults that arise during the design, training, or implementation of an AI system is known as error handling in AI. Mistakes can occur at different stages of the AI development process and can be brought on by a variety of things, such as poor data quality, algorithmic mistakes, or system malfunctions.

3.2 Error Handling Techniques

Control over data quality: Ensuring that the information used to train AI models is true, full, and appropriate for the environment in which the model will be employed. Verifying the model's performance and that its outputs are compatible with the input data is known as model validation.

Error analysis: Determining the root causes of errors and creating mitigation or elimination measures.

Testing and evaluation: The AI system will be put through thorough testing and evaluation to find any problems or mistakes.

Creating methods for real-time monitoring and feedback that inform developers and data scientists about the performance of the AI system.

Together with these methods, it's crucial for developers and data scientists to adhere to best practices for handling errors [9], such as creating concise documentation, performing regular code reviews, and putting in place reliable mechanisms for error reporting and logging. Developers may make sure AI systems are dependable, strong, and successful in reaching their intended goals by adhering to these best practices.

To speed up learning for problems involving two-class imbalance, the use of neural networks technique had been suggested. This entails figuring out a weight-space direction that reduces errors by an equal amount for both majority and minority classes. Subsequently, the class imbalance might be addressed despite the correction of the clustering class placement faults. To solve this issue, the updated learning algorithm has been suggested. In addition to speeding up the convergence of the neural network, a modified back propagation had been suggested to prevent the ignoring of minority classes during a training process. Deep neural network training is used in more recent research to handle imbalance datasets [10]. Both majority class and minority class classification errors were equally caught by this strategy.

4. Conclusion

In conclusion, business forecasting and error handling in AI are crucial components for successful decision-making and planning in any organization. By utilizing sophisticated machine learning algorithms, businesses can generate accurate forecasts and predictions to help them anticipate changes and make informed decisions. However, even the best AI systems can produce errors, and it is essential to understand how to identify and manage these errors to avoid costly mistakes. By implementing robust error handling procedures, businesses can reduce the risks associated with AI-generated forecasts and predictions, improving their ability to respond to changing market conditions and achieve their strategic objectives. Ultimately, by embracing the power of AI while also recognizing its limitations, businesses can gain a competitive advantage and thrive in an ever-evolving landscape.

References

[1] Gupta, C.B., Gonsalves, T.A., Usha Rani, N., and Safari. *Principles and Practice of Management.* (10th[h] Edn.) Mayoor Publishers, 485 pp., 1997.

[2] Gilliland, M., Tashman, L., and Sglavo, U. *Business Forecasting: The Emerging Role of Artificial Intelligence and Machine Learning.* John Wiley & Sons, 2021. Accessed: Mar. 17, 2023. [Online]. Available: https://books.google.co.in/books?hl=en&lr=&id=s_ArEAAAQBAJ&oi=fnd&pg=PR1&dq=related:1AL0rLAh-V8J:scholar.google.com/&ots=WQd8n1dcGH&sig=w-0QAVwJGBt0eJ5fNE9_RoZVxyY&redir_esc=y#v=onepage&q&f=false.

[3] New developments in business forecasting – ProQuest., *www.proquest.com.* https://www.proquest.com/docview/226920994?pq-origsite=gscholar&fromopenview=true. Accessed: Dec. 4, 2022.

[4] Zhang, G.P. *Neural Networks in Business Forecasting.* Idea Group Inc (IGI), 2004. Accessed: Mar. 17, 2023. [Online]. Available: https://books.google.co.in/books?hl=en&lr=&id=T4vkIKbSAykC&oi=fnd&pg=PR6&dq=business+forecasting+neural+networks&ots=dK7hmRjUsU&sig=Jw_hcn4os9AVp1BBWBw-rTuFlVk&redir_esc=y#v=onepage&q=business%20forecasting%20neural%20networks&f=false.

[5] Hinman, J., and Hickey, E. Modeling and Forecasting Short-term ELECTRICITY Load Using Regression Analysis. 2009. Accessed: Aug. 20, 2019. [Online]. Available

[6] Shim, J.K. *Strategic Business Forecasting: The Complete Guide to Forecasting Real World Company Performance.* (Revised Edn.). CRC Press, 2000. Accessed: Mar. 17, 2023. [Online]. Available: https://books.google.co.in/books?hl=en&lr=&id=tVtdbIHTy-QC&oi=fnd&pg=PA1&dq=business+forecasting+regression+analysis&ots=a2oZrhp40m&sig=3kcQO0yziiZ6aDEUwmGwUP1BeXc&redir_esc=y#v=onepage&q=business%20forecasting%20regression%20analysis&f=false.

[7] Yuan, F.C. Parameters optimization using genetic algorithms in support vector regression for sales volume forecasting. *Applied Mathematics*, 3(10): 1480–1486, 2012. doi: https://doi.org/10.4236/am.2012.330207.

[8] Jana, S., Kang, Y., Roth, S., and Ray, B. Automatically Detecting Error Handling Bugs Using Error Specifications Automatically Detecting Error Handling Bugs using Error Specifications. 2016. Available: https://www.usenix.org/system/files/conference/usenixsecurity16/sec16_paper_jana.pdf.

[9] Hou, R. Optimization Model of K-Means Clustering Using Artificial Neural Networks to Handle Class Imbalance Problem. You may also like Melanoma detection using adversarial training and deep transfer learning Hasib Zunair and A. Ben Hamza–Imbalanced fault identification via embedding-augmented Gaussian prototype network with meta-learning perspective Detection of Radio Pulsars in Single-pulse Searches Within and Across Surveys Optimization Model of K-Means Clustering Using Artificial Neural Networks to Handle Class Imbalance Problem. doi: https://doi.org/10.1088/1757-899X/288/1/012075.

[10] Md. Tabrez Quasim1, Rupak Chattopadhyay. Artificial Intelligence as a Business Forecasting and Error Handling Tool. p. 3, Feb. 2015. Available: https://www.researchgate.net/publication/315459176_Artificial_Intelligence_as_a_Business_Forecasting_and_Error_Handling_Tool.

CHAPTER 16

Practical Benefits of Using AI for More Accurate Forecasting in Mental Health Care

D Swainson Sujana[1,2] and *D Peter Augustine*[1,2]

1. Artificial Intelligence: An Introduction

1.1 The Timeline

Artificial Intelligence (AI) is the general term for being able to make computers do things that require human-like intelligence. AI is the novel idea of the computer pioneers like Alan Turning and John von Neumann in the 1940s. Their novel intuition towards making machines think is the key start for this AI technology evolution. As shown in Fig. 1, the first milestone of AI happened in the year 1956 when it was proved by a group of researchers that a machine could solve any problem with the use of an unlimited amount of memory. Here they named this program General Problem Solver (GPS).

The next phase in AI evolution is more focused on real-world problems. This led to the development of expert systems, which allow machines to learn from past experiences and make predictions based on the given data. The aim of AI is to imitate the human brain, here the expert system is not as much as complex as the human brain. But they can be used to identify patterns and make decisions by training them with enough data. These systems can be used in medical and manufacturing fields.

1 Department of Computer Science.
2 Christ University, Bangalore.
Emails: d.sujana@res.christuniversity.in; peter.augustine@christuniversity.in

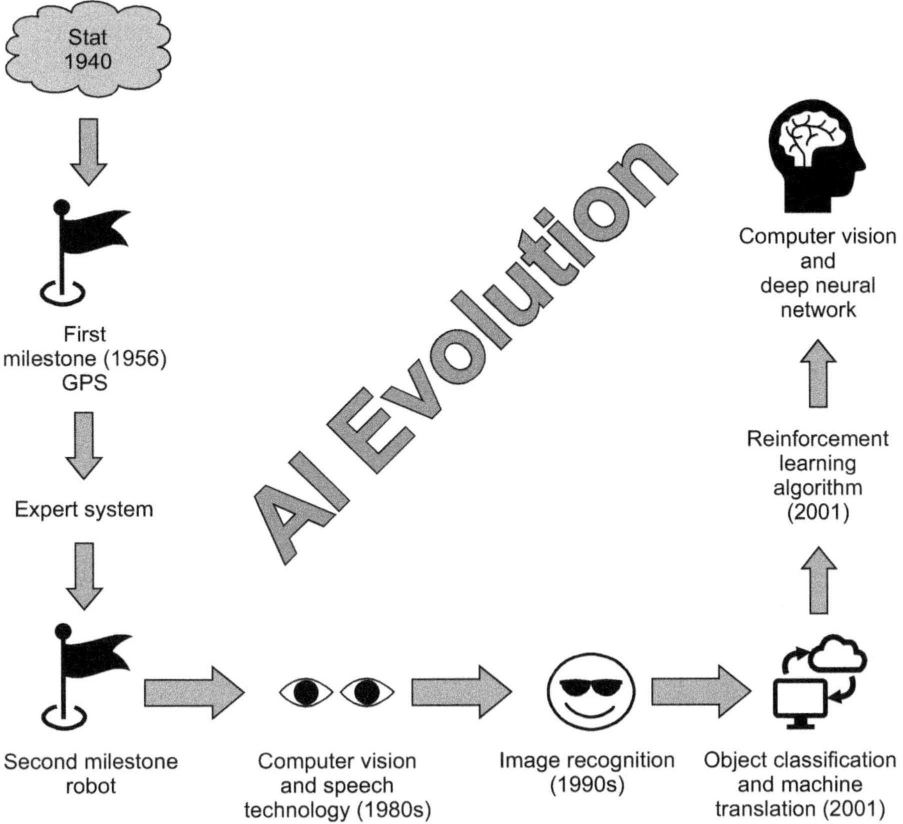

Fig. 1: The Evolution of AI.

In 1965, the development of automated programs like Shakely the robot and ELIZA had simple conversations with humans which paved a way for speech recognition technology which led to the development of the modern-day Siri and Alexa. After a long period, AI interest revived again in the 1980s. This revival led to some advances in computer vision and speech recognition. During this period the focus was more on making the system learn and understand real-world data with less human intervention.

The technological advances in terms of computing power, information storage, and advances in computer hardware brought a major boom in AI in the mid-1990s. As a result, machines proved their capability with humans in image recognition tasks. AI had seen many notable innovations in the initial years of the 21st century. In the year 2001, it outperformed humans in the areas like object classification and machine translation. Another major advancement in these years is the development of generative model-based reinforcement learning algorithms. The present emphasis is highly focused on using deep neural networks for computer vision tasks, leading to significant advancements in many fields with an effective forecast.

AI forecasting is a field in AI used to make scientific predictions about the future based on past data without requiring oversight. Here in this chapter, we are going to see some of the fields where AI can be used to predict the future which is far better than the traditional way of doing things.

1.2 Types of AI

AI can be classified into 4 categories based on its functionalities [2]:

1. Reactive Machine
2. Limited Memory
3. Theory of mind
4. Self-aware AI.

1. Reactive machine

A reactive machine is the basic form of AI. It works with present data and does not store memories or use past experiences to predict future actions. They don't have competencies beyond the assigned tasks. They observe the world and react to it.

IBM's Deep Blue is an example of the reactive machine that beat chess grandmaster Garry Kasparov. Deep Blue can make predictions about the next move for it and its opponent. But it doesn't have the knowledge of prior experiences, it ignores past actions and at the present moment it chooses from the possible next moves.

2. Limited memory

This type of AI learns from past data and decisions will be made based on that. But the memory of this system is only for a short time period. It can use past data for a specific period of time but cannot last for a long time and cannot be added to a library of their experiences. One of the major application areas of this Limited Memory AI is self-driving cars.

3. Theory of mind

Theory of mind AI is an advanced technology that requires a thorough understanding of people and things and can vary the state of mind and behaviors within the environment. This type of AI exists only as a concept. It is still improving but not completed yet. It can understand people's feelings, sentiments, and opinions. Kismet, a robot head designed in the 1990s, is a good example of this AI. It can mimic and recognize human emotions but it cannot understand gazes and convey attention to humans. The robotics field is gaining a peak in several healthcare areas including diagnosis, simulation, and mental health support.

4. Self-aware AI

This type of AI only exists as a hypothesis. The aim of this AI is to understand the states, internal characters, and situations, and perceive human feelings. This type of AI can understand and induce emotions in those it interacts with and also has its own needs, emotions, and beliefs. Affective computing is a technology that focuses on emotion understanding by machines and expressing emotions through robots and virtual agents.

2. Global Pandemic and Mental Health

The global pandemic pushed the healthcare field to find novel channels to pick swiftly and safely with the help of technology. The mental health field has seen an increased need for treatment from the very beginning of the global pandemic. Conferring to a survey taken by the American Psychological Association, there's been 84% of psychologists treat anxiety disorders which were 74% a year earlier [3].

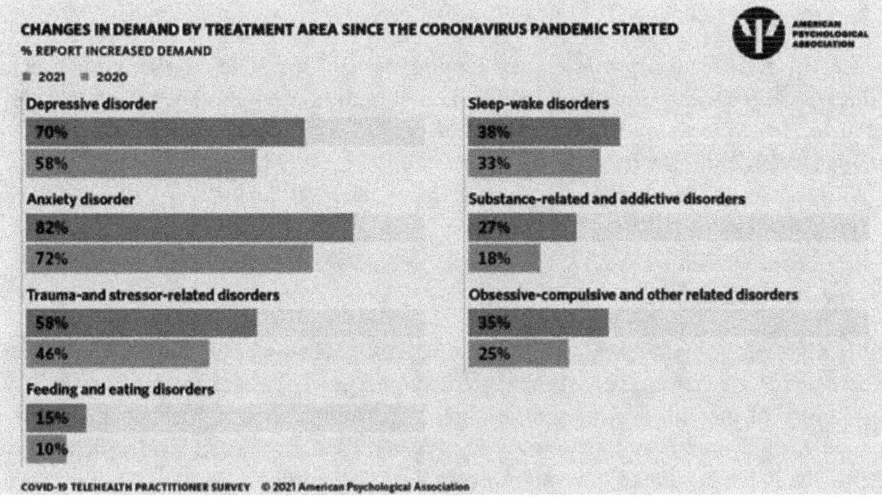

Fig. 2: Demand in the mental health sector due to Covid-19 [4].

2.1 Barriers to Accessing the Mental Health Services

Before discussing how AI can help in mental health services, let's see some of the barricades that prevent the approachability of mental health services, shown in Fig. 3.

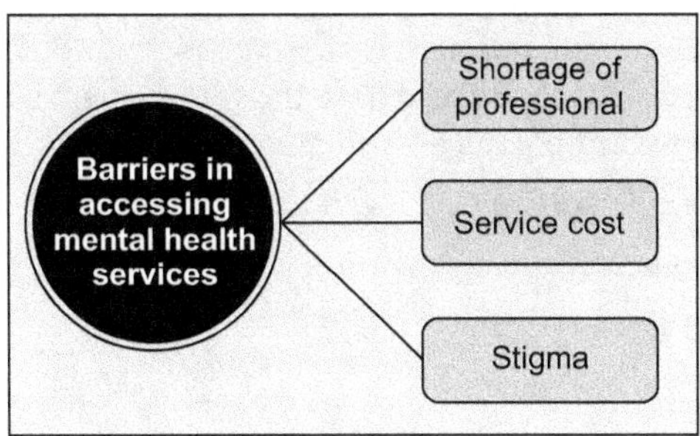

Fig. 3: Barriers to mental health services.

1. **Shortage of mental health professionals**

The global pandemic (Covid-19) caused a pressing need for more health services and support. A substantial shortage of psychiatrists causes long wait times of 6 months to 1 year. This is a major cause of the lack of access. Counselors and therapists also face the same problem, 1 among 10 patients waiting up to four months to access counseling service according to the Canadian Institute of Health Information on mental healthcare access.

2. Service cost

The risk of mental health problems is highest among young people aged 15–24 in Canada according to Canada's CAMH (Centre for Addiction and Mental Health). Cost is a significant barrier for the youth to access mental health services.

3. Stigma

A set of beliefs and wrong understanding which surrounds those with mental illnesses avert them from looking for mental health services. According to WHO (World Health Organization), 30–80% of people with mental health issues do not pursue treatment.

Because of the above-mentioned issues, mental health professionals have opened the doors for AI in their practice.

3. AI in Healthcare

In healthcare, AI is helping doctors to diagnose diseases by taking data from past health records, scanning reports, and medical images. The excellent decision-making capability of machine learning algorithms takes more accurate decisions based on past data. This helps doctors to make faster diagnoses at the right time which in turn helps the patients to take the timely treatment. According to Forbes, 7 million lives would be out of danger in 2035 due to the AI forecast in the healthcare system.

3.1 AI in the Mental Health System

Mental health is significant for the well-being of life at all stages. It includes emotional, psychological, and social well-being. It determines how we handle stress and make healthy decisions. A person's mental health may change at any point over a period of time. When the demands placed on a person exceed their abilities and resources then mental health could be impacted. Also, some physical health problems such as diabetes, heart disease, and stroke also pave the way to mental health problems.

The following are the benefits of AI in the mental health field:

1. Better personalized care
2. Remove barriers to treatment
3. Boost efficiency
4. Shorter wait times

As shown in Fig. 4, with AI in the mental health field more personalized care can be given to each patient and also there are no barriers in terms of place and time to access the right treatment. The waiting time for the clinician can be shortened and thus ensures efficiency by assigning the exact therapy at the correct time by the right therapist.

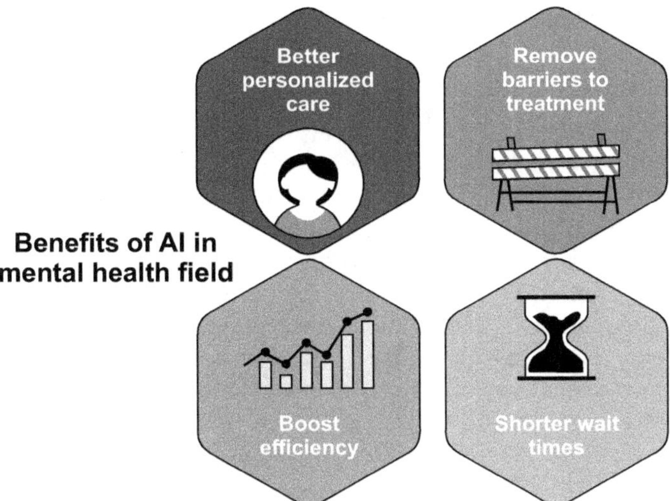

Fig. 4: Benefits of AI in mental health.

3.2 AI and Neurodevelopmental Disorders

Autism Spectrum Disorder (ASD), and Attention Deficit and Hyperactive Disorder (ADHD) are some of the neurodevelopmental disorders that are more prevalent among children.

People who have neurodevelopmental disorders have differences in brain development and brain activity. Another important truth is that these disorders do not have any permanent cure. But one consoling fact is that early prediction and identification of brain anomalies can reduce the core symptoms of these disorders and the medication intake in case of severe levels can be significantly reduced. So early diagnoses play a vital role in treating neurodevelopmental disorders.

1. Medical brain imaging and AI's machine vision

The human brain is one of the most complex structures consisting of numerous spatially distributed but functionally connected brain regions. Researchers trying to understand the underpinnings of the brain structure to address neurodevelopmental disorders. Advancement in this effort led to the invention of many medical imaging tools such as Computational Tomography (CT), EEG (Electro Encephalography), NIRS (Near-infrared spectroscopy), and Magnetic Resonance Imaging (MRI). This medical imaging can be arranged into two broad categories: (1) Invasive method, (2) Non-invasive method. Computational Tomography (CT) is a form of X-ray to capture cross-sectional images of the human body. CT scans offer superior precision than X-rays with more detailed images of the bones, blood vessels, and soft tissue within the body. But it is an invasive method of imaging that require a higher dose of radiation which is not good for the patients.

The MRI is a non-invasive method of medical imaging that does not require any radiation but uses superconducting magnets and radio waves to capture images that do not affect the patient by any means. Strong magnetic fields and radio waves are used to generate images of organs and tissues. It plays a major role in diagnosing brain disorders.

2. Early detection with AI's machine vision

Medical imaging integrates AI to distinguish, identify, and recognize early traces of abnormalities where the human eye cannot. By timely detection, an appropriate diagnosis can be made and treatment can begin before something goes serious and unrecoverable. Early detection increases the chances of successful treatment.

Medical professionals turn to AI to identify the most minor abnormalities in medical images and brain signals which cannot be seen through the naked eye. A version of the technology in AI called machine vision which is used in medical image analysis focused on machine-based image processing [6]. Machine vision also called computer vision enables systems to extract in-depth information from digital images. Computer vision works more like human vision, but it takes much less time than humans. As shown in Fig. 5, with cameras, data, and algorithms it can quickly surpass human capabilities.

Computer vision requires lots of data. It examines the data repeatedly until it discerns some differences and recognizes images.

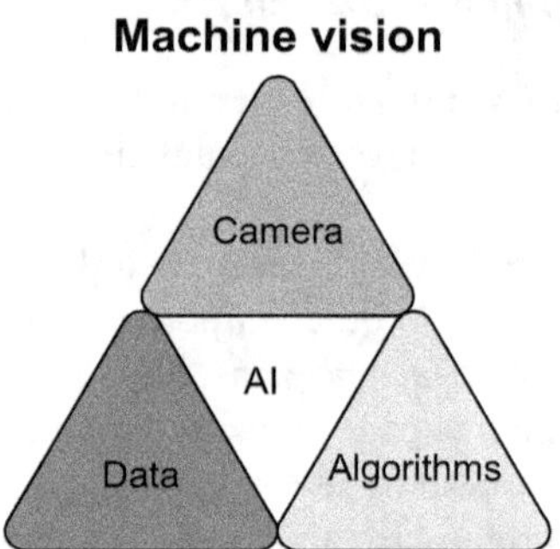

Fig. 5: Elements of machine vision.

3.3 Brain-computer Interface (BCI)

The vision of controlling one's environment through thoughts has come true through the advancement in technology. This is possible through BCIs that obtain brain signals, process them, and interpret them into commands. These commands can be fed to output devices which then bring out the desired operations. A BCI is a collaboration between a brain and a device that reads the electrical signals from the brain and uses them for further assistance. It acts as an interface between the brain and the object to be controlled by the brain. BCI technology can be used to help people with severe motor impairments by artificially augmenting or re-exciting synaptic plasticity in affected neural circuits. Plasticity is an intrinsic characteristic of the brain. Recent technological advances induce the plasticity of the brain by decoding neural activities and delivering external signals into targeted brain areas. BCI can be useful for people who are badly captivated by neuromuscular disorders to replace or restore their neuro-ignited functions. It can also be used as a therapy treatment for people with strokes, head trauma, and other disorders.

The brain signals activated through some external stimulus actions were collected through the electrodes. These electrodes will be placed on the top of the scalp and some of them placed on the cortical surface of the brain. The pertinent signals were collected and processed for feature extraction. Feature extraction analyzes the digital signals and separates the pertinent signals which are the signal features that carry the person's intent from the peripheral content. Then the extracted signal features are then fed into the AI translation algorithms which then translate the signals into commands for the connected output device. The connected output device may need functions in the form of letter selection, cursor control, moving or lifting of a robotic arm, etc.

As humans learn from past experiences, AI systems make predictions about future states by taking large amounts of labeled data, analyzing them by finding correlations and patterns, and finally, it makes predictions based on the patterns identified. Cognitive skills are the essential skills the human brain uses in the day-to-day life to think, read, learn, remember, reason, and pay attention. Cognitive skills are crucial in processing new information. Merging AI technologies with the human brain will help one day to improve core cognitive skills in humans or to restore certain brain areas which are inadvertently affected by strokes, aging, and traumatic brain injuries.

Cognitive skills, such as learning, reasoning, and self-correction are the primary skills that are focused on by AI programming. With this, it can predict the next or future action to be performed. So, with AI, the machine can replicate the cognitive capabilities of the human brain.

3.4 Brain Implants and AI

Another highly advanced technology called Deep-brain stimulation involves brain implants. In this procedure, certain specific areas of the brain can be stimulated through electrodes which are implanted into the brain through neurosurgery. This technique yields some therapeutic profits for issues such as depression, chronic pain, Parkinson's disease, and Tourette's syndrome. Deep-brain sensors can act as an interface between the external sensors and the AI-enabled systems that activate the needed responses to create initiations in a person's atmosphere.

3.5 The Super Clinician

Integrated AI technologies are able to deliver a simulated practitioner with the abilities of human practitioners which are called a "Super Clinician". These super clinicians are built on advanced technologies such as infrared imaging, optical sensors, and machine olfaction. Infrared imaging technology detects body temperature changes which indicate the changes in the internal states. Optical sensors help in facial expression, eye blinking, and vocal characteristics to provide some clinically relevant information. The simulation of sensing the smell of alcohol can be done through Machine olfaction technology. The super clinician can conduct consulting sessions with patients with complete autonomy or it can serve as an assisting practitioner during clinical assessment and treatment. An example of this super clinician is, a project developed by DARPA (Defense Advanced Research Projects Agency) and DCAPS (Detection and Computational Analysis of 1psychological Signals). It encompasses the development of an AI system that uses machine learning (ML), computer vision, and Natural Language Processing (NLP) to analyze language and physical gestures to detect psychological distress traces in humans [29].

4. AI and Mental Health Therapy

Integration of AI and mental health therapy sessions can offer more operative and customized treatment options. AI technology also provides more intuitions into patients' needs and supports therapists to create new practices and training in a more personalized manner. AI can improve mental health therapy in various ways (Fig. 6).

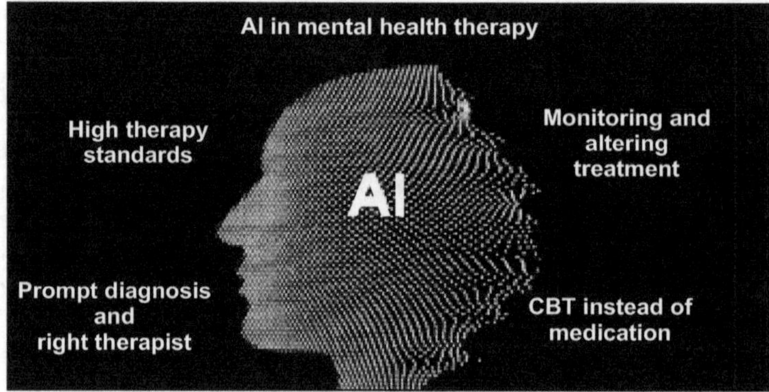

Fig. 6: AI in mental health therapy.

1. Maintaining high therapy standards with better quality control

Due to amplified demand for services and overextended workloads, now mental health clinics are looking for automated and better ways to monitor good quality control among therapists. NLP is a technique where the machine process transcripts. Computers can understand the text and spoken words of a person with the help of NLP in the same way as human beings do. NLP combines computational linguistics rule-based modeling of human language with machine learning, deep learning, and statistical models which are all part of AI. With the help of these technologies, computers can process and understand human language in the form of either text or voice data to understand its actual meaning, complete with the utterer's purpose and sentiment.

These technologies also offer some interactive tools to clinics for a fine understanding of the words spoken between therapists and clients. **ieso** is an example of a mental health clinic that uses an AI technology called NLP to scrutinize the language used in therapy sessions. ieso therapy service captures with permission the exchange between a therapist and patient. With session-by-session progress measures, ieso has a globally unique, de-identified dataset which is essential to increase the quality and effectiveness of therapy. Monitoring quality with the help of AI eases clinicians, so they can attend to more clients because it reduces unproductive sessions through better therapy.

In UK and USA, **Lyssn** the software company runs technology-enriched clinics and universities with the intention to help child welfare, human services, wellness, and behavioral health organizations reduce costs, and improve quality.

2. Precision in Diagnosis and Assigning the Right Therapist

Prompt and timely diagnosis and early intervention can do significant wonders and results in some life-changing significances for a person's mental health. Early intervention can lead to more precision in diagnosis which then leads to the right treatment through targeted referrals to specialist services.

AI is helping doctors to find mental illness in its early stages and to make precise choices in treatment strategies. With the help of AI, the insights from clients' data can be used to design more fruitful therapy sessions, assign the right clients to the right therapist, and suggest a more suitable therapy for an individual that would yield better results. AI translates the natural language spoken during the therapy sessions into a kind of barcode or fingerprints. These translated therapy sessions tell the roles played by different words during the session. Each barcode or fingerprint of a therapy session can show the statistics of how much time was spent in useful therapy versus general chat. Seeing these statistics can help the therapist to focus more on future sessions. Figure 7 shows some of the AI-empowered mental health diagnosis tools and therapy providers.

ieso is one of the largest therapy providers over the Internet which uses AI technology. It provides Cognitive Behavioural Therapy (CBT) through text or video over the internet. ieso addresses various mental health problems like depression, mood and anxiety disorder, and PTSD (post-traumatic stress disorder). Also, it has delivered more than 460,000 hours of CBT to 86,000 clients over the

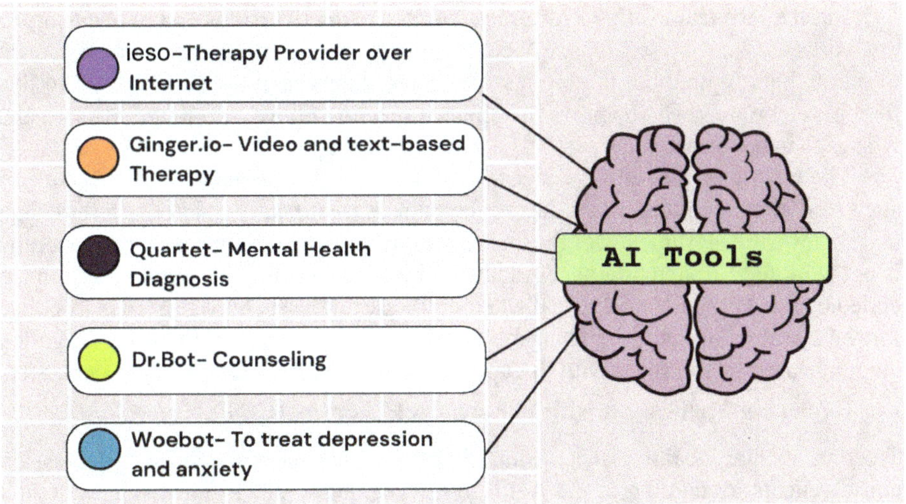

Fig. 7: AI-empowered diagnosis and therapy provider.

internet. According to ieso, its recovery rate among all disorders is 53%. ieso's focus is more on depression and generalized anxiety disorder and it also includes some data-driven techniques. The recovery rate of ieso boosts drastically because of the adaptation of NLP which is its key feature. According to ieso, the recovery rate in 2021 for depression is 62% while the national average is 50% and 73% for anxiety disorder compared to a national average of 58%.

Collecting and analyzing the data are two basic-level but time-consuming operations that can be carried out by any AI-based solutions more quickly than humans. Thus, AI-based solutions help the mental health professionals in terms of data analytics. It also suggests effective treatment plans based on data analytics. **Ginger.io** virtual mental health services are a good example that provides therapy and coaching sessions through video and text. The real-time data collected using mobile devices have been analyzed, the Ginger.io app helps specialists to track patients' progress, identify the time of crisis, and develop personalized care plans.

Mental health diagnosis also can be done through machine learning tools. It can automatically predict the issues based on the past experience and new data, an example of this is **Quartet Health**. It analyzes patients' medical histories and behavioral patterns to reveal unpredicted mental health complications. For example, Quartet can flag possible anxiety issues based on their repeated visits and tests for a non-existent cardiac problem.

3. Watching Patient Progress and Altering Treatment

The next step after assigning the right therapist is to monitor the patient's progress and track down the improvements. Here AI can play a vital role to identify if there is any treatment alteration that needs to be done or if there is a need to change the therapist.

An algorithm used by Lyssn analyzes the utterance between therapist and clients during a therapy session to calculate the time spent on constructive therapy or general chitchat to make some effective changes in the therapy session.

But the ieso team follows different monitoring methods; instead of focusing on the words used during sessions, it focuses more on clients. In recent research, a team of researchers figured out the replies voiced by clients such as "I don't want to live like this anymore," which shows the client is reacting on ways to move forward and wishes to make a change.

The team concluded that not reaching out to such responses or the absence of such responses from the client during the path of treatment would be a failure sign and it shows the inefficiency of the therapy and it has been considered that the therapy is not working. So such utterances made by the clients during the therapy session are coded by the AI models and reveal whether the particular therapy session is effective for the client or needs to be changed. AI also opens doors to find an investigation language used by the successful therapist, to train another therapist in the same area.

4. Replacing Medication with Behavioural Therapy (CBT)

Treating mental health issues using drugs has witnessed increased depression among clients. According to the NHS, in the year 2020–2021 in England the count of patients who were suggested antidepressants rose compared to the past years.

According to the UK's National Institute for Health and Care Excellence (NICE), updated guidelines the use of CBT as advisable instead of medication for cases with notable depression levels.

According to recent research, AI can help with an effective CBT treatment. An AI model was trained through 90,000 hours of internet-enabled CBT to classify therapist utterances. From this research, it has been found that usage of an increased quantity of therapy content showed significant improvement in patient symptoms, but the greater quantity of non-therapy content showed less improvement in patients. CBT identifies negative thought patterns and finds ways to break them by use of phrases to discuss means of change and preparation for the future. Finally, the conclusion made by the research team is that higher levels of CBT chat than general chat in the sessions lead to a better recovery rate. Thus, AI helps to optimize the therapy sessions and thus ensures the low intake of medicines by the clients.

ieso is one of the largest therapy providers in the UK over the internet by text or video. It uses NLP as a core part of treating depression and generalized anxiety disorder. As shown in Fig. 8, its recovery rate across all mental health disorders is 53%, compared with the national average of 51%. Focusing on depression and generalized anxiety disorder, the recovery rate for depression is 62% compared to the national average of 50%, and 73% for generalized anxiety disorder, as compared to 58% of the national average.

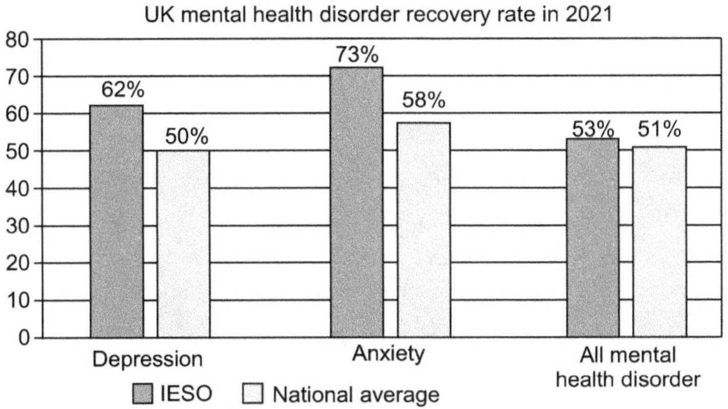

Fig. 8: Mental health disorder recovery rate using IESO.

4.1 Wearable Technologies and Mental Therapy

AI and the Internet of Things (Iot) are the two most popular wearable technologies in the healthcare sector. Some IoT wearable devices help to locate the nearest physician in case of an emergency. This enables the patient to receive treatment on time [10]. Some wearables like glucose monitors are linked to inhalers, which helps the doctor to diagnose and treat the patient accurately.

4.1.1 Fitbit

Another track where AI is proving its efficiency in mental health therapy is through wearable technologies. In addition, with clinical sessions, therapists use these wearable technologies to improve the treatments instead of counting on the patients to give accurate reports. One application of this wearable technology is Fitbit. Fitbit is a lightweight, portable, and inexpensive wearable sensor that is usually worn on the wrist to obtain the client's passive sleep data [11]. According to research, 90% of patients with major depressive disorder have sleep disturbances. It can be measured through contextual factors like fear, worry, or life stressors due to poor mental health. Fitbit cannot be used to identify mental health issues, but it can act as a prominent device to unveil the early symptoms of mental health disorders.

4.1.2 ACTi watch

Actiwatch models are used to provide sleep schedule variability, sleep quantity, and sleep quality statistics. This helps the therapist to better identify the therapeutic options for the particular patient and helps them to understand how the patient responds to the treatment. This Actiwatch model uses actigraphy principles and it is equipped with a highly sensitive acceleromete [12]. This is ideal for those who prefer to wear a wristwatch.

4.1.3 Thync

Thync is a wearable patch and an innovative approach that improves brain functioning through electricity, the natural language of the nervous system. This device sends electric stimulation programs to specific areas of the brain and peripheral nervous system to instigate the desired positive effect. It increases energy, clarity, and alertness for a few hours through neurostimulation. The basic motive of this wearable device is to make people happier and more energetic without drugs, or unnatural chemicals which affect the parts of the body [26].

4.1.4 Muse

Muse is a headband plus brain-sensing tool that helps users to improve their focus. It works by translating the brainwaves making the people calmer and more settled. This brain-sensing tool is suitable for people who are lazy and not able to focus [13].

4.1.5 The OURA ring

The OURA ring tracks our sleep patterns and measures our body's signals and guides us in making optimal choices in our daily life. It basically believes in the fact that better health, brighter moods, and more energy depend on nice and sound sleep. So, it tracks sleep patterns and guides us to take better decisions. It provides a complete picture of our health using 20 biometric signals. It can also tell you when you might be sick even before you fall sick [27].

4.1.6 Apollo

Apollo is a touch therapy wearable device that helps to calm down our nervous system. It not only tracks our health but also it improves it. It was developed by neuroscientists and physicians for better sleep, focus, and relaxation. It can be worn on a band around our wrist or ankle or waistband. The soothing waves of vibration from the device help to restore the body's natural rhythms that help to fall asleep, relax, or focus. Studies show wearing the Apollo for three hours a day, five days a week will strengthen our nervous system and build our resilience to stress [28].

4.1.7 AI-powered screening tool

START (Symptom Tracker and Resources for Treatment) is an early warning system developed by Newyork-Presbyterian (NYP). This AI-enabled digital tool is designed for healthcare workers to warn them when they are at near risk of depression, anxiety, or trauma symptoms [14]. This is an intelligent tool (Fig. 9) that immediately provides feedback to the participants about the severity of their symptoms. Also, it connects them to the available mental healthcare resources. The procedure goes like this, every two weeks, participants are asked to fill in a set of short questions that evaluate the symptoms of their mental stress which leads to poor mental health.

Here the AI algorithm captures the most relevant clinical symptom data in two ways: First, it mechanically selects the most informative questions based on the earlier responses given by the participant. Second, it focuses on questions that are highly associated with specific connectivity of the brain network pattern. Finally, symptoms and behaviors in depression can be identified based on the disruptive connections between the brain hubs.

Fig. 9: AI-powered screening tool.

5. AI Chatbots: Transforming Mental Health Diagnosis

A Chatbot is a computer program that imitates conversation with the user through a text-based or voice-based chat interface [15]. The underlying system of a basic healthcare chatbot is based on a variety of foundations including the sophisticated features of AI as shown in Fig. 11:

1. Natural Language Processing (NLP)—assesses and understands a patient's queries.
2. Knowledge Management—Keeps track of all the records of patients.
3. Deep Learning—Deep Learning (DL) helps the bot to give content-based replies.
4. Sentiment Analysis—The bot learns from the past mistakes and improves through sentiment analysis.

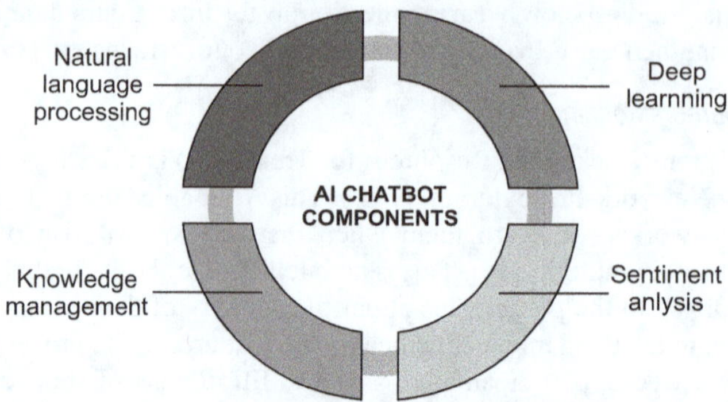

Fig. 10: AI Chatbot Components.

Talk to Eliza

> Hello, I am Eliza.
* Hello, Eliza
> How are you today.. What would you like to discuss?

Input:

Fig. 11: Eliza chatbot screen.

The following are some of the healthcare chatbots developed in the mental healthcare field.

5.1 Eliza

The first well-programmed chatbot was invented in the year 1996. It is programmed to simulate Rogerian psychotherapy. This type of therapist is known for reflecting some information back to the user. It asks people to talk to the bot as if it was a therapist. It operates by recognizing phrases and keywords from the input and using those keywords to produce a response from pre-programmed responses [1]. For example, as shown in Fig. 12, if we type "Hello, Eliza," immediately it will respond as "How are you today? What would you like to discuss?"

By its response, it creates an illusion of having an interaction with a real human being, though the responses are pre-programmed.

5.2 Wysa: A Mental Health Chatbot for the World

Wysa is an AI-safe space where we are led by an "AI penguin". Instead of talking to another human being, we can talk to this AI penguin which helps to release us

Fig. 12: Woebot chatbot screen.

from our stress by its empathetic and patient listening. It is making people more open to AI than other human beings. Wysa offers mental health support through text-based sessions. Users can choose to remain anonymous and the coach can refer them through their chosen nicknames. Wysa also offers audio-based emotional well-being sessions for 30–45 minutes and maintains complete confidentiality by recording nothing [5]. Research points out that people will be more comfortable and open to AI three times faster than a therapist. An AI bot can do a cognitive behavioral therapy exercise in 10 minutes which can be done in three therapy sessions by a therapist. Wysa1.0 can detect some suicidal thoughts and escalate them to a helpline. Wysa has an overseeing therapist, it maintains a log for sessions which will be referred in case of any threat. But all this is done in an anonymous manner. Wysa has more than 100 different AI models and can detect over 70 distinct emotion types.

5.3 Dr. Bot

Another AI tool used in counseling is called Dr. Bot. It is a chatbot that uses a text or voice-based AI interface to simulate human conversation [17]. These chatbots are greatly beneficial for employers and health insurers to identify individuals who might be stressed with depression, or anxiety and offers access to appropriate and cost-effective care.

5.4 Woebot

In the year 2017 Stanford University has developed a chatbot with the name Woebot with the help of some psychologists. It uses the digital version of cognitive behavioral therapy technique to treat depression and anxiety which is of 40 years old [18]. This digital version of CBT is highly organized and structured talk psychotherapy that attempts to alter or remove a patient's adverse thought patterns in a few number of sessions. It is a free-to-use app available in the app store on a smartphone device (Fig. 13).

5.5 Tess

Tess is a chatbot that offers mental health support to people through textual conversations [19]. Similar to texting a friend, people can text Tess to get through their

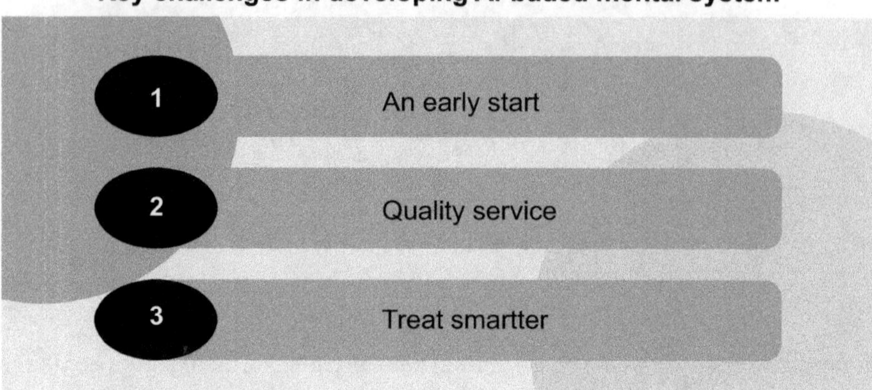

Fig. 13: Key challenges in developing an AI-based mental health system.

tough times. Having text message conversations with Tess is completely confidential and it builds resilience. The user can stop their conversation at any point in time. Research findings show that Tess is reducing depression by 28% and anxiety by 18% [3]. Using Tess is so easy and there is no need of downloading any application. The user can start chatting with a simple 'Hi' to the IMPACT phone number.

With expert guidance from psychologists and with the help of AI, Tess works in the same way a therapist would. While providing the training Tess also learns about specific concerns and interventions. This helps it to develop in order to provide better support. Tess is different from other mental health chatbots by providing "Integrative mental health support", that is, Tess can provide a variety of interventions according to each person's unique need. When it hears some strong emotions such as suicide or self-harm, Tess either recommends a crisis line to the person or it will send a notification to a crisis counselor.

5.6 BlueSkeye AI

Blueskeye is an AI-based mental healthcare approach that can understand mood and mental well-being [20]. It is an effort taken with the vision to improve people's quality of life and personal safety by blending AI technologies like computer vision, and ML with human behavior. BlueSkeye AI was developed with a vision to ensure that people suffering from mental health issues get the right care that is needed and that the diagnosis happens at the right time.

5.7 Deliberate AI

Deliberate AI develops and offers a data analytics platform which provides deeper data insights for mental health care and neurological clinical trials [21]. It uses AI-based algorithms to analyze a constellation of behavioral and physiological biomarkers across multiple modalities like audio, video, and NLP to accurately diagnose and assess neurological conditions. The specialty of this AI model lies in multimodality diagnosis. Sometimes individual biomarkers fail to achieve clinical relevance. But Deliberate AI is supported by multimodal ML, which uses combinations of biomarkers that can better understand, measure, and develop predictions for psychiatric and neurological disorders.

5.8 Upheal: For Mental Health Therapists

Upheal is an AI-powered notepad for mental health professionals. The mental health profession is where documentation plays a crucial role. Often this documentation limits the clinicians to provide quality therapy sessions because of the rigid time schedule. Also, therapists find this documenting process burdensome and inefficient. But the regulatory perspective documenting patient interactions is used for legal protection and to monitor the progress of the patient.

The AI-powered notebook Upheal brought a solution to the documenting process in a therapy session. The online therapy sessions are recorded and transcribed to text using NLP. Upheal then creates summaries of these sessions with different insights and analytics which helps the therapist to access the relevant information without missing any details [22].

5.9 Clare&me: Audio Bot

Clare&me is an automated AI-driven audio bot for mental illnesses such as anxiety and mild depression [23]. It is a conviction-free, unidentified safe space available anytime and anywhere for its users. Based on the information shared by the users about their current well-being and concerns, Clare can guide the users with a self-reflective exercise with an emphasis on trying to help them feel better about themselves. The user can share their information through calls or text messages.

6. AI and Mental Illness

Mental illness treatment incorporates imagination, insight, and empathy. These traits can only be pretended by AI. An algorithm that analyzes patient records may not have an interior understanding of human beings but it can identify psychiatric problems.

6.1 AI and the Language of Suicide

John Pestian was a computer scientist who started using ML to study mental illness. He wanted to confirm that computers can figure out the persons with the intention of self-harm [24]. He worked with a clinical psychologist and collected hundreds of suicidal notes from different families and analyzed them. Pestian and his colleagues fed the notes into an AI system language model which learned words and phrases. Then the AI model was tested for its ability to recognize suicidal thoughts in statements that people made. The results have shown that the algorithm could identify "the language of suicide".

The next level Pestian collected the audio recordings taken from patients who attempted suicide. He developed software to analyze their spoken words and the sounds of their speech. Their research revealed that people experiencing suicidal thoughts had more sighs but less laughs. Their speech included longer passes and words with short vowels that were less intelligible. Their speech expressed more anger and less hope. As a trial, Pestian's team recorded the speech of hundreds of patients and used algorithms to classify them as suicidal, not suicidal but mentally ill, or neither. For almost 85%, the AI model gave the same conclusion as clinicians. Thus, it proves its efficiency and it is useful for inexperienced, uncertain, and overbooked clinicians [24].

7. AI Utilization in Mental Health

Mental health is a peculiar field that involves empathy, perspective-taking, and emotional intelligence. The growing need for mental health can be successfully managed by states and companies by adapting the technologies like AI, ML, and DL. These technologies not only support the existing mental health workforce but also improves the present mental health service. There are some dimensions of AI solutions that positively impact the mental health industry [9]. They are:

Availability

The most important advantage of AI-powered mental health solutions is the feature of availability: 24/7-hour availability, rapid response time, and ease of communication are the main factors that increased access to AI-powered mental health services.

Personalized treatment

AI-driven apps and wearable sensors can be used to individualize treatment plans by collecting, tracking, and monitoring behavior as well as bodily functioning. This information supports the communication between the patients and the health professionals.

Safe space

AI-powered mental health solutions available in the form of bots provide a safe space for people who feel ashamed or insecure. Bots acts as virtual therapist, to whom people were more willing to express their feelings knowing that it is anonymous and they won't be judged.

Data processing

AI's great potential lies in gathering lots of data and analyzing the datasets quickly and efficiently. With the help of automatic learning algorithms, it filters, determines, and identifies useful information, and thus assesses huge datasets. This data assessment technique is helpful in the prediction, prevention, and treatment of patients in the mental health sector [25].

Speech pattern recognition

Researchers discovered that differences in speech patterns could be used to detect mental health problems. Depressed people talk in a monotonous and quiet voice with many breaks in between but anxious people speak much faster. Using ML, some solutions are already able to recognize vocal patterns to detect signs of mental health issues with over 80% clinical accuracy.

7.1 Practical Benefits of AI in Mental Health

More Empathic conversation using human-AI collaboration

One of the key supportive skills and strategies in providing mental healthcare services online is empathy. Expressing empathy more effectively in online mental health support conversations helps the users to feel the real-time experience. Behavioural Data Science at the Allen school in collaboration with clinical psychologists from different medical schools developed AI tools that can identify empathy in conversations and give more intelligent and actionable real-time feedback to users. They also developed Hailey, a peer-to-peer, AI text-based mental health support, which provides timely feedback to help participants to receive mental health support on time and respond more responsibly to those seeking help. The human-AI collaboration approach has seen a 19.60% increase in informal empathy which is a core essential skill in providing online mental health services.

Virtual world and artificial companion

Virtual reality is an evolving application of AI. This virtual reality technology is used by clinical virtual reality in the mental healthcare sector to treat a variety of psychological disorders. These virtual environments make intelligent agents that can listen and talk with users. This increases flexibility and realism in the interaction. These agents have the capability to express emotions with human users. Also, virtual companions, such as virtual pets, promote the mental well-being of people who are struggling with loneliness. AI makes these virtual companions more interactive, realistic, and flexible to a patient's needs. An example of this virtual companion is animal robots that are created with the intention to offer therapy for dementia patients [30].

7.2 Key Challenges of the Mental Health Support System

There are three key challenges considered before developing an AI-based mental health support system (Fig. 10).

1. **An early start**

An early diagnosis of mental health issues can lead to faster and more efficient recovery. Accessing psychological therapy sooner is associated with better mental health outcomes and it also lowers healthcare costs and the usage of medications. An AI-based therapy service provided by ieso allows patients to access treatment faster and makes therapy delivery more efficient. The efficiency can be achieved in 6.1 ieso sessions versus the average of 7.5 sessions.

2. **Quality service**

AI-based monitoring tools can check the quality of the therapy delivered to the patients. It can be done by measuring patient progress at each session and analyzing the core elements of psychological therapy and highlighting the active ingredients of therapy to ensure the patient receives exactly the right service.

3. **Treat smarter**

The treatment of mental health disorders will be more effective if the diagnosis is very specific instead of broad where very different things are grouped together. Different subtypes of mental health conditions should be treated separately instead of treating them all with solitary therapy. AI-based therapy services offer new ways to personalize treatments based on the different subtypes of mental health conditions which aid the quality and faster recovery.

8. Current AI Trends in Mental Health

In the year 2022, many startup companies raised a notable amount in AI and mental healthcare. Among them is the AI chatbot Wysa raised 20$ million, the startup BlueSkeye, which is focusing on early diagnosis of mental health issues raised £3.4 million, the Upheal smart notebook for mental health professionals raised €1.068 million and Clare&me, the AI-based mental health companion got a deal of €1.2 million.

Researchers at Vanderbilt University medical center have developed an ML algorithm that makes 80% accurate prediction of whether an individual is likely to take his/her life based on the person's hospital admission data, age, gender, and past medical diagnosis. The University of Florida researchers developed a new AI platform to diagnose early Parkinson's disease.

9. Future of AI in Mental Healthcare

The interference of AI and ML in the mental health sector has the ability to transform the diagnosis and treatment procedures which were followed in the long past years. In the near future, algorithms will be the front-line warriors of defence against the mental health attacks that are devastating the lives of many people so far. The following things will be possible with AI in the near future [8]:

1. **Integrated physical and mental health care**

In the future, ML algorithms are expected to alert doctors and surgeons about patients who are at peril of dangerous mental disorders based on their medical history. For example, in the opioid crisis, according to data 10% of people who take opioids (a drug used to reduce moderate to severe pain) for 90 days after cardiac surgery will become too dependent on them. Here, the AI system can alert the prescribing doctor about the end of the prescription window and the patient who is under the clutch of opioid dependence could be referred to a therapist. So, patients can come to know about the techniques available to overcome their pain and added symptoms.

2. **Reducing bias and human errors**

Many ML algorithms have already established their efficiency in recognizing speech patterns and identifying facial expressions to detect the signs of depression and other stress disorders like post-traumatic stress. These tools can be used by mental and physical health providers in patient intake meetings to observe the signs of trouble exhibited by the patients. For example, certain medicines with benzodiazepine components make people more dependent on it and makes them as high functioning in specific areas of their lives. An AI tool can help a physician to dig deeper into the problems that are on the way instead of looking at the patient's surface-level appearance.

3. **Early warning flagging**

Timely interventions are vital in many zones of mental health. Algorithms based on language analysis proved their 100% efficiency in identifying teens who were at risk of developing psychosis. AI models built on these algorithms are used to monitor the patients who are under treatment and they can alert the doctor when they enter the crucial zone. The early sign of disorder can be detected by answering a few questions online every day and thus limiting the in-person visit to the doctor every day. These answers could help the app to detect the early symptoms of trouble and thus create checkpoints to identify the exact point of deterioration before it changes into serious trouble.

4. **24/7 support: A lifeline**

Chatbots and apps are accessible all around the world and all the time. They are always awake and affordable. People feel easier to share their struggles with unidentified chatbots than human beings. These tools are more useful to track and improve the mood of people who have trouble getting mental healthcare. The chatbots sometimes act as lifelines for patients who are besieged in the middle of the night.

9.1 Will AI Replace Therapists?

AI cannot replace human therapists, but it can support them. AI aims to improve care in therapy and fill the gaps. It assists the therapist in monitoring patients' health, does preliminary assessments with more accuracy, and helps the therapist deliver fruitful treatments. AI also acts as a tool to analyze a huge amount of data used by mental health professionals to provide better services. With AI's remote therapeutic sessions, people living in rural places also can get better mental health services at a minimal cost [31].

9.2 Pitfalls of Using AI in Mental Healthcare

The potential bias in AI systems may increase due to insufficient and poor-quality databases. They often require large amounts of sensitive patient data, and security and data privacy become a great concern here. Integrating AI tools into the existing healthcare systems requires training for medical professionals to use AI-based tools effectively, and it is more time-consuming and difficult [9].

Technological barriers

Currently, we are in a period experiencing rapid advancements in technology. But still, multiple technological barriers remain a challenge in the field of AI. In general, AI technology is good at the things that humans find difficult to do but not so good at things that are natural for humans. Designing intelligent machines with human-like intelligence and common sense still remains a challenge for the field of AI.

Ethics and AI

According to AI, the more data the algorithm process, the more accurate the results will be. The algorithm becomes better if it learns more. The real risk is around the data used by the technology. The risk of privacy and ineffective care is still to be considered as a crucial factor when designing ML algorithms. So, the design and development of ML algorithms should strictly avoid the replica of human biases and stigmas. The privacy aspects in data storage, processing, and usage also should be considered more seriously. AI systems can be developed under the assistance of the ethicists to make sure they would behave in an ethical manner. The teamwork of ethicists and mental healthcare professionals can address the ethical requirements in the practical mental healthcare.

References

[1] http://psych.fullerton.edu/mbirnbaum/psych101/eliza.htm. n.d.. http://psych.fullerton.edu/mbirnbaum/psych101/eliza.htm.

[2] Lateef, Z. What are The Types of Artificial Intelligence? | Branches of AI. 2019, June 18. Edureka. https://www.edureka.co/blog/types-of-artificial-intelligence/.

[3] 4 ways artificial intelligence is improving mental health therapy. World Economic Forum. n.d.. https://www.weforum.org/agenda/2021/12/ai-mental-health-cbt-therapy/.

[4] https://www.apa.org/pubs/reports/practitioner/2022-covid-psychologist-workload. (n.d.).

[5] Everything you need to know about Wysa's coaching services 2020, January 30. Wys https://blogs.wysa.io/blog/most-read/everything-you-need-to-know-about-wysas-coaching-services.

[6] https://www.utoledo.edu/offices/rocketwellness/docs/Tess%20Program.pdf.

[7] What is Computer Vision for Medical Imaging? 2022, July 18. Premio Inc. https://premioinc.com/blogs/blog/what-is-computer-vision-for-medical-imaging.

[8] Asar, A. Council Post: Five Ways AI Can Help Revolutionize Mental Healthcare. Forbes. 2020, August 19. https://www.forbes.com/sites/forbestechcouncil/2020/08/19/five-ways-ai-can-help-revolutionize-mental-healthcare/.

[9] AI in Mental Health –Examples, Benefits & Trends &mdash. 2022, December 13. ITRex. https://itrexgroup.com/blog/ai-mental-health-examples-trends/.

[10] Can, Y.S., Arnrich, B., and Ersoy, C. Stress detection in daily life scenarios using smart phones and wearable sensors: A survey. *Journal of Biomedical Informatics*, 92: 103139. 2019, April. https://doi.org/10.1016/j.jbi.2019.103139.

[11] 3 ways to take better care of your mind and body in 2023. 2023, January 5. Google. https://blog.google/products/fitbit/fitbit-pixel-watch-nest-better-sleep-stress-less-recovery-2023/.

[12] Actigraphy monitoring devices. n.d.. Philips. https://www.usa.philips.com/healthcare/sites/actigraphy/solutions/motion-biosensors.

[13] Neurofeedback EEG Device –How it Works. n.d. Muse. https://choosemuse.com/how-it-works/.

[14] This tool helps health workers look after their mental health. World Economic Forum. n.d. https://www.weforum.org/agenda/2020/09/ai-tool-mental-health-frontline-workers/.

[15] What is a Chatbot? n.d. IBM. https://www.ibm.com/in-en/topics/chatbots.

[16] Everything you need to know about Wysa's coaching services. 2020, January 30. Wysa. https://blogs.wysa.io/blog/most-read/everything-you-need-to-know-about-wysas-coaching-services.

[17] Merchant, S. The Best ChatBots for Behavioral Health. AIM. 2021, March 2. https://www.aimblog.io/2021/03/02/these-chatbots-are-helping-with-mental-health-right-now/.

[18] What Powers Woebot? 2022, June 1. Woebot Health. https://woebothealth.com/what-powers-woebot/.

[19] Mental Health Chatbot. n.d. X2. https://www.x2ai.com/.

[20] World Leaders in Behavioural Understanding –Ethical AI. n.d.. Blueskeye AI. https://www.blueskeye.com/.

[21] Multimodal assessment of psychiatric and neurological health. n.d.. https://www.deliberate.ai.

[22] Upheal · Portfolio · Calm/Storm Ventures. n.d.. https://www.calmstorm.vc/portfolio/upheal.

[23] clare&me. n.d. https://www.clareandme.com/post/current-ai-utilization-in-mental-health.

[24] Nast, C., and N. Can A.I. Treat Mental Illness? *The New Yorker*. 2023, February 27. https://www.newyorker.com/magazine/2023/03/06/can-ai-treat-mental-illness.

[25] Bell, I.H., Lim, M.H., Rossell, S.L., and Thomas, N. Ecological momentary assessment and intervention in the treatment of psychotic disorders: A systematic review. *Psychiatric Services, 68*(11), pp. 1172–1181, 2017, November. https://doi.org/10.1176/appi.ps.201600523.

[26] Science of Neurostimulation. n.d. Thync Global, Inc. https://thync.com/pages/neurostimulation.

[27] Accurate Health Information Accessible to Everyone. n.d. Oura Ring. https://ouraring.com.

[28] Wearable Touch Therapy for Stress Relief. n.d.. Apollo Neuro. https://apolloneuro.com/.

[29] Luxton, D.D. Artificial intelligence in psychological practice: Current and future applications and implications. *Professional Psychology: Research and Practice*, 45(5): 332–339. 2014, October. https://doi.org/10.1037/a0034559.

[30] Shibata, T., and Wada, K. Robot Therapy: A New Approach for Mental Healthcare of the Elderly: A Mini-Review. *Gerontology*, 57(4): 378–386. 2010, July 15. https://doi.org/10.1159/000319015.

[31] Mazza, G. AI and the Future of Mental Health –CENGN. 2022, August 29. https://www.cengn.ca/information-centre/innovation/artificial-intelligence-ai-and-the-future-of-mental-health/.

Chapter 17
Predicting Stock Market Indexes with Artificial Intelligence

Yasmeen Fatima,[1] *Mohammad Asif,*[1] *Preethi Nanjundan*[2,*] and *Jossy P George*[2]

1. Introduction

The forecasting of the Share market has been a popular research area, involving the analysis of input and output stock data using computer technology and algorithmic knowledge. This involves building unpredictable relationships among the data and analyzing the stock market trends to provide a reference for investors. The inception of artificial intelligence (AI) technology, blended with the web, immense data, and cloud computing has provided technical support for various industries. AI technology is employed to scrutinize and predict the equity market, exploring curvilinear associations amid stock market information, and furnishing a foundation for investors to formulate investment determinations. Predicting equity prices is a demanding undertaking due to diverse factors like governmental happenings, fiscal circumstances, business resolutions, investor mentality, and overseas currency hazards. The securities exchange is a vastly active and disordered framework, and producing precise projections of the securities exchange is of paramount significance.

[1] School of Commerce, Finance and Accountancy, CHRIST (deemed to be university), Pune Lavasa. Emails: yasmeen.fatima@bcomfan.christuniversity.in; md.asif@bcomfan.christuniversity.in

[2] Department of Data Science, CHRIST (deemed to be university), Pune Lavasa. Email: frjossy@christuniversity.in

* Corresponding author: preethi.n@christuniversity.in

2. Artificial Intelligence

2.1 Brief Description about AI

AI pertains to the cognitive abilities displayed by machinery that has been created by humans. It is commonly defined as computer programs that simulate human intelligence and act as intelligent agents. These agents are able to interpret and utilize external data to complete tasks assigned to them by humans. The field of AI is complex and requires expertise in various areas such as reasoning, knowledge, planning, learning, perception, communication, and manipulation of machines. There are many different tools and algorithms that utilize AI, ranging from mathematical optimization to logical deduction. As AI continues to develop, it has the potential to become a smarter system that is superior to humans in many ways.

AI is increasingly being applied in various fields, such as software and robotics, to make decisions related to business and politics. It involves simulating human behavior, such as learning, planning, and decision-making, to improve performance in completing tasks. The field of AI is interdisciplinary; combining knowledge from both natural sciences (such as physics, biology, and chemistry) and social sciences (such as psychology, sociology, and economics), one can gain a more comprehensive understanding of complex issues and phenomena. The relationship between AI and other areas, such as mathematics and linguistics, is still being explored [3].

There are different perspectives on the capabilities of AI. The concept of a strong AI suggests that it is feasible to develop machines that possess consciousness, introspection, and proficiency in logic and analytical thinking. On the other hand, a weak AI, believes it is difficult to invent machines that can handle problems like humans. Despite the advancements in AI, machines still lack the ability for independent thinking and self-awareness.

AI is an exciting field that has gained significant attention. In the past few years, there has been a growing focus on the advancement of clever computer systems that can undertake activities commonly linked with human intelligence, such as acquiring knowledge, logical thinking, and making judgments. Although AI has diverse uses, it is frequently employed in disciplines like computing, cognitive science, and other areas related to the natural and social sciences.

The study of AI is complex and requires specialized knowledge. It encompasses a wide range of topics, from mathematical optimization and deductive reasoning to bionics, cognitive science, statistical inference, and the study of economics. Researchers use various algorithms and techniques to develop intelligent systems that can simulate human behavior and decision-making. As such, AI has the potential to evolve into a smarter system that surpasses human intelligence.

The development of AI technology is also being promoted in the field of software, with the emergence of computers embedded with AI features and robots capable of making decisions related to business or political issues. The AI simulation process involves training machines to simulate human behaviors, such as learning, understanding, planning, and decision-making. Through this process, computers can perform tasks with greater efficiency and accuracy, and become better equipped to handle complex tasks.

Despite its many benefits, the development of AI also presents challenges. AI systems have a limited capacity for logical thinking, which means they may struggle with tasks that involve abstract thinking or visual perception. Moreover, the development of AI requires a deep understanding of various disciplines, including mathematics, computer science, and linguistics. However, with continued research and development, it is possible that AI systems will become increasingly sophisticated and capable of handling even more complex tasks.

2.2 AI and Its Principles

Machine Learning (ML)/AI is a field that has been developing for numerous decades, with various methods and strategies aimed at simulating human intelligence. The ultimate goal of AI is to enable machines to understand, think, and analyze like humans, allowing computers and other devices to mimic human intelligence. One popular approach to AI involves training neural networks to perform specific tasks by learning from sets of data. This is achieved through supervised learning, where the neural network improves its algorithm by processing pairs of input and output values. Unsupervised learning, also known as clustering, involves training the neural network with input values only. The rise of the internet has led to the collection of vast amounts of data, while the increasing power of computers and advancements in algorithms have facilitated the development of deep learning. Logical regression is the basis for neural network prediction, with weights (w) and thresholds (b) used to represent the Sigmoid function. The Sigmoid function transforms calculation results into values between 0 and 1, with closer values to 1 indicating more accurate predictions (Fig. 1).

The AI model depends on the cost function to evaluate the correctness of its forecasts by calculating the disparity between the projected and real values. A smaller difference between these values indicates higher prediction accuracy. The neural network learns by attempting to minimize the cost of the mislaying task. The weights (w) and thresholds (b) play an essential role in regulating the cost of the loss function, so the challenge is to search for the essential merit of w and b. This is where gradient descent algorithms come into play, as they adjust w and b iteratively to decrease the cost of the mislaying role, ultimately finding the minimum or a near-minimum value.

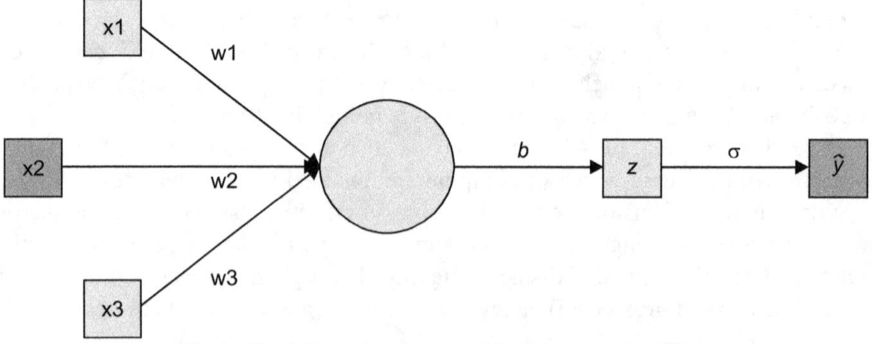

Fig. 1: The delineaetive diagram of logistic regression [1].

AI has emerged as a powerful tool for cognitive processes, managing, and responding. The human psyche may struggle with complex medical and engineering calculations, but computers empowered with AI can perform these tasks faster and more accurately. As a result, what were once considered complex problems are now routinely solved with the help of AI. AI programs cover an extensive range of topics, from machine learning and machine vision to smart robots and deliverance or intelligent retrieval. They also include applications such as reading contexts and linguistics, controlling gestures, and many others. AI has become an integral part of modern technology, making it possible for computers to emulate human intelligence and perform tasks that were previously thought to be beyond their capabilities [7].

Deep learning is an important field in AI that involves building networks with multiple hidden layers. These networks are initialized randomly and trained through a process of adjusting the weights of connections based on the network's output. Over time, the network learns to recognize patterns and make predictions based on input data.

The theory of AI is based on the analysis of simulated neural networks. One common type of AI design is the complex perceptron, which uses multiple layers between the input layer and output layer to create a more abstract representation of input data.

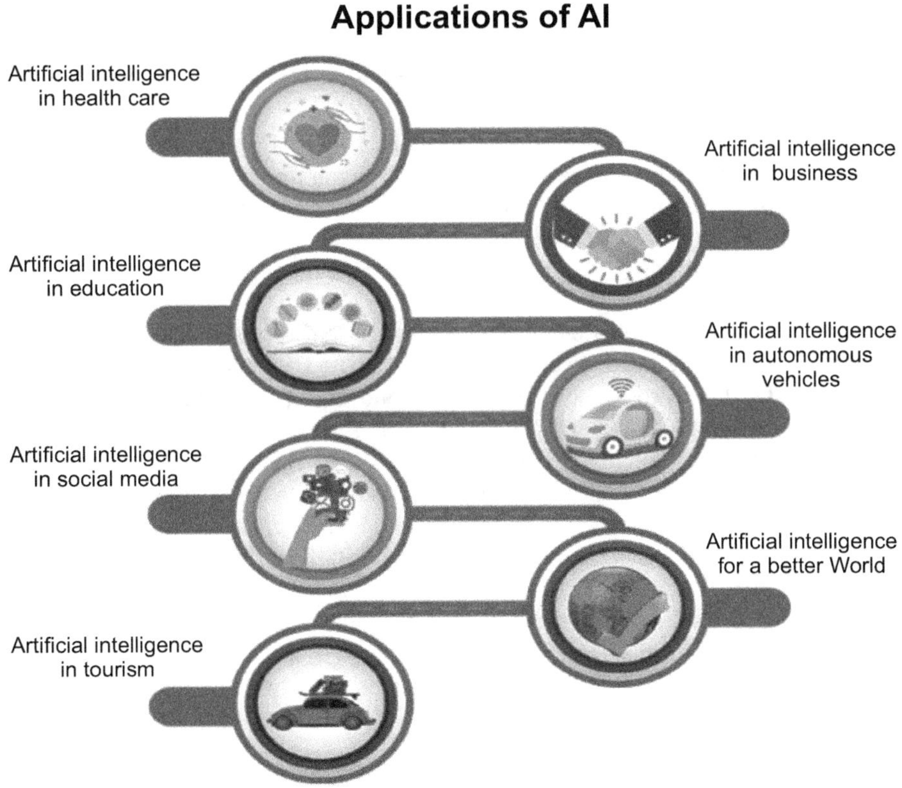

Fig. 2: The applications of AI technology [2].

AI unites subsidiary features to form predominant representations of data. In 2006, [9] an unattended intemperate stratified training algorithm called the Deep Belief Network (DBN) to resolve reform problems in AI. Another AI form is the complex autoencoder, which can be used for unsupervised learning.

One of the most well-known deep learning models is the shift-invariant, which was discovered by [11]. Shift invariant or convolutional neural networks (CNNs) use the relative situation between input data to decrease the number of criteria and sharpen training performance.

Deep learning has applications in various fields, including computer vision. Computer vision involves teaching machines to recognize patterns, demonstrations, and pictures. This has a variety of utilization, together with medical tableau analysis, which can enhance disease prediction, verification, and cure. AI techniques can be classified as supervised or unsupervised learning, similar to machine learning methods. The specific learning models used can vary between different frameworks and platforms.

Top of Form

Pictorial representation techniques and other technical mechanisms are used to size up figure analysis tasks into smaller and more viable parts. One example of a widely used technology is speech recognition, which involves converting speech into text and processing it for various applications such as medical transcription, voice writing, voice command in information processing systems, and after-sale service.

Another useful technology is the recommendation engine, which analyzes a user's attitude and allocates to actively identify their current or potential needs. Through algorithmic analysis and processing, the engine can push relevant information to the user's network and immediately provide them with the most useful information, thereby improving their browsing experience and increasing the rate of conversion.

3. Predicting the Stock Market with Artificial Intelligence

AI is a rapidly growing field that is revolutionizing the way we think about predicting the stock market. One key concept in this field is genetic algorithms, which use the principles of natural progress to develop descendants of software integral that are growingly well-suited for their goal or aim. These algorithms have proven to be highly effective in predicting stock market trends, as they are able to learn from past data and adapt to changing market conditions. By processing large amounts of historical stock market data and analyzing trends and patterns, genetic algorithms are able to generate highly accurate predictions about future market movements (Fig. 3). This makes them an invaluable tool for investors looking to make informed decisions about buying and selling stocks, as well as for financial analysts and economists studying the nature of the share market over time. As the field of AI continues to accelerate, we can expect to see even more modern and powerful tools for predicting stock market trends in the years ahead.

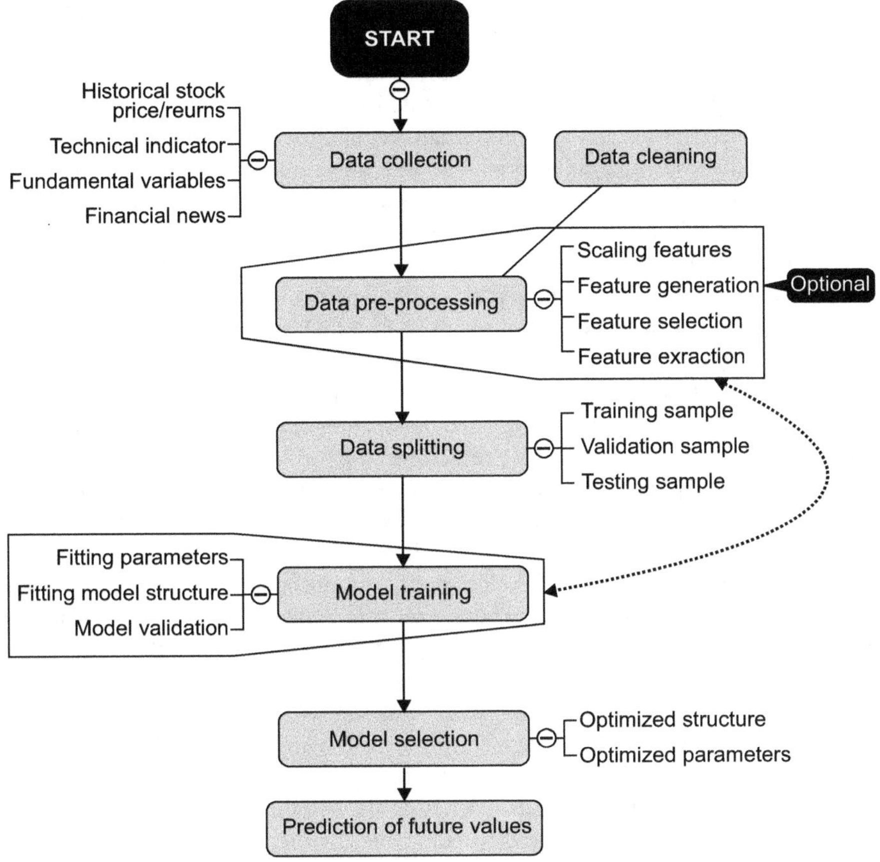

Fig. 3: Process of prediction using AI [10].

3.1 Stock Prediction Concept

The field of predicting stock market trends involves predicting the future prices of stocks or the overall behavior of the stock market. Predictions can be of two types: dummy and real-time. In dummy predictions, rules are defined and the future prices of stocks are calculated by finding the average price. In real-time predictions, developers use the latest market data to analyze the current prices of shares. To achieve greater accuracy in predicting stock prices, a unique algorithm may combine past data with various parameters and analyze market trends.

AI plays a crucial game in share market prediction, and there are various AI-based applications available in the market that provide daily and weekly predictions. Walnut Algorithms is a French start-up that uses evolved AI algorithms and monetary data to develop unconditional gains and investing strategies. The company's trading model constantly learns and adapts to improve its predictions. Alpaca, a California-based Fintech startup, is developing an AI-powered platform for financial trading. The platform provides an API that allows users to trade using algorithms, connect with different programs, and create various services. Sentient Technologies is focused on building a trading platform that uses AI to create individual traders for different companies. The company has received $174M in funding from startups. Binatix is an AI trading organization that swear to be gainful, having used its approach for over 3 years. The company is also functioning with hedge funds to develop and implement investing strategies dependent on their technology.

AI has become increasingly important in the stock market prediction world. AI-based applications and algorithms are being developed to predict market trends and stock prices using various techniques such as machine learning, deep learning, and genetic algorithms. There are two ways of forecasting: dummy and real-time. Dummy forecasting uses predefined rules to predict future stock prices, while real-time predictions use current data to predict stock prices.

One of the most unique algorithms used for stock market prediction is genetic algorithms. These algorithms generate successive generations of software entities by processing natural evolution, which are increasingly better suited for the intended purpose. These algorithms use historical data and other parameters to analyze market trends and improve the accuracy of predictions.

Several AI-based programs and applications are currently available in the market for stock market prediction. For example, Walnut Algorithms is a French startup that uses modern AI algorithms and monetary data to develop unconditional gains and investing strategies. They have developed a trading model that continuously learns and improves in a timely manner. The platform is created to recognize meaningful patterns in monetary markets and scale over various assets globally, adapting to changes in market conditions.

Another notable example is Alpaca, a California-based Fintech startup that is developing AI and database software for monetary trading. The company's platform is an API (Application Programming Interface) for stock trading, allowing users to trade with algorithms, connect with other systems, and build services.

Sentient Technologies is another company that is working on an AI-based trading platform. The company has developed billions of AI-based traders for various companies, using a combination of deep learning, Bayesian methods, and evolutionary algorithms at a very large scale to create products that drive significant ROI. They have received significant funding from startups, with total investment reaching $174M.

Binatix is a trading firm that utilizes deep learning and emerged from stealth mode in 2014, declaring significant profitability through the use of their methodology for more than three years. Limited information is available regarding the company, beyond what was disclosed in a Recode article from 2014. Binatix is reputedly collaborating with hedge funds to create and execute investment strategies founded on their technology.

3.2 Forecasting Stock Market by Machine Learning

In the world of international finance, staying up-to-date on trading activity is of paramount importance. One crucial aspect of this activity is predicting future prices of financial instruments traded on exchanges. This involves using technical and fundamental analysis techniques to make predictions about the direction in which prices will move. Many brokers rely on time series analysis to make these predictions, using programming languages like Python to help them do so.

In recent years, machine learning (ML) techniques have become increasingly popular for making these predictions. One such method is the Support Vector Machine (SVM), which can be trained on historical data of price to forecast future movement in price. This study focuses on using SVM to predict stock prices in three different markets, for both large and small capitalizations. By analyzing daily price data, the model can make forecasting about the direction of coming price movements.

It's important to note that predicting the stock market is a complex and uncertain process, and there is no guarantee that any model or approach will be accurate. However, using ML algorithms like SVM can help traders make more informed decisions based on historical data, allowing them to potentially improve their overall trading performance.

3.3 Stock Market Forecasting Using Intelligent Retrieval on Tick Data

The Indian Institute of Technology in Delhi, located in Hauz Khas, has conducted research on predicting the Indian share market utilizing intelligent retrieval networks and tick records. The stock market is a platform where stocks and derivatives of a company are traded at a mutually agreed-upon price. The stock market is a rapidly growing sector in most countries, and a large number of people are indirectly or directly involved in this area. Therefore, it is crucial to understand the trends in the market, which has led to an increasing interest in forecasting share prices. However, owing to the dynamic nature of the share market and its susceptibility to rapid changes in stock prices, forecasting share prices is a challenging role.

Previous studies have suggested constructive methods for studying event depictions that represent linguistic and semantic information from a text collection, illustrating their usefulness for predicting script events. However, in the end, take out from underdone texts lack prudent knowledge, such as the intentions and sentiment of the event content, which can be functional for a well-known event match with minute dissimilarities in their aspect understanding. To inscribe this topic, this paper presents leveraging external prudence knowledge about the intention and attitude of the event.

The utilization of simulated neural networks on tick records to forecast the Indian share market is a unique appeal that combines both technology and finance. This method is different from traditional approaches such as fundamental analysis and technical analysis. The utilization of simulated neural networks allows for more accurate forecasting of stock prices by analyzing large amounts of data and identifying patterns.

The research conducted by the Indian Institute of Technology in Delhi provides new insights into the utilization of simulated neural networks for forecasting share prices. The study also highlights the importance of understanding market trends and the dynamic nature of the stock market. It emphasizes the need for accurate predictions to make informed investment decisions.

Furthermore, the paper proposes a novel approach to enhance event representations by leveraging external common-sense knowledge. This approach can improve the accuracy of event predictions and aid in distinguishing event pairs with subtle differences in their surface realizations.

Overall, the research conducted by the Indian Institute of Technology in Delhi is a significant contribution to the field of finance and technology. It provides a unique and innovative approach to predicting stock prices, which can potentially lead to better investment decisions and increased profitability.

3.4 Support Vector Machines Use in Forecasting Stock Price Direction

The work of Support SVM is to predict the direction of stock prices. It is a popular method employed by financial organizations and traders. The goal of these models is to outperform the market for their clients, but achieving consistently high levels of profitability is a challenging task. The accuracy of stock price forecasting is critical because even a slight improvement in forecasting accuracy can result in a significant increase in profitability for these organizations.

SVMs are a type of AI algorithm that has been widely utilized for share price forecasting due to their ability to handle high-dimensional data and nonlinear relationships. SVMs have been shown to outperform traditional statistical models in predicting stock price direction.

In SVM-based stock price forecasting, historical stock price data is used to train the algorithm to search out different patterns and trends in the data. Once trained, the algorithm can be utilized to predict the future direction of stock prices. The accuracy of the forecasting hangs on the standard of the data utilized to train the algorithm and the selection of appropriate features.

While SVM-based stock price forecasting has shown promise, it is prime to note that no model can consistently predict the direction of stock prices with perfect accuracy. However, accurate predictions can still provide valuable insights for financial organizations and traders to make informed investment decisions.

Overall, the use of SVMs for stock price forecasting is a valuable tool for financial organizations and traders. With the increasing availability of high-quality data and advances in AI algorithms, the accuracy of share price forecasting is expected to continue to improve in the future.

3.5 Prediction Accuracy

Market predictions can be successful, but they are not always accurate, and sometimes the right decision is not made at all. According to economist Friedrich Hayek and journalist James Surowiecki, prediction markets are useful for gathering and synthesizing vast amounts of data held by individuals. However, the wisdom of crowds can also lead to failure in prediction markets when individuals in the crowd are wrong in their unbiased judgments due to peer pressure, panic, and bias, leading to a lack of diversity. In some cases, the lack of knowledge in the crowd can also contribute to the failure of crowd predictions.

The marginal traders' hypothesis suggests that individuals are always looking for opportunities where the crowd is incorrect, and these specials can put the forecasting market back on track when the prediction of the crowd is incorrect. Prediction markets have been triumphant in business and constitution, where questions require a quick turnaround for the correct solution. However, they have failed to gain traction with researchers and the general public.

In 2020, the stock market witnessed a wild ride due to outside events that overshadowed important investment factors like company earnings. Despite this, the stock market managed to gain a strong benefit, with major market indices reaching all-time highs in the last weeks of the year. Investment leaders from US financial institutions reviewed policy positions laid out by the new leadership as Joe Biden became President on January 20, 2021, and a new Congress was seated. Crowd predictions for stock market gains became a real event, but accuracy was not always achieved.

In conclusion, prediction markets can be useful tools for gathering data, but their accuracy is not guaranteed. The wisdom of crowds can lead to successful predictions, but it can also lead to failures. It is important to consider all factors when making investment decisions, including company earnings, policy positions, and outside events that may impact the market

4. The Application of Random Forests in Predicting Stock Prices and the Influence of Fiscal Performance and Trend Analysis on this Prediction

The application of ML and AI techniques for the prices of stock forecasting has become a popular research area. As researchers continue to work on developing approaches to ameliorate the exactness of stock price forecasting models, there is a wide range of options available, but not all methods work equally well. Different methods may produce varying results even when the same dataset is used. The paper being discussed employs the RFA to forecast stock prices with the use of certain ratios of finance from the preceding quarter, offering one approach to the problem of stock price prediction through a predictive model. However, other factors involving the opinion of investors, public opinion, news reports, and market-wide events can also impact stock prices. By combining financial ratios by utilizing a proficient analysis tool for evaluating the opinion of investors, it is possible to enhance the precision of the stock price prediction model. This can lead to better results in forecasting the stock market.

In recent years, the financial industry has witnessed a surge in the utilization of ML and AI techniques. These advanced technologies have gained significant momentum in the sector. These techniques have been used to develop predictive models that can assist in making investment decisions, particularly in the stock market. However, accurately predicting stock prices is a complex and challenging task, as there are numerous factors that can influence stock prices, including economic indicators, political events. One of the most commonly used ML algorithms to forecast stock prices is the random forest algorithm. The algorithm is a process of group learning method which combines countless decision trees to produce a more precise and vigorous forecasting model. Random forest models are particularly useful for handling large datasets with many variables, as they can automatically select the most relevant features for prediction.

In addition to financial ratios, investor sentiments and public opinion have been shown to be important factors in predicting stock prices. Sentiment analysis, that includes analysis of news articles, social media posts, and other sources of information to determine the overall opinion towards a particular company or industry, can be used to supplement financial ratios in a predictive model. By combining these factors, a more accurate and comprehensive stock price prediction model can be developed.

Overall the usage of ML and AI can be used, such as the random forest algorithm, has the potential to revolutionize the stock market and improve investment decision-making. However, it is necessary to continue to investigate and refine these techniques to improve their correctness and effectiveness in predicting stock prices.

5. Module Identification

Module identification is a crucial step towards the successful completion of a project, and the data collection module serves as the primary step in this regard. The data collection module focuses on collecting the right dataset, which is essential for market prediction. The dataset should be filtered based on various factors, and additional external data must be added to enhance the dataset's quality. In our case, stock prices of the preceding year are being used as our primary source of data. Initially, we will be examining the dataset available on Kaggle and then proceed to select the appropriate model for accurate prediction analysis. A robust data collection module forms the foundation for building an effective market prediction system.

In the data collection module, the quality of the data plays a crucial role in the accuracy of the predictions made. The dataset should be relevant and up-to-date to ensure that the market trends are accurately reflected in the predictions. Additionally, external data such as news articles, economic indicators, and social media sentiment can also be incorporated to enhance the dataset's quality and make more accurate predictions.

Post data collection, it needs to go through the process of data cleaning, and structuring so as to ensure that it is in a format that can be used for analysis. This includes removing any irrelevant or incomplete data, dealing with missing values, and normalizing the dataset to guarantee consistency.

Selection of appropriate models for analysis is also an essential step in the process of collecting data. The selection process is on the basis of the standard of dataset available and the accuracy required. The chosen model should be able to handle the dataset complexity and provide accurate predictions.

In summary, the data collection module is a critical component of market prediction, and its success relies on the standard of the dataset collected, the insertion of external data sources, and the selection of an appropriate model for analysis.

5.1 Captivating Data

It is a crucial step in the investigation of the data that helps to clean and transform raw data into a more suitable format for further analysis. In many cases, raw data contains errors, missing values, and inconsistencies that can lead to inaccurate analysis results.

Preprocessing involves a series of techniques that can help to address these issues and prepare the data for modeling.

One of the key tasks in preprocessing is handling missing values, which can transpire under numerous scenarios such as data entry errors or incomplete records. Techniques such as mean, mode, and regression ascribes can be used to fill in the missing values based on the characteristics of the dataset.

Another important task is outlier detection, where anomalies or outliers in the data are identified and treated accordingly. A heretic can have a notable effect on the analysis outcomes, so it is essential to recognize and address it before modeling.

Feature scaling is another crucial step in pre-processing, which involves transforming numerical features to a similar scale to avoid bias in the modeling process. Feature scaling techniques such as standardization and normalization can help to achieve this.

Overall, captivating data is a critical step in data analysis to assist in ensuring the accuracy and reliability of the modeling results. It involves a series of techniques that can help to clean and transform raw data into a more suitable format for further analysis.

5.2 Training the Machine

Training the system involves providing the algorithm with the necessary data to refine the model's performance. The training datasets are used for the adjustment of the models, while the test sets are left untouched to ensure that the model's performance is evaluated solely on the basis of concealed data. The training of the model includes cross-estimation, which helps in obtaining an accurate estimate of the model's performance using the training data. In Financial market forecasting, ML algorithms are employed to make predictions based on historical data, and the models are trained on this data.

The process of tuning models involves optimizing meta parameters similar to the count of trees in the random forest. The entire univariate kink is performed for every set of hyperparameter values, and a cross-validated score is calculated for each set of hyperparameters. The best hyperparameters are then selected based on this score. The purpose of tutoring the model is to start with few Opening values with the dataset and optimize the parameters required in the model. This process is repeated until the optimal values are obtained. Finally, the predictions are made by the trained model on the inputs from the test dataset.

The dataset is distributed into a ratio of 4:1, with 3/4th of the data used for tutoring the model, while the remaining 1/4th is kept for the trial of the model's performance. By training the model on a part of the data and experimenting on the left data, we can evaluate the model's accuracy and ensure that it can make accurate predictions on unseen data. The training process is critical in developing a robust model that can accurately predict stock prices.

5.3 Data Scoring

It is the act of using a forecasting model to analyze a dataset. The Random Forest Algorithm is utilized to process the dataset, which is an ensemble technique commonly used for classification and regression. The results achieved are based on the learning models applied. The last module of this process presents the outcome of the model, which is used to forecast the likelihood of a stock to rebel or fall on the basis of particular criteria. It helps in identifying the sensitivity of a specific entity or stock. To ensure the security of the data, a user authentication system control is implemented to restrict access to authorized entities only.

To elaborate further, the RFA is a supervised learning algorithm that uses decision trees to classify data. It creates several decision trees and merges them for overall improvement and accuracy of the model. The result of the model is used to forecast the stock's future on the basis of the given parameters, such as the current market trends, company financials, and previous stock prices. This helps investors make informed decisions about buying or selling stocks. The vulnerabilities identified by the model can aid in managing investment risks and avoiding financial losses.

The user authentication system control is a security measure that ensures the confidentiality of the data generated by the model. It restricts access to the model's results to only authorized entities, such as investors or financial advisors. This prevents unauthorized access to sensitive data that could result in financial harm to investors or other parties

5.4 Experimental Results

The data for our findings is available in an xls file, which is in its raw format. The file contains 11 columns, also known as attributes, that help us to understand the uptrend and downtrend of stock prices. Among the 11 attributes, some important ones include HIGH, LOW, OPENP, and CLOSEP. The HIGH attribute illustrates the uppermost value of the stock in the preceding year, whereas the LOW attribute represents the bottommost value of the stock in the preceding year. The OPENP attribute showcase the worth of the stock at the inception of the trading day, and the CLOSEP attribute represents the value of the stock at the end of the trading day. However, the 4 attributes mentioned above are critical in our analysis as they provide insights into the performance of the stock over a specific period.

Figure 4 is a graphical view of the raw statistics which is present in our xlsx file. The dataset consists of 121,608 data points. However, few statistics contain irrelevant information that cannot be used to train our system. Thus, it is necessary to operate the raw data to obtain a filtered set of data to train our system.

Figure 5 portrays the output of applying the head () function, which returns the first 5 rows of the dataset. The pandas (Python Data Analysis) library is used to perform this analysis, and the inbuilt value for the no. of rows to return is 5. The code of trading attribute in the treated dataset is insignificant, so in order to eliminate it strip() function is used. Additionally, altogether trading codes are replaced with the 'GP'.

Figure 6 portrays a broadcast of the stock attribute 'CLOSEP' against the 'DATE' attribute, generated with the use of "matplotlib.pyplot" library. The plot illustrates the variation in the stock's closing price over a 2-year period. Additionally, a candlestick plot is provided, generated using the "mpl_finance" library, which depicts the certain

DATE	TRADING CODE	LTP	HIGH	LOW	OPENP	CLOSEP	YCP	TRADE	VALUE (mr	VOLUME
28-12-2017	1JANATAMF	6.4	6.5	6.4	6.4	6.4	6.5	79	1.888	2,94,720
27-12-2017	1JANATAMF	6.5	6.5	6.4	6.5	6.5	6.5	73	1.295	2,00,062
26-12-2017	1JANATAMF	6.5	6.6	6.4	6.5	6.5	6.5	103	4.119	6,30,548
24-12-2017	1JANATAMF	6.6	6.6	6.4	6.5	6.5	6.5	46	0.654	1,01,104
21-12-2017	1JANATAMF	6.6	6.6	6.4	6.4	6.5	6.4	24	0.241	37,098
20-12-2017	1JANATAMF	6.4	6.5	6.4	6.4	6.4	6.4	37	0.296	45,885
19-12-2017	1JANATAMF	6.4	6.6	6.4	6.5	6.4	6.5	55	1.387	2,16,529
18-12-2017	1JANATAMF	6.4	6.5	6.4	6.4	6.5	6.4	36	0.141	21,817
17-12-2017	1JANATAMF	6.5	6.5	6.4	6.5	6.4	6.6	118	2.904	4,52,125
14-12-2017	1JANATAMF	6.5	6.6	6.5	6.6	6.6	6.6	36	0.596	90,597

Fig. 4: Raw Data [2].

	DATE	TRADING CODE	LTP	HIGH	LOW	OPENP	CLOSEP	YCP	TRADE	VALUE (mn)	VOLUME
0	2018-08-16	1JANATAMF	6.2	6.3	6.1	6.2	6.2	6.2	56	0.757	122741
1	2018-08-16	1STPRIMFMF	11.2	11.2	10.9	11.0	11.1	10.9	145	2.640	238810
2	2018-08-16	AAMRANET	80.1	80.4	78.5	78.5	79.7	78.3	545	15.488	195035
3	2018-08-16	AAMRATECH	30.8	31.6	30.7	31.0	30.9	31.0	195	5.100	164899
4	2018-08-16	ABB1STMF	6.1	6.1	5.9	6.0	6.1	6.0	109	11.214	1857588

Fig. 5: HEAD() [6].

Fig. 6: Time Series Plot of Gp.

attributes such as date of the trade, when it was open, what was the opening value topmost and bottommost, and 'CLOSEP' (Fig. 7). The broadcast represents the same stock's trading information and is useful for identifying price trends and patterns.

Figure 8 is a histogram that displays the distribution of the differences between the CLOSEP and OPENP attributes. The histogram is created using the matplotlib. pyplot library and shows the frequency of occurrence for the differences between the two attributes. The graph shows that most of the differences fall between the range of –0.2 and 0.2.

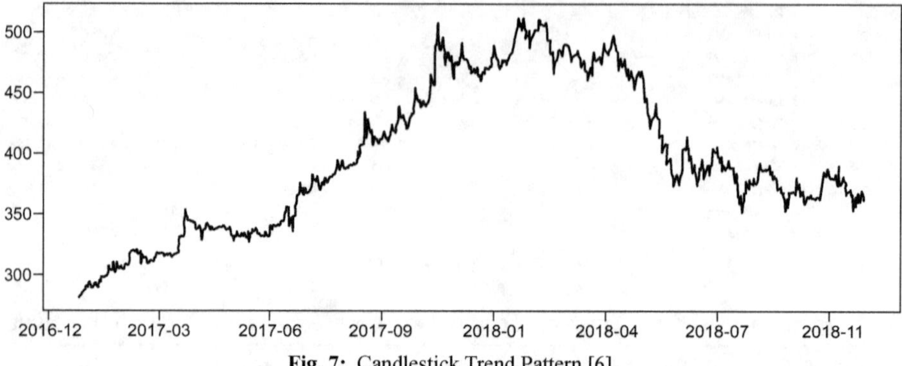

Fig. 7: Candlestick Trend Pattern [6].

Figures 8 and 9 show histograms of the attributes 'CLOSEP-OPENP' and 'high-LOW'. The purpose of plotting these histograms is to study the effect of today's closing and opening prices, as well as the topmost and bottommost prices of the stock over the past year, on the prices of the stock at a future date. On the basis of reasoning, a logic is developed to assign a value of 1 to DEX if today's CLOSE is higher than yesterday's, else a negative value of –1 is assigned. The entire dataset is then processed, and using the top() function, a glimpse of the obtained data is obtained. The next step involves setting the feature and target variables, along with the training size. Using the sklearn libraries, the SVC classifier is imported and fit with the training data. After training the model and running the test data through the trained model, the confusion matrix obtained is shown in Fig. 9.

Additionally, another model is trained using the same dataset. It utilizes the Random Decision Forest, which belongs to the aggregate approach. Post connecting the model with the data and running it against expected data, it is found to have an accuracy rating of 0.808. In conclusion, the SVC model has an accuracy of 0.787 in the test set, on the other hand, the correctness rating of the RFC is calculated to be 0.89.

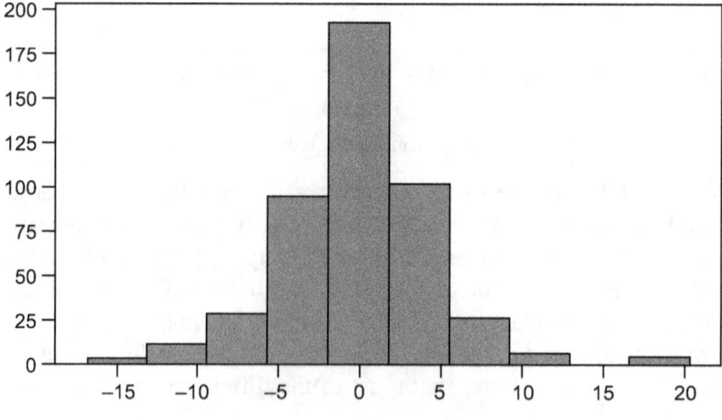

Fig. 8: Histogram Of Closep-Openp [6].

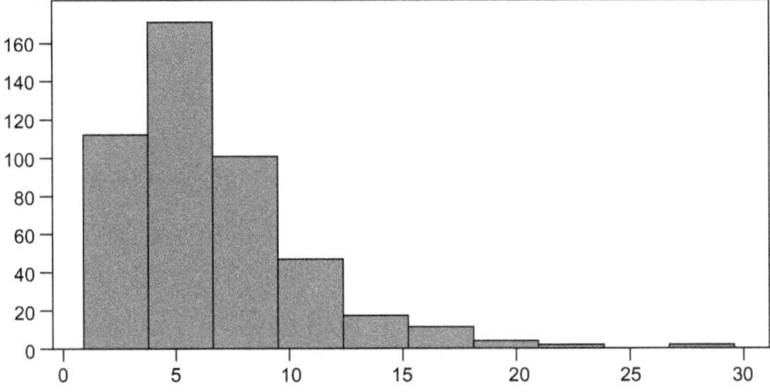

Fig. 9: Histogram Of High-Low [1].

6. Future Scope

To enhance our trading prediction utility, we plan to incorporate sentiment analysis for a more comprehensive evaluation in the field of cryptocurrency trading. In the future, we aim to expand this project by including additional parameters such as financial ratios, multiple timeframes, and more. By considering more parameters, we can increase the accuracy of our predictions. We can also apply the algorithms to analyze public comments and identify patterns and relationships between customers and company employees. With the use of data mining techniques along with the algorithms, the overall structural performance of the company can be enhanced..

7. Conclusion

After evaluating the transparency of the algorithms used, we have determined the algorithm, namely random forest, which is the most appropriate in the prediction of the stock market using past data from various data points. The algorithm would be highly beneficial for players in the financial market or stock market who are looking to invest in the stock market since it has been trained on a huge collection of collected data and has been selected after testing on a specimen data set. This extrapolate showcases the model of ML that predicts keep in prices with more transparency compared to previous ML models.

References

[1] Stock market prediction using machine learning algorithms. *International Journal of Recent Technology and Engineering*, *8*(2S4): 280–283, 2019. https://doi.org/10.35940/ijrte.b1052.0782s419.

[2] Chopra, R., and Sharma, G.D. Application of Artificial Intelligence in stock market forecasting: A critique, review, and research agenda. *J. Risk Financial Manag.*, 14: 526, 2021.

[3] Karimuzzaman, M., Islam, N., Afroz, S., and Hossain, M.M. Predicting stock market price of bangladesh: A comparative study of linear classification models. *Annals of Data Science*, 8(1): 21–38. 2021.4.

[4] Duong, D., Nguyen, T., and Dang, M. Stock market prediction using financial news articles on ho chi minh stock exchange. *In: Proceedings of the 10th International Conference on Ubiquitous Information Management and Communication*. 2016, January, pp. 1–6.

[5] Marc-André Mittermaye. Forecasting intraday stock price trends with text mining techniques. *In: 37th Hawaii International Conference on System Sciences*, 2004.
[6] Sachin Sampat Patil, Prof. Kailash Patidar, and Asst. Prof. Megha Jain. A Survey on Stock Market Prediction Using SVM. *IJCTET*, 2016.
[7] Hakob Grigoryan. A Stock Market Prediction Method Based on Support Vector Machines (SVM) and Independent Component Analysis (ICA). *DSJ*, 2016.
[8] Nam, K., and Seong, N. Financial news-based stock movement prediction using causality analysis of influence in the Korean stock market. *Decision Support Systems*, 117: 100–112, 2019.
[9] Devi, K.N., Bhaskaran, V.M., and Kumar, G.P. . Cuckoo optimized SVM for stock market prediction. *In: 2015 International Conference on Innovations in Information, Embedded and Communication Systems (ICIIECS)*, pp. 1–5. 2015, March. IEEE.
[10] Nti, I.K., Adekoya, A.F., and Weyori, B.A. A novel multi-source information-fusion predictive framework based on deep neural networks for accuracy enhancement in stock market prediction. *Journal of Big Data*, 8(1): 1–28. 2021.
[11] Raut Sushrut Deepak, Shinde Isha Uday, and Dr. D. Malathi. Machine Learning Approach in Stock Market Market Prediction. *IJPAM*, 115(8): 71–77. 2021.

Chapter 18

The Intellectual Structure of Application of Artificial Intelligence in Forecasting Methods

A Literature Review using Bibliometric Thematic Analysis

Anshul Saxena,[1,*] *Jayant Mahajan,*[2] *Vandana Bhagat,*[2] *Subha MV,*[3] *Binu P Paul*[4] and *Vaishali Jain*[5]

1. Introduction

Crude oil is a valuable asset class which forms the nucleus of the energy core of the transport sector for any country. According to report [1], crude oil helps in meeting 93% of energy needs for the transportation sector globally. It has been projected across various forums that crude oil along with coal and natural gas is going to satisfy world energy needs for the forthcoming years. Consequently, it has been observed that fluctuations in crude oil prices tilt the economies of scale across the world.

Crude oil prices in the international market affect the price of its derivatives in the form of petrol and diesel in the domestic market. To get an accurate forecast

[1] Research Scholar, Christ University, India.
[2] Associate Professor, Christ University, India.
[3] Associate Professor, Anna University, India.
[4] Associate Professor, Christ Academy, India.
[5] Associate Professor, Marwadi University, India.

of the crude oil price is one of the biggest problems faced by global oil markets. Reliable forecasting of crude oil prices will be a boon both for producers and consumers. Oil-producing countries can determine the production level based on predicted rates. Countries like India, China, who are dependent on imports to fulfill their energy needs, might be able to manage their deficit in a better way. Industries who are dependent on oil can determine their production and inventories in advance with the help of an accurate forecast. Substantial fluctuations in crude oil prices [2] have resulted in reduced economic output and price asymmetry.

Crude oil is the largest actively traded commodity across the globe [3]. Price discovery for crude oil takes places across multiple stock exchanges (like IMEX and NYMEX) on which variety of financial instruments (like spot, derivatives) trade, thus making it a complicated job to accomplish. Also, the oil price is influenced by a multitude of long- and short-term factors. The volatility of crude oil prices in recent years has made it a difficult task to achieve. Short-term volatility, in combination with geographical and political factors, has changed the dynamics of crude oil pricing. Therefore, accurate forecasting of the crude oil prices is of utmost importance.

The history of crude oil prices is a fascinating tale of economic and geopolitical factors that have shaped the global energy landscape (Fig. 1). Since the inception of the modern oil industry in the mid-19th century, crude oil prices have experienced numerous fluctuations, influenced by shifting supply and demand dynamics, geopolitical events, and technological advancements.

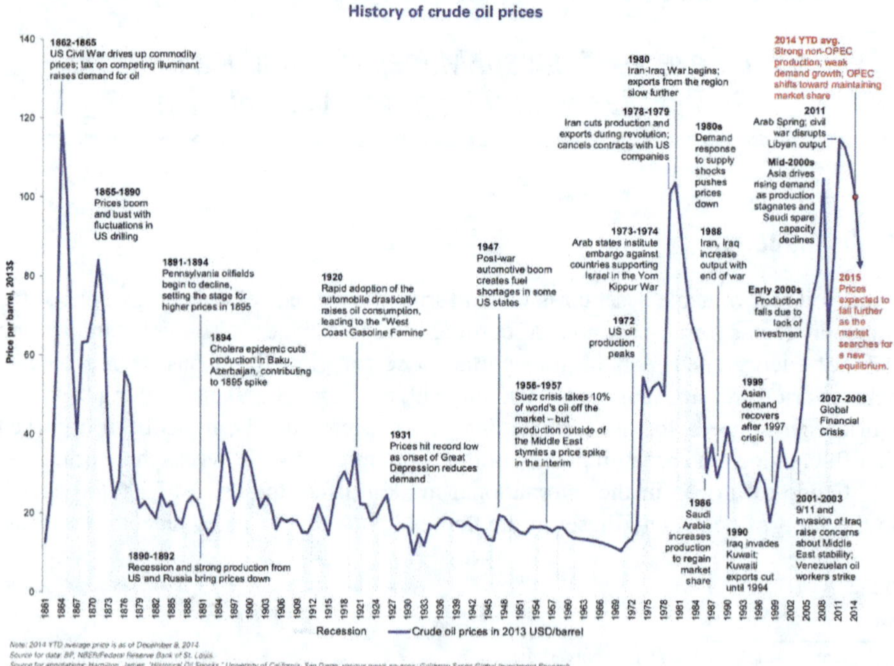

Fig. 1: History of crude oil prices (*Source*: BP Website).

Early Years: 1860s–[1940s]

The birth of the modern oil industry can be traced back to the drilling of the first commercial oil well in 1859 by Edwin Drake in Titusville, Pennsylvania. In the early years, the U.S. dominated global oil production, and prices were relatively stable. The market was driven primarily by local factors, with limited international trade.

Post-WWII: 1940s–[1970s]

After World War II, the global oil industry began to change dramatically. The Middle East emerged as a major oil-producing region, and multinational oil companies secured long-term contracts with countries like Saudi Arabia, Iran, and Iraq. During this period, the oil market experienced relative stability, with prices regulated by large oil companies and the Seven Sisters (a group of multinational oil companies). Prices remained relatively low, averaging around $1.80–$2.50 per barrel in the 1940s and 1950s, and gradually increasing to around $3 per barrel by the early 1970s.

Oil Crises and Volatility: 1970s–[1990s]

The 1970s witnessed the first major disruption in crude oil prices, with two significant oil crises. The 1973 oil crisis, precipitated by the Yom Kippur War, resulted in the Arab members of OPEC (Organization of Petroleum Exporting Countries) imposing an oil embargo on the U.S. and other Western countries. Consequently, oil prices quadrupled, reaching nearly $12 per barrel by 1974. The second oil crisis occurred in 1979, following the Iranian Revolution, when prices spiked again, reaching nearly $40 per barrel by 1980. The 1980s and 1990s were marked by price volatility and periods of oversupply. Oil prices plummeted in 1986 due to increased production from non-OPEC countries, notably the North Sea and Alaska. Prices remained relatively low throughout the 1990s, averaging around $20 per barrel.

2000s: Surging Prices and Financial Crisis

In the early 2000s, oil prices began a steady climb, driven by growing demand from emerging economies, particularly China and India. Prices reached an all-time high in July 2008, with the U.S. benchmark (WTI) hitting $147 per barrel. However, the global financial crisis soon followed, leading to a sharp decline in oil prices, which bottomed out at around $30 per barrel in December 2008.

2010s: Shale Revolution and Price Collapse

The 2010s were characterized by the rise of U.S. shale oil production, which significantly altered the global oil market. Increased U.S. output contributed to an oversupply of crude oil, culminating in a price collapse in 2014–2016. Prices dropped from over $100 per barrel in mid-2014 to below $30 per barrel in early 2016.

2020s: COVID-19 Pandemic and Market Recovery

The COVID-19 pandemic had a dramatic impact on oil prices in 2020. As global demand plummeted due to lockdown measures, oil prices experienced an unprecedented crash, with the U.S. benchmark (WTI) even dipping into negative territory in April 2020. However, as the pandemic subsided and demand recovered, oil prices began to rebound, rising steadily through 2021.

Need for Study

Crumbling Crude Oil Future Prices (Have touched –ve price zone in May 2020 (Fig. 2)) have resulted in a sharp drop in the revenue for Oil Exploration and production companies. Record fall has been observed in the demand for crude oil due to the recent lockdown imposed in various countries of the world as a result of Covid-19 restrictions [4]. High Production cost for US Shale companies has resulted in closure of oil production companies owing to the current pandemic. (Concho Resources, Encana, Diamondback Energy, Devon Energy, Pioneer Natural Resources, QEP Resources, and WPX Energy). In an unprecedented turn of events, global oil demand is projected to plummet by a staggering 9.3 million barrels per day (mb/d) in 2020, on a year-on-year basis. As the world grapples with the ongoing pandemic, the International Energy Agency (IEA) has highlighted critical concerns to the G20 energy ministers regarding the future of the oil market. While low oil prices may seemingly offer an enticing prospect to consumers, the IEA underscores that such advantages are, in reality, of minimal value. The reason? Approximately 4 billion individuals across the globe currently live under some form of Covid-19 lockdown. This unparalleled situation has resulted in far-reaching consequences for both the oil market and those affected by the pandemic. In the midst of these extraordinary circumstances, it becomes crucial for the G20 energy ministers to

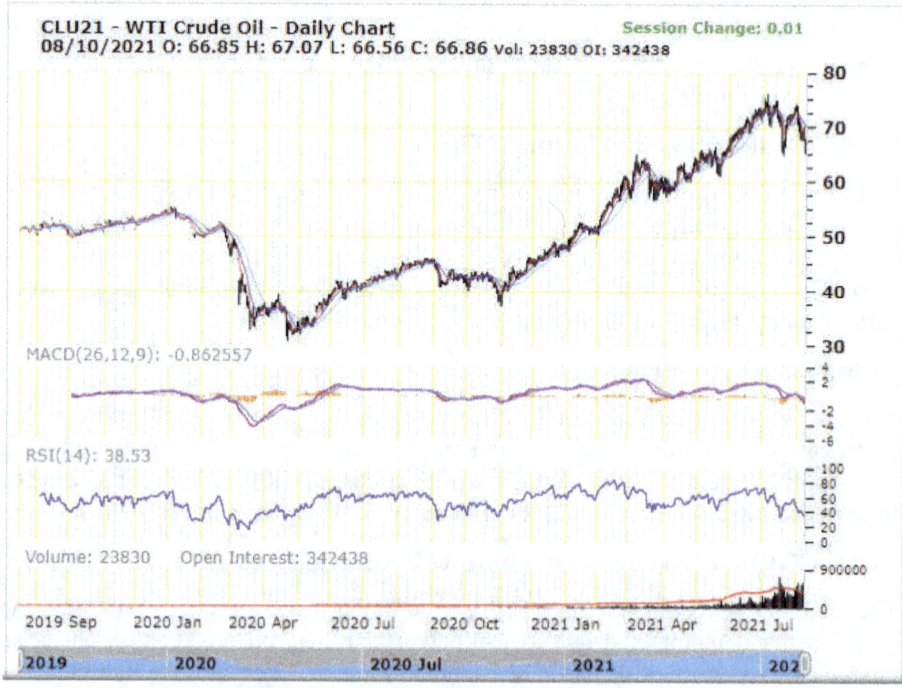

Fig. 2: Swings in crude oil prices (*Source*: www.tradepolicy.com).

take into account the broader implications of the oil market's instability. Despite the allure of inexpensive oil, the IEA emphasizes that it offers little consolation to those who are restricted in their daily activities due to the pandemic. The ability to capitalize on low prices is significantly hindered for a substantial portion of the world's population. Therefore, it is essential that the G20 energy ministers look beyond the immediate imbalances plaguing the market. In doing so, they can develop a more comprehensive understanding of the challenges that lie ahead for the global oil industry. In conclusion, the record-setting drop in global oil demand, coupled with the minimal benefits of low prices for billions of individuals, necessitates a thorough examination by the G20 energy ministers. By contemplating the broader ramifications of these market imbalances, the ministers can work towards crafting effective policies that address the complexities of this unprecedented scenario. All the factors mentioned above have contributed to making the sharp fall in crude oil prices a tail risk event, resulting in a black swan event for the oil and petroleum industry. The list of event signals towards volatile markets shortly, thus signaling a need for a robust and dynamic forecasting framework for crude oil prices, which can adapt themselves swiftly based on market conditions (Fig. 2).

2. Bibliometric Analysis

Bibliometric analysis offers a credible approach for comprehending the temporal progression of a research domain from a multidisciplinary standpoint. This technique has established itself as an invaluable tool in obtaining an all-encompassing insight into the extent and limits of a particular area of study. It also aids in pinpointing influential contributors and unearthing novel avenues warranting further exploration. A wide range of scholars from various disciplines have embraced this methodology. By assessing the quantitative aspects of research publications, this analysis delves into the intricate web of knowledge production and dissemination. It maps the intellectual landscape, shedding light on the intricate connections between ideas, researchers, and institutions [5–8]. Bibliometric studies have proved instrumental in unveiling the evolutionary trajectory of research fields, highlighting the emergence of new themes, their interaction, and convergence. Through this lens, one can discern how academic disciplines have adapted to changes, merging or branching out as needed. Moreover, this approach allows for the identification of influential authors who have shaped the research landscape. By tracing their impact, we can better understand the genesis of ground-breaking ideas and their subsequent development. Bibliometric analysis also unveils emerging trends in research, providing invaluable foresight for scholars and policymakers alike [9–11]. This empowers them to make well-informed decisions about future investments and collaborative efforts in research. In conclusion, the versatility and comprehensiveness of bibliometric analysis have rendered it a favored method among scholars seeking a holistic view of their research domain. By illuminating the past, present, and potential future of a field, this technique fosters a more profound understanding of the intricate tapestry of knowledge production.

Methodology

This study endeavors to offer a comprehensive perspective on research within the crude oil price forecasting models domain. The paper briefly outlines prevalent keywords, prominent publication outlets, authors, countries, and organizations by scrutinizing the existing literature.

Database Curation

To fulfill the study's objectives, a two-step evaluation and review strategy was employed. The initial step entailed defining suitable search terms to mine the Scopus database for pertinent publications. The choice of this database hinged on its extensive collection of peer-reviewed research articles in esteemed journals, as well as its wide accessibility, recognition, and trustworthiness in academia. In the subsequent phase, we pinpointed publications for bibliometric and content analysis, adhering to specific criteria.

Identifying Keywords

After an initial review of existing research, keywords were selected to capture synonymous or frequently used terms in the literature. A Google Scholar search for "Crude oil price forecasting" was executed, with the first 50 works analyzed to discern commonly employed phrases. Additionally, articles in *ABS3* (or above) journals and previously published SLRs on peer-to-peer lending were scrutinized to pinpoint apt terms. "Econometrics", "Artificial Intelligence" and "Machine Learning" emerged as viable keywords based on this review. To further verify the selected keywords, we convened a panel of experts from the realms of information systems and financial technologies—one each from industry and academia—with extensive research experience in these areas. Upon their recommendations, the search scope broadened to include "Hybrid Models" and "Learning Process". This selection aligned with the keywords employed by Thayyib et al., 2023 and Zhang et al., 2022 [13,12].

Thematic Analysis

Thematic analysis is a research method employed for identifying, analyzing, and reporting patterns or themes within data. This approach allows for the extraction of meaningful insights from qualitative information, facilitating a comprehensive understanding of the subject matter. It can be applied to various data types, such as interviews, observations, or documents. The process encompasses several stages, starting with familiarization of the data, followed by the generation of initial codes. Next, researchers search for and review themes by examining the relationships between codes. The identified themes are then refined and defined, ensuring their relevance to the research question. In essence, thematic analysis offers a systematic yet flexible means of exploring and interpreting complex data, enabling scholars to uncover rich and nuanced understandings of their subject matter. This method is particularly valuable for shedding light on the intricacies of human experiences and perceptions.

Over the years, the domain of crude oil price forecasting has experienced a remarkable transformation, marked by distinct methodological shifts. In the 1970s [12],

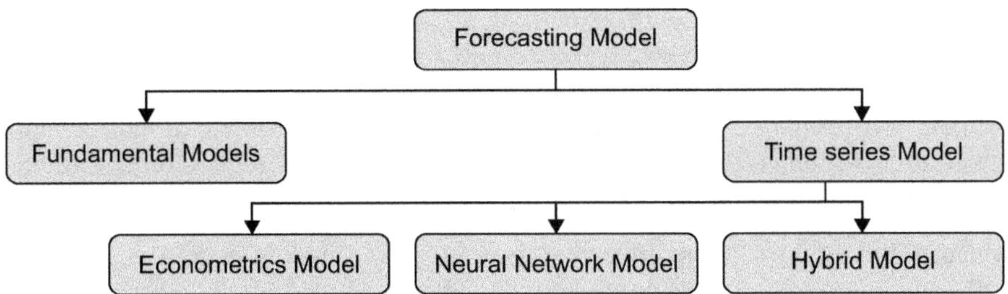

Fig. 3: Classification of oil prices: analysing-forecasting methods.

the fundamental model held sway, laying the groundwork for understanding and predicting oil price trends. As the 1990s unfolded, the focus shifted to econometric models, which provided a more refined approach to analyzing oil market dynamics [13]. The early 2000s [13] saw the rise of machine learning and neural networks, revolutionizing the forecasting arena with their data-driven prowess and adaptability. These cutting-edge techniques harnessed data to anticipate price changes with heightened accuracy. Since 2015 [13], the advent of hybrid models has further enriched the field, fusing the strengths of diverse methodologies to yield superior forecasting capabilities. Consequently, the realm of crude oil price forecasting continues to mature, benefiting from the integration of various approaches and ongoing innovation. Subsequent sections (Fig. 3) discuss the various theoretical, time series, and artificial intelligence (AI) model used for crude oil price forecasting.

3. Fundamental Models

Fundamental models estimate crude oil price based on it is the relationships with economic, financial, social, and political signals. These signals and their relevance in determining the crude oil prices are captured through various theoretical and structure models. Crude oil prices reflect the influence of various long- and short-term factors in the form of news. Above mentioned models attempt to make use of these key oil price drivers in predicting the crude oil price.

- Theoretical analysis model
- Structure Models

3.1 Theoretical Analysis Model

Theoretical analysis model deploys economic theories specific to demand and supply and production cost to determine the oil price. Each factor was modeled in as the mathematical problem or a dilemma situation which further helps in understanding the dynamics of crude oil prices about factors influencing it. The different theories used in theoretical analysis model is discussed further in the underlying paragraph

1. Exhaustible Resource Theory (ETR): This theory uses the Hoteling rule as an underlying assumption. Hotelling rule puts forth a dilemma situation where an oil producer has to evaluate the opportunity cost of extracting the oil from the ground for the current year or use the reserve the oil reserve next year when the oil price is supposed to peak. In other words, Hotelling's Theory elucidates the optimal price for extracting and selling a non-renewable resource, such as oil, in contrast to preserving it for future use. In the context of crude oil prices, this can be perceived as a price increase that results from the interplay between time and the interest rate of capital associated with oil reserves held by oil-producing nations. ETR has been applied in the development of the following structural model. *(*Invetopedia.com).

2. Cost Theory: Cost theory infers that "In a perfectly competitive market, an increase in the marginal cost of consumption or an anticipated future reduction in marginal costs could result in a decrease of the presently produced quantity." This relationship highlights the intricate dynamics that influence production levels under competitive market conditions (Han and Li, 2015). This theory assumes that production cost can influence the crude oil prices in the long term future. Recently this phenomenon has been observed when due to decrease in the crude oil consumption which was the result of the lockdown in 133 countries following the spread of coronavirus has compelled shale oil-producing countries like USA, Russia, and Canada to cut the oil production in order to sustain the marginal cost of production. Application of cost theory is discussed below under the aegis of the structure model.

3. Market Structure Theory (MST): Crude oil industry is recognized by its concentrated market structure, where it has been observed that competitive countries or enterprises have shown oligopoly markets characteristics. Since oil production is concentrated in the hands of few countries like USA and Russia or cartels like Organization of petroleum exporting countries (OPEC) (as shown in Fig. 4), these production centers tend to influence the oil prices.

Effects of influence of production center on crude oil price have been observed recently when once spot price of crude oil touched 22$/barrel due to competition between OPEC and Russia, the major oil-producing countries decided to cut down its production in order to stabilize the crude oil prices (Bloomberg, 2020).

Fig. 4: Major production centres of crude oil (*Source*: OPEC).

Table 1: Summary of structure model using Fundamental theory.

Author	Theoretical and Conceptual Framework	Assumption
[14]	**Strategy Selection**	Total demand is modeled as a function of price/ production level determined by OPEC and differentialy.
[15]		Simulation of price path defined by Gatley and Huntington (2004) for Non- OPEC countries or by OPEC production level.
MacAvoy (1982)	**Econometric Fit**	Oil supply level is estimated for 5 groups of oil producing countries (3 OPEC + USA + Others) using income, lagged demand and price (log linear function).
[16]		Estimation of demand with the help of economic activity and price of oil using three actors: OCED, Non-OCED and OPEC countries.
[17, 18]		This analysis delves into the connection between oil prices and capacity utilization levels, with the assumption of a nonlinear relationship. By examining these variables, the study seeks to uncover the underlying complexities that exist between oil prices and the degree to which production resources are being utilized.
Dees et al. (2007)		In estimating demand, it is considered a function of various factors, such as income, domestic prices, exchange rates, and technological progress. By examining the interplay of these elements, a more comprehensive understanding of the forces driving demand can be achieved.
[15]	**Capacity Utilization**	Demand of crude oil based on the production capacity utilized by OPEC countries.
[19]		Demand estimation is approached as a log-linear function incorporating factors such as income, price (adjusted for exchange rate), and technological advancements. By considering the relationships between these elements in a log-linear manner, a more nuanced understanding of demand can be obtained.
[20]	**Optimization Models**	Optimization of the price by assuming weighted average of marginal costs for short term.
[21]		The optimization of crude oil prices is achieved through the analysis of 4 key cost components: production, royalty, transportation, and processing. By employing an intertemporal model, a more comprehensive understanding of the factors influencing oil prices can be developed over time.
[22]		Sensitive opportunity cost, substitution cost and coalition structure of OPEC.
Ezzati (1976)		Estimation of crude oil price using consumption and GNP growth rate.

Table 1 contd. ...

...Table 1 contd.

Author	Theoretical and Conceptual Framework	Assumption
[23]		Demand as a liner function which grows constantly.
Daly et al. (1982)		Demand is characterized as a log-linear function that considers both economic activity and the lag distribution of previous oil prices relative to other goods. This approach offers a more comprehensive understanding of demand dynamics by taking into account the interplay of these factors.
[24]		Forecasts the NPV of cash flow and look to simulate the condition in which it can be maximized.
[25]	**Game Theoretic Models**	Oligopoly players will be competitive in nature and set the prices according to competition in market and other set of will be conservative in nature and put a premium in preserving the depleted resources.
[26]		OPEC members set the future price ensuring maximization of present value of net income and Non-OPEC players maximize price by controlling the production level.
[27]		Consider linear function of price and lagged demand.

Many economists have proposed to reach an optimal crude oil price after assuming Nash–Cournot game theory approach (treating OPEC countries and dominant producers as price taker or price maker) [25]. Other approaches treat OPEC countries as the single monopolistic player [23] or by considering the capacity utilisation goals whereas discussed above oil-producing countries try to influence the oil price through controlling the production rate.

This study would like to conclude that in comparison to Hotelling rules or cost theory, models based on market structure theories have practical significance for long-term oil-price volatility analysis.

3.2 Structural Models

Informed by the principles of microeconomics, these mathematical models endeavor to optimize or simulate crude oil prices, aiming to establish a framework that incorporates specific constraints. In practice, these models serve to imitate the behavior of key market players, who play a crucial role in determining the price of crude oil. The objective is to create a representation that mirrors the real-world dynamics that influence oil prices. These models gained popularity in the early 1970s and remained influential throughout the 1980s, a period during which OPEC was the primary force in determining oil prices. The OPEC cartel's influence on the market made understanding and predicting their behavior essential for those interested in the oil industry. Economists designing these models have focused on several primary factors, which they believe most significantly impact crude oil prices. These factors include global oil demand, OPEC production, and non-OPEC production. By incorporating these elements into their models, economists aim to paint a comprehensive picture of the forces driving oil prices. The intent behind developing these models is to provide valuable insight into the complex relationships that exist between various market forces. By simulating the behavior of market players and the consequences of their actions, these models can help stakeholders make more informed decisions and navigate the often-turbulent oil market. While the effectiveness of these models may have varied over the years, their importance cannot be understated. As the global oil market continues to evolve and new players emerge, refining and adapting these models becomes increasingly necessary.

4. Econometrics Models

This section focusses on models which were built to analyze the historical behavior of crude oil prices. The fundamental model has been found to be good at identifying the factors which can influence the crude oil prices in the long and short term. To further understand the behavior of crude oil prices, time series and econometrics models have been applied to the commodity exchange related data. It has been observed that stochastic models have been used to understand the mean reverting properties of the crude oil prices (Chan, 2015).

To forecast the crude oil prices based on its past behavior, time-series analysis has been used afterwards. In the beginning, time-series properties were assumed to follow the Gaussian distribution; however, later this theory has been discarded since it was found that crude oil prices follow non-Gaussian distribution (Andrew et al., 2001; Kim, 2012). In further studies, it has been discovered that in fact, crude oil prices show nonlinear and complex properties which need specialized treatment of variables to obtain the accurate results. Researchers have proposed three categories (Fig. 2) of models to overcome the problems highlighted above (Table 2 exhibits the underlying study for a model of each category).

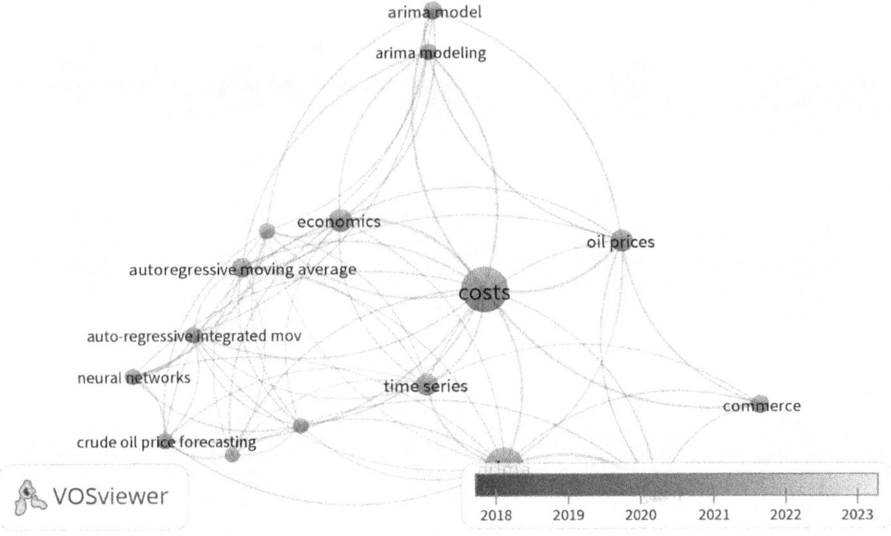

Fig. 5: Thematic analysis of Econometrics model.

Table 2: Summary of implementation of time series model to predict crude oil prices.

Model Type	Author	Assumptions
ARMA Model	[28]	Reviewed models for crude oil price in log levels and in log differences.
Random Walk Model and Mean-reverting Model	[29,28]	A stochastic model has been utilized to examine the time series properties of crude oil. By employing this approach, a deeper understanding of the underlying patterns and trends in crude oil prices can be achieved, providing valuable insights into the market's behavior.
Asymmetric models	[30,31,32,33]	Total output is modelled as function of oil prices, macroeconomic effect of oil price increase and decrease.
Regime-switching models	[34, 30]	Measure the effect of oil production relative to oil prices of G7 countries and output of Greek industries relative to oil price movement.
Time-varying Volatility and GARCH Model	[35,36,37,38, 39,40]	Appraise the influence of oil price volatility on real stock returns.
Vector Autoregressive Model (VAR)	[41,42,43]	interdependencies between multiple time series.
Cointegration Model	[44,45,46]	Key areas of interest when examining crude oil prices include: (1) the relationship between oil prices, economic activity, and inflation, (2) the impact of oil prices on real exchange rates, (3) the correlation between oil prices and GDP, (4) the influence of OPEC's behavior and the coverage rate of OECD's anticipated future oil demand, and (5) the connection between spot prices, futures prices, and stock indices related to crude oil.
The Factor-augmented VAR model (FAVAR)	[47]	The dynamics of oil futures have been investigated utilizing a comprehensive panel dataset, which encompasses global macroeconomic indicators, financial market indices, and a variety of data on energy product quantities and prices. This approach offers a holistic perspective on the factors influencing oil futures.
Structural Vector Autoregression Models (SVAR)	[48]	1. The data used in the study comprises oil production, inventory, price, and an indicator of real economic activity. 2. The researchers discovered that the relationship between oil prices and the US business cycle is contingent upon the type of fundamental shock being examined.

Stochastic Models: Several researchers have studied the time series properties of crude oil price assuming it follows the random walk and mean reverting properties.

Random Walk: model implies that to best forecast for tomorrow's price is today's price. Mean reverting models assume that oil price in the longer run will revert to long term equilibrium price.

Regression Models: Regression models are used to find the relationship between the crude oil prices and exogenous variables. According to the review carried out as part of this study, most of the researchers deployed the ARIMA (autoregressive integrated moving average) model and GARCH (generalized autoregressive conditional heteroskedasticity) family models to forecast the crude oil prices. ARIMA models have helped the researcher in capturing the uncertainties related with oil price volatility. The ARMAX model has been modified to capture the volume and price related to crude oil price. Crude oil price distribution often results in fat tails signifying the shock in the underlying market. To capture the uncertainties prevailing in the market, RSM (response surface methodology) models are used.

Crude oil prices are correlated with multiple commodities market, and different types of crude oil derivatives instrument traded across the globe. To account for the complex interactions between multiple time series generated from the observed interdependencies among various markets that influence crude oil prices, univariate autoregressive models are adapted into vector autoregression (VAR) models. Which models enable the estimation of multiple time series using a single unified model, thus providing a more comprehensive understanding of the relationships at play. The combination of VAR and cointegration models has been utilized to study the intricate connections between macroeconomic, industrial, and financial market variables. This approach allows for a more nuanced analysis of how these factors influence each other and, in turn, crude oil prices. One specific model that leverages large panel datasets to forecast crude oil prices is the Factor-Augmented Vector Autoregression (FAVAR) model. This model incorporates a diverse range of data, including energy product prices, financial market indices, and macroeconomic indicators. The FAVAR model serves as a powerful tool for understanding the interplay between these variables and their collective impact on oil prices. By employing a series of structural equations, the FAVAR model is able to model the relationships between oil prices and other variables. This allows analysts to gain a deeper understanding of how individual factors interact and influence one another. Once these relationships are established, structural parameters can be used to derive economic insights and predictions. Structural vector autoregression (SVAR) models prove invaluable in explaining both observed and predicted fluctuations in the oil market. By incorporating multiple time series and analyzing the relationships between various macroeconomic, industrial, and financial market variables, these models provide a robust framework for understanding the complex factors that drive crude oil prices. As a result, analysts can make better-informed decisions and forecasts based on the insights derived from these sophisticated models.

Regression models has helped in identifying the crude oil price drivers. On further analysis, it has been found that when crude oil prices have been forecasted with the help of fundamental models, they have displayed uncertainty, noisiness, and non-stationary characteristics inherited from factors underlying.

5. Artificial Intelligence Models

The dynamics of crude oil prices are characterized by both linear and nonlinear properties, making traditional time series forecasting methods insufficient for accurately predicting oil prices. In contrast, Artificial Intelligence (AI) models have demonstrated the potential to forecast crude oil prices with higher precision. The present study has focused on analyzing these AI models as a means of circumventing the complex and varied characteristics of crude oil prices and their constituent factors. AI models are capable of capturing complex and nonlinear relationships, making them well-suited for modeling crude oil prices as nonparametric models that map input-output relationships without needing to explore the underlying processes. These models employ sophisticated algorithms that enable them to learn from historical data, extract patterns, and generate accurate forecasts. One of the primary advantages of using AI models is their ability to effectively account for a wide range of factors that may influence crude oil prices. By taking into consideration numerous variables such as geopolitical events, supply and demand dynamics, and global economic conditions, these models can provide a more comprehensive understanding of the forces driving crude oil prices. Furthermore, the use of AI models has demonstrated significant promise in forecasting future trends and potential changes in crude oil prices. As such, they have become an increasingly popular tool for investors, traders, and policymakers seeking to make informed decisions in the highly dynamic and complex crude oil market. In conclusion, the application of AI models provides a powerful means of analyzing crude oil prices, offering more accurate forecasts and a deeper understanding of the market's underlying dynamics. Given the highly volatile and unpredictable nature of crude oil prices, the use of these models is likely to become even more essential in the future for ensuring informed decision-making in the oil market.

Existing AI models such as Artificial Neural Networks (ANNs) and Support Vector Machines (SVMs) (Fig. 3) provide powerful solutions to nonlinear petroleum price prediction. Details related to various AI models has been discussed below.

1. **Neural Network:** Academic researchers have employed standalone models such as Recurrent Neural Networks (RNNs), Long Short-Term Memory (LSTM) models, and their variations to better understand the nonlinear properties of time-series data, including changing variances, asymmetric cycles, and thresholds. Additionally, neural networks have increasingly been used in conjunction with

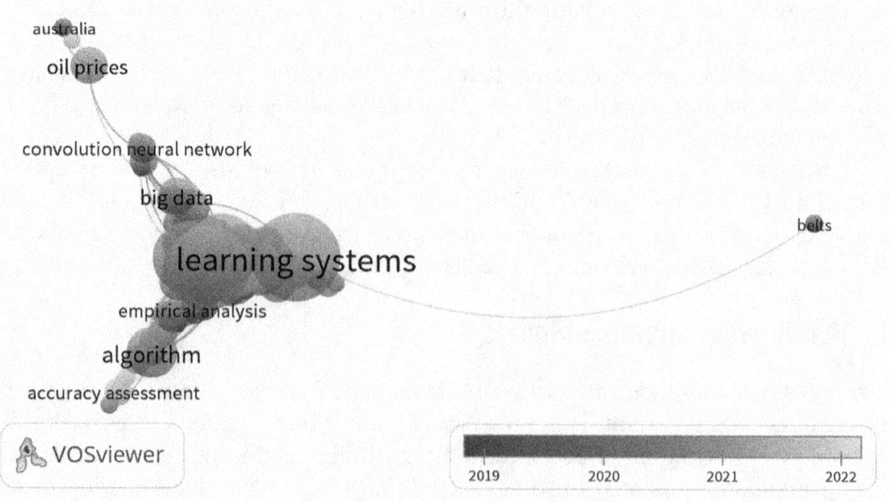

Fig. 6: Thematic analysis of neural networks.

other AI models such as wavelets, genetic algorithms, and SVMs to achieve even greater accuracy.

2. **Support Vector Machine:** SVM is a supervised machine learning technique that can effectively identify intricate patterns within noisy and complex datasets [49]. The algorithm was initially proposed by Vapnik in 1994 and utilizes rule-based methods to determine decision boundaries or hyperplanes based on feature engineering. SVM generates hyperplanes in boundless dimensional spaces and then utilizes them as classifiers or regressors. One of the key advantages of SVM is its ability to handle high-dimensional data and nonlinear relationships, making it particularly useful for studying crude oil prices. This technique can identify the key influencers of crude oil prices by analyzing complex interactions between variables such as global demand, production levels, geopolitical events, and economic factors. SVMs use a set of support vectors or data points to identify the decision boundaries, which are further used to classify or regress the input data. SVMs work by finding the optimal hyperplane that maximizes the margin between the classes. The hyperplane is defined as the decision boundary that separates the two classes with maximum distance. SVM techniques have gained significant traction in recent years due to their effectiveness in handling complex datasets and providing accurate predictions. This makes them an increasingly popular tool for analysts and investors looking to understand and anticipate the dynamics of the crude oil market.
3. **Fuzzy Logic:** "Fuzzy logic is a methodology used for reasoning about ambiguous concepts, making it particularly useful in situations where precise definitions are difficult to establish. This approach is commonly used as a lower-level machine control algorithm and is often combined with neural networks to form ensemble models. In this context, the fuzzy logic parameters are fine-tuned using approximation techniques derived from neural networks, a process known as fuzzy neural networks" [50]. Fuzzy logic algorithms are well-suited for handling complex and dynamic data, particularly in applications where real-world data is noisy or incomplete. The technique is particularly useful in the study of crude oil prices, where numerous factors can influence price fluctuations, including supply and demand dynamics, geopolitical events, and macroeconomic trends. When coupled with neural networks, fuzzy logic models can provide a powerful tool for accurately predicting crude oil prices. The combination of these two methodologies enables analysts to extract meaningful insights from complex data sets and identify patterns that would be difficult or impossible to discern using traditional statistical methods. Overall, fuzzy logic models are becoming an increasingly popular tool for studying complex systems such as crude oil prices, offering a more nuanced and dynamic approach to forecasting that is better suited to the challenges of real-world data analysis.
4. **Wavelet Algorithms**: Wavelet analysis is a widely used technique in financial time series analysis, where it is used to decompose complex data into different scales. This decomposition is achieved by identifying oscillations or wavelets of different frequencies within the data, which can then be used to derive scale-based estimators for variance and correlation. The wavelet approach is

particularly useful for identifying economic shocks that are localized in time, as the different scales act as lagging and leading oscillators that can detect even small fluctuations in the data. By analyzing the properties of wavelet coefficients, analysts can identify significant differences in the data and better understand the underlying dynamics of the system. Wavelet analysis provides a powerful tool for studying financial time series data, allowing analysts to identify complex patterns and relationships that may not be apparent using traditional statistical techniques. This approach is particularly useful for studying crude oil prices, where a range of economic, political, and environmental factors can influence price fluctuations. By using wavelet analysis, analysts can better understand these complex dynamics and make more accurate predictions about future price movements.

6. Hybrid Models

In recent years, the forecasting of crude oil prices has witnessed significant advancements with the emergence of hybrid models. These models synergistically combine various methodologies, such as econometric techniques, machine learning algorithms, and other advanced methods, to capture the intricacies of oil markets [51,52]. One of the main strengths of hybrid models lies in their ability to address the limitations of individual methods. By integrating diverse techniques, they ensure a more comprehensive understanding of market dynamics and offer improved forecasting capabilities [53]. For instance, Zhang et al. [54,55] demonstrated that a hybrid model, combining wavelet analysis and ANNs, significantly outperformed single-method models in predicting crude oil prices. Researchers [56] have explored various combinations of methodologies to develop more accurate and robust hybrid models. [57] proposed a hybrid model based on empirical mode decomposition (EMD) and SVMs, which proved effective in capturing both linear and nonlinear relationships within crude oil price data. Similarly, employed a hybrid model utilizing

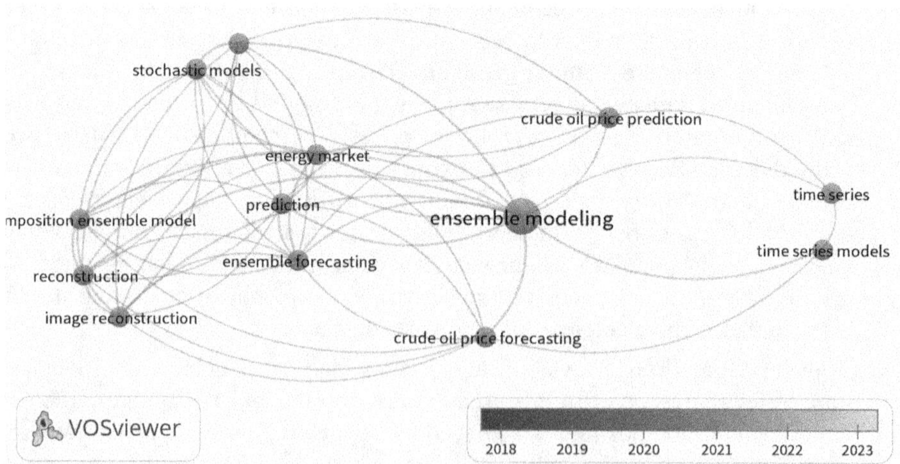

Fig. 7: Thematic analysis in hybrid model.

genetic algorithms and ANNs, achieving promising results in forecasting crude oil prices. These innovative hybrid models have garnered interest from both academia and industry, as they provide valuable insights into the factors influencing oil prices and contribute to more informed decision-making processes (Yüksel et al., 2017). For example, Chen et al. (2020) highlighted the role of geopolitical risk in oil price fluctuations, using a hybrid model that incorporated wavelet-based techniques and machine learning algorithms (Fig. 3).

7. Comparative Analysis

There is no standalone model which is able to forecast the crude oil price with better accuracy. Although AI and hybrid models have shown promises in tackling the nonlinearity, non-stationary and time-varying structures of crude oil prices, they are yet to be accepted as the standard model among the economists (Table 3). Similarly, recent trends in crude oil prices have widened the spread between WTI and Brent crude oil class thus making it difficult to conclude products from which market can be for the price discovery among both.

Hybrid models enhance forecasting performance by merging multiple techniques, addressing the limitations of individual models. These models combine diverse approaches that strongly disagree with one another, effectively reducing

Table 3: Comparative analysis of forecasting models.

Model Name	Model Type	Assumptions	Limitations
Fundamental Model	Mathematical	Probabilistic models based on the underlying economic theories	- Theoretical assumptions and condition proposed as a part of model
AR, MA and ARMA Model	Stochastic	Random walk, White noise and stationary, linear properties	- Presence of trend, non-stationary, and violation of Gaussian Distribution [64]
ARIMA and GARCH Models	Regression	Gaussian Distribution, linear properties	- Obtaining the reliable historical data related to explanatory variables, - nonlinearity and violation of Gaussian distribution [58]
Neural Network	Artificial Intelligence	Non Parametric, Mimicking the behavior of neurons and perceptons	Over fitting of data., Optimization of model related to no. of hidden layers and activation function [59]
Support Vector Machine	Artificial Intelligence	Division of vectors in hyper planes based on structural risk minimization principle, convex optimization	Poised problem in training large data set, training of convex optimization problem [64]
Fuzzy Logic	Artificial Intelligence	Reinforcement Learning and Learning Rate	Optimization of learning velocity of model [58]
Wavelet	Artificial Intelligence	Morlet Function, Learning Rate,	Vanishing and Exploding gradient descent problem [60]

generalization variance or error. The goal is to minimize the risk of relying on a single, inadequate model, and to achieve higher accuracy in predictions. When the underlying process is difficult to determine or a single model fails to capture all-time series characteristics, hybrid models become especially valuable. The assumption driving this approach is that the true data-generating process may remain unidentified, or a solitary model might not fully represent complex time-series dynamics. Practically, hybrid models can be constructed by blending models founded on distinct principles, such as statistical, machine learning, or econometric models. This fusion leads to more precise and dependable forecasts, as well as a deeper comprehension of underlying patterns and relationships within the data. In summary, hybrid models serve as a crucial instrument for elevating forecasting accuracy and reducing risk, specifically in the realm of crude oil price analysis. By uniting diverse models, these hybrids successfully address individual limitations and provide a more comprehensive understanding of data trends.

8. Conclusion

Determination of crude oil prices hinges on two tenets. The first tenet is Price discovery mechanism. It states that price discovery for crude oil is related to the stochastic discovery factor, which can affect crude oil prices. Crude oil often jumps up and down based on the cue received from the global markets—the infusion of the news helps in generating the momentum and volatility at the exchanges. Two class of models have been proposed for price exploration in the crude oil market: Permanent Transitory Models and Information Share Model which focus various components of Vector error correction model (VECM). Based on the extensive literature review conducted, it has been discovered that the majority two-market plays helps in crude price discovery: WTI and Brent. It was evident that till 2010, WTI and Brent crude oil market was highly cointegrated. However, afterwards the WTI market has contributed majorly in understanding demand (recently it has been observed that WTI has reached 10$/barrel when demand for crude oil has gone drastically down due to global lockdown enforced due to spread of coronavirus).

The second tenet is related to the estimation of prices based on mean reversion and time-series momentum displayed by crude oil prices. Chronologically, Fundamental models were favored till 1990 to understand the contribution of various actors like OPEC and world events towards crude oil prices. They were mathematical models based on microeconomics theory. Upon financialization of crude oil prices due to the collapse of the OPEC-administered pricing system in 1985, econometrics models have come to the fore. Techniques like ARIMA, GARCH, and ARCH have helped the economists in understanding the linear properties of time series in a better way. However, these models fail to comprehend the nonlinear properties of time-series like time changing variance, asymmetric cycles, and higher-moment structures. Afterwards, due to the rise in computational and data storage capacity of the systems, researchers have started using AI family of models. It has helped them in capturing the nonlinear aspects of crude oil price movements. Contemporary studies, nonetheless, have made ensemble techniques like fuzzy network or wavelets as the cornerstone of crude oil price estimation techniques which were successful in trapping the mean revision and momentum properties of time series.

This review examines the prevailing trends in crude oil price estimation techniques used by various researchers. A comparative analysis of the performance of different models has been conducted. Researchers have suggested integrating neural networks with traditional methods, such as support vector regression and wavelets, through an ensemble approach to overcome the limitations of single models. Furthermore, the impact of various factors, including geopolitical and economic events, is emphasized as a constraint to understanding crude oil prices. The study identifies a research gap in developing an intelligent system that can comprehend the dynamics of crude oil prices and optimize model performance. This system should be based on fundamental principles to address the current limitations of crude oil price estimation techniques.

References

[1] U.S. Energy Information Agency. Annual Energy Outlook 2019 with projections to 2050. *Annual Energy Outlook 2019 with Projections to 2050*, 44(8): 1–64, 2019. doi: DOE/EIA-0383(2012) U.S.

[2] Bassam. Fattouh. *An Anatomy of the Crude Oil Pricing System*. Oxford Institute for Energy Studies. 2011, 83pp.

[3] A Ranking of Traded Commodities by Liquidity. https://www.thebalance.com/the-most-actively-traded-commodities-809314 (accessed May 29, 2020).

[4] Failler, P., and Liu, Y. The complexity of energy price fluctuations. *Energies*. 16(5): 2354, Mar. 2023. doi: 10.3390/EN16052354.

[5] Donthu, N., Kumar, S., Mukherjee, D., Pandey, N., and Lim, W.M. How to conduct a bibliometric analysis: An overview and guidelines. *J. Bus. Res.*, 133: 285–296, Sep. 2021. doi: 10.1016/j.jbusres.2021.04.070.

[6] Goyal, K., and Kumar, S. Financial literacy: A systematic review and bibliometric analysis. *International Journal of Consumer Studies*, 45(1). Blackwell Publishing Ltd., pp. 80–105, Jan. 01, 2021. doi: 10.1111/ijcs.12605.

[7] Donthu, N., Kumar, S., and Pattnaik, D. Forty-five years of Journal of Business Research: A bibliometric analysis. *J. Bus. Res.*, 109: 1–14, Mar. 2020. doi: 10.1016/j.jbusres.2019.10.039.

[8] Donthu, N., Kumar, S., Sahoo, S., Lim, W.M., and Joshi, Y. Thirty years of product and brand management research: A retrospective review of the Journal of Product and Brand Management using bibliometric analysis" *Journal of Product and Brand Management*, 2022. doi: 10.1108/JPBM-02-2022-3878.

[9] Tandon, A., Kaur, P., Mäntymäki, M., and Dhir, A. Blockchain applications in management: A bibliometric analysis and literature review. *Technol. Forecast Soc. Change*, 166, May 2021. doi: 10.1016/j.techfore.2021.120649.

[10] Kumar, S., and Goyal, N. Behavioural biases in investment decision making: A systematic literature review. *Qualitative Research in Financial Markets*, 7(1): 88–108, Feb. 2015. doi: 10.1108/QRFM-07-2014-0022.

[11] Kumar, S., Colombage, S., and Rao, P. Research on capital structure determinants: A review and future directions. *International Journal of Managerial Finance*, 13(2). Emerald Group Publishing Ltd., pp. 106–132, 2017. doi: 10.1108/IJMF-09-2014-0135.

[12] Thayyib, P.V. State-of-the-Art of Artificial Intelligence and big data analytics reviews in five different domains: a bibliometric summary, *Sustainability (Switzerland)*, 15(5). MDPI, Mar. 01, 2023. doi: 10.3390/su15054026.

[13] Zhang, Q., Hu, Y., Jiao, J., and Wang, S. Exploring the trend of commodity prices: a review and bibliometric analysis. *Sustainability (Switzerland)*, 14(15). MDPI, Aug. 01, 2022. doi: 10.3390/su14159536.

[14] Blitzer, C., Meeraus, A., and Stoutjesdijk, A. A dynamic model of OPEC trade and production. *J. Dev. Econ.*, 2(4): 319–335, 1975. doi: 10.1016/0304-3878(76)90006-7.

[15] Gately, D. OPEC: Retrospective and prospects 1973–1990. *Eur. Econ. Rev.*, 21(3): 313–331, 1983. doi: 10.1016/0014-2921(83)90095-8.

[16] Amano, A. A small forecasting model of the world oil market. *J. Policy Model*, 9(4): 615–635, 1987. doi: 10.1016/0161-8938(87)90016-0.

[17] Kaufmann, R.K., Dees, S., Gasteuil, A., and Mann, M. Oil prices: The role of refinery utilization, futures markets, and nonlinearities. *Energy Econ.*, 30(5): 2609–2622, 2008. doi: 10.1016/j.eneco.2008.04.010.

[18] Kaufmann, R.K., Dees, S., Karadeloglou, P., and Sánchez, M. Does OPEC Matter? An econometric analysis of oil prices. Published by: international association for energy economics stable URL: http://www.jstor.org/stable/41323358 REFERENCES Linked references are available on JSTOR for this article: You may need. 25(4): 67–90, 2004.

[19] Baldwin, N., and Prosser, R. World oil market simulation. *Energy Econ.*, 10(3): 185–198, 1988. doi: 10.1016/0140-9883(88)90002-3.

[20] Deam, R.J. World energy modelling: The development of Western European oil prices. *Energy Policy*, 1(1): 21–34, 1973. doi: 10.1016/0301-4215(73)90019-0.

[21] Nordhaus, William, D. The of energy allocation resources. *Brookings Pap. Econ. Act, 2*, January, pp. 1–22, 1972.

[22] Kalymon, B.A. Economic incentives in OPEC oil pricing policy. *J. Dev. Econ.*, 2(4): 337–362, 1975. doi: 10.1016/0304-3878(76)90007-9.

[23] Cremer, J., and Weitzman, M.L. OPEC and the monopoly price of world oil. *Eur. Econ. Rev.*, 8(2): 155–164, 1976. doi: 10.1016/0014-2921(76)90010-6.

[24] Horn, M. OPEC's optimal crude oil price,. *Energy Policy*,. 32(2): 269–280, 2004. doi: 10.1016/S0301-4215(02)00289-6.

[25] Salant, S.W. Imperfect competition in the world oil market: A computerized Nash-Cournot model. 3(2): 223–414, August 2015, 1982.

[26] Alsmiller, R.G., Horwedel, J.E., Marshalla, R.A., Nesbitt, D.M., and Haas, S.M. A model of the world oil market with an OPEC cartel. *Energy*, 10(10): 1089–1102, 1985. doi: 10.1016/0360-5442(85)90024-6.

[27] R. April. "Pricing Policies for a Two-P.4Rt Exhac'Stible Resource Cartel." no. April, 1976. AQ: Please check. This ref. is as follows:. Robert. S. Pricing policies for a two-part exhaustible resource cartel: The case of OPEC. *European Economic Review*, 8(2): 139–154, April 1976.

[28] Alquist, R., Kilian, L., and Vigfusson, R.J. *Forecasting the Price of Oil*, Vol. 2. Elsevier B.V., 2013. doi: 10.1016/B978-0-444-53683-9.00008-6.

[29] Maslyuk, S., and Smyth, R. Unit root properties of crude oil spot and futures prices. *Energy Policy*, 36(7): 2591–2600, 2008. doi: 10.1016/j.enpol.2008.03.018.

[30] Papapetrou, E. Oil Price asymmetric shocks and economic activity: The case of greece. *Austrian Association for Energy Economics 10th IAEE European Energy Conference from 7–10 September 2009 in Vienna*, pp. 1–21, 2009.

[31] Hamilton, J.D. What is an oil shock? *J. Econom.*, 113(2): 363–398, 2003. doi: 10.1016/S0304-4076(02)00207-5.

[32] Hamilton, J.D. W15002. 2009.

[33] Kilian, L., and Vigfusson, R.J. Pitfalls in estimating asymmetric effects of energy price shocks. *International Finance Discussion Paper*, 2009(970): 1–56, 2009. doi: 10.17016/ifdp.2009.970.

[34] Cologni, A., and Manera, M. Oil prices, inflation and interest rates in a structural cointegrated VAR model for the G-7 countries. *Energy Econ.*, 30(3): 856–888, 2008. doi: 10.1016/j.eneco.2006.11.001.

[35] Sadorsky, P. Oil price shocks and stock market activity. *Energy Econ.*, 21(5): 449–469, 1999. doi: 10.1016/S0140-9883(99)00020-1.

[36] Sadorsky, P. The macroeconomic determinants of technology stock price volatility. *Review of Financial Economics*, 12(2): 191–205, 2003. doi: 10.1016/S1058-3300(02)00071-X.

[37] Sadorsky, P. Modeling and forecasting petroleum futures volatility. *Energy Econ.*, 28(4): 467–488, 2006. doi: 10.1016/j.eneco.2006.04.005.

[38] Morana, C. A semiparametric approach to short-term oil price forecasting. *Energy Econ.*, 233: 325–338, 2001. doi: 10.1016/S0140-9883(00)00075-X.

[39] Adrangi, B., Chatrath, A., Dhanda, K.K., and Raffiee, K. Chaos in oil prices? Evidence from futures markets. *Energy Econ.*, 23(4): 405–425, 2001. doi: 10.1016/S0140-9883(00)00079-7.

[40] Sadeghi, M., and Shavvalpour, S. Energy risk management and value at risk modeling. *Energy Policy*, 34(18): 3367–3373, 2006. doi: 10.1016/j.enpol.2005.07.004.

[41] Cunado, J., and Perez de Gracia, F. Oil prices, economic activity and inflation: Evidence for some Asian countries. *Quarterly Review of Economics and Finance*, 45(1): 65–83, 2005. doi: 10.1016/j.qref.2004.02.003.

[42] Chen, S.S., and Chen, H.C. Oil prices and real exchange rates. *Energy Econ.*, 29(3): 390–404, 2007. doi: 10.1016/j.eneco.2006.08.003.

[43] Lardic, S., and Mignon, V. Oil prices and economic activity: An asymmetric cointegration approach. *Energy Econ.*, 30(3): 847–855, 2008. doi: 10.1016/j.eneco.2006.10.010.

[44] Chevillon, G., and Rifflart, C. Physical market determinants of the price of crude oil and the market premium. *Energy Econ.*, 31(4): 537–549, 2009. doi: 10.1016/j.eneco.2009.01.002.

[45] Gülen, S.G. Efficiency in the crude oil futures market. *Journal of Energy Finance & Development*, 3(1): 13–21, 1998. doi: 10.1016/s1085-7443(99)80065-9.

[46] Henriques, I., and Sadorsky, P. Oil prices and the stock prices of alternative energy companies. *Energy Econ*, 30(3): 998–1010, 2008. doi: 10.1016/j.eneco.2007.11.001.

[47] Zagaglia, P. Macroeconomic factors and oil futures prices: A data-rich model. *Energy Econ.*, 32(2): 409–417, 2010. doi: 10.1016/j.eneco.2009.11.003.

[48] Lippi, F., and Nobili, A. Oil and the macroeconomy: A quantitative structural analysis. *J. Eur. Econ. Assoc.*, 10(5): 1059–1083, 2012. doi: 10.1111/j.1542-4774.2012.01079.x.

[49] Bao, Y., Yang, Y., Xiong, T., and Zhang, J. A comparative study of multi-step-ahead prediction for crude oil price with support vector regression. *Proceedings of the 4th International Joint Conference on Computational Sciences and Optimization, CSO* 2011, pp. 598–602, 2011. doi: 10.1109/CSO.2011.70.

[50] He, K., Xie, C., Chen, S., and Lai, K.K. Estimating VaR in crude oil market: A novel multi-scale nonlinear ensemble approach incorporating wavelet analysis and neural network. *Neurocomputing*, 72(16–18): 3428–3438, 2009. doi: 10.1016/j.neucom.2008.09.026.

[51] Dai, Z., and Kang, J. Bond yield and crude oil prices predictability. *Energy Econ.*, 97: 105205, May 2021. doi: 10.1016/J.ENECO.2021.105205.

[52] Tiwari, A.K., Mishra, B.R., and Solarin, S.A. Analysing the spillovers between crude oil prices, stock prices and metal prices: The importance of frequency domain in USA. *Energy*, 220: 119732, Apr. 2021. doi: 10.1016/J.ENERGY.2020.119732.

[53] Gong, X., Guan, K., Chen, L., Liu, T., and Fu, C. What drives oil prices?—A Markov switching VAR approach. *Resources Policy*, 74: 102316, Dec. 2021. doi: 10.1016/J.RESOURPOL.2021.102316.

[54] Zhang, Y.J., and Zhang, L. Interpreting the crude oil price movements: Evidence from the Markov regime switching model. *Appl. Energy*, 143: 96–109, Apr. 2015. doi: 10.1016/j.apenergy.2015.01.005.

[55] Zhang, P.G. Time series forecasting using a hybrid ARIMA and neural network model. *Neurocomputing*, 50: 159–175, Jan. 2003. doi: 10.1016/S0925-2312(01)00702-0.

[56] Alkhammash, E.H. An optimized gradient boosting model by genetic algorithm for forecasting crude oil production. *Energies (Basel)*, 1517, Sep. 2022. doi: 10.3390/en15176416.

[57] Dang-Ha, T.H., Bianchi, F.M., and Olsson, R. Local short term electricity load forecasting: Automatic approaches. *Proceedings of the International Joint Conference on Neural Networks*, Vol. 2017 May, pp. 4267–4274, Jun. 2017. doi: 10.1109/IJCNN.2017.7966396.

[58] Xie, W., Yu, L., Xu, S., and Wang, S. A new method for crude oil price forecasting based on support vector machines. *Lecture Notes in Computer Science (including subseries Lecture Notes in Artificial Intelligence and Lecture Notes in Bioinformatics), 3994* LNCS-IV, pp. 444–451, 2006. doi: 10.1007/11758549_63.

[59] Pang, Y. Forecasting the crude oil spot price by wavelet neural networks using OECD Petroleum Inventory Levels. *New Mathematics and Natural Computation*, 7(2): 281–297, 2011. doi: 10.1142/s1793005711001937.

CHAPTER 19

Effective Temperature Prediction for An Enhanced Climate Forecast System

Mohammed Mueen Pasha M[1,*] and *Sibu Cyriac*[2]

1. Introduction

Ever since the first industrial revolution, there has been a subtle temperature change. The transition to new manufacturing processes in conjunction with the surge in population has a negative consequence on the earth's atmosphere. Climate change has been identified as the most crucial environmental issue of this century, and it has sparked heated discussions [1]. Temperature is the most common metric to evaluate the change in climate/global warming. It is anticipated that climate change will result in an adverse and enduring impact on the ecosystem. Weather forecasting today extensively depends on conventional methodologies and requires complex and complicated infrastructure [2]. Prime problems concern quality of acquired data, timeliness, availability, reliability, and usability constraints on forecast preparation.

In a recent interview, the Commonwealth Secretary-General—Baroness Patricia Scotland has warned the world that about 42 countries across the globe are on the verge of disappearance if no stern measures against climate change are put to practice [15]. Similarly, a report by the United Nations and global scientific partner organizations provides details on the rapid and large-scale rise in climate change factors [12].

[1] Assistant Professor, St. Joseph's University, Bengaluru, India.
[2] Department of Computer Science and Applications, CHRIST University Bengaluru, India.
* Corresponding author: mohammedmueen7@gmail.com

Currently the data captured from the satellite, Radiosondes, super computers are fed into a complicated computational system. Though the data at times isn't quite reliable but we have been trying to improve the input mechanisms. Less effective methodologies are deployed to handle the dynamicity of the temperature in the weather data. Millions of calculations need to be performed and often forecasters get blamed for different forecast. Furthermore, inter-comparison between studies is difficult due to the lack of a standard dataset and appraisal metrics [9–13]. The time and expense incurred is exponentially high when compared to artificial intelligence-(AI) based predictions. Current advancements in machine learning (ML), especially deep learning (DL) pave a way for researchers to build and deploy models that would assist weather forecast. To propose a strong data-driven approach, appropriate feature engineering techniques are to be used. A specific set of supervised and unsupervised algorithms will be applied to the pre-processed dataset. In this work, we seek to consider the performance of our models for forecasting of the weather [6–8]. The model works on the assumption that the air temperature of tomorrow is dependent on today's air temperature.

2. Methods

2.1 Dataset

The Berkeley Earth Surface Temperature study dataset comprises of 1.6 billion temperature records. It very well may be appropriately bundled and has invigorating subsets (like nations, urban communities, and so on). They have posted the stockpile data for the changes. They have shielded procedures which have environment perceptions from a brief time interval to be secured in this dataset. There various records include Worldwide Land/Sea and Land Temperatures document from 1750–2015.

The record contains a short rundown of the progressions in Earth's worldwide average surface temperature assessed by joining the Berkeley Earth land-surface temperature field with an interpolated rendition of the HadSST4 sea temperature field. Other archives incorporate—Global Average Land Temperature record for Country, Global Average Land Temperature record for State, Global Land Temperatures record for Major City, Global Land Temperatures record for City [2]. The dataset has time series data which is acquired consistently at an equal time space.

2.1.1 Algorithm, methods, and models

(a) Linear Interpolation

Interpolation is a technique in python to impute missing values between any two given data points. This is a simple and effective method to use on series data, particularly linear interpolation is the default method used. In this approach, the missing data is increasingly plotted sequentially in a straight line.

(b) KNN

KNN (K-Nearest Neighbor) is a widely used imputer for missing values. The missing values are filled by taking the mean of the nearest neighboring data points. It uses a distance measure, usually Euclidian distance to locate the neighboring data points in a dataset. The 'k' samples determine how close the features are in a dataset by using the mean value.

(c) Cubic Spline

Our dataset is coarsely equally spaced and would result in Runge's phenomena. Polynomial interpolation is not a good choice here, thus spline is used to reduce the oscillation between data points. The resultant interpolating polynomial is smoother and has minimized error [13].

(d) Lag features

Lag feature is a powerful feature engineering technique which takes a point in time t and considers previous points like t-1, t-2, etc. The lag could be one step or even multiple, t-1 (lag 1), t-2 (lag 2), the seasonality of features can be determined with autocorrelation and fed as feature set. The feature set reveals new patterns to learn since the lagged features are of future data points.

(e) Rolling window features

This is a statistical technique for times series forecast. It is pretty much similar to expanding window features but here the window size remains fixed and allows reverse estimation from the recent observations. It can be used in addition with lag features due to its favorability with the old lag values are higher. Thus the feature set generated looks like a 'rolling window' [6].

(f) Auto Correlation Function (ACF) and Partial Auto Correlation Function (PACF)

Correlation helped us understand the relationship between the variables within the dataset. Correlation can be positive, negative or even neutral [3]. With the help of methods like ACF and PACF we can determine the temporal dynamics of single time series. Different lag lengths are considered in ACF whereas PACF uses previous lag lengths (shorter) or, in other words, the intervening observations are removed.

(g) Approximate Entropy (forecasting ability of a time series)

This can be used to quantify the regularity and unpredictability of fluctuations in a time series [14].

(h) SARIMAX

SARIMAX (Seasonal Autoregressive Integrated Moving Average with eXogenous factors) is a commonly used algorithm to operate on time-series data. Based on the nature of the data we chose SARIMAX to operate on data for forecasting. The ARIMA model does not support seasonality and is preferred for univariate dataset [12]. This is a class of models that comprehend its own lags and the lagged forecast errors which help us to determine the forthcoming values [4].

The Model can be dissected into 2, namely, Auto Regression (AR) and Moving Average (MA). Seasonal differencing is like standard differencing, yet, rather than deducting sequential terms, we deduct the value from past season. The model is denoted as ***SARIMA(p,d,q)x(P,D,Q)***, where, P, D and Q are Seasonal Auto Regression (SAR), order of seasonal differencing and Seasonal Moving Average(SMA) terms, respectively, and 'x' is the count of occurrences of the time series. p = non-seasonal autoregressive (AR) order, d = non-seasonal differencing, q= non-seasonal moving average (MA) order, P = seasonal AR order, D = seasonal differencing, Q = seasonal MA order, and S = length of repeating seasonal pattern. By adding those seasonal AR and seasonal MA components, SARIMA solves the seasonality problem [13]. Hyperparameter optimization or simply hyperparameter tuning is used to pick the best among the parameters to train the algorithm. We chose **Grid Search** to hyper tune the parameters. In grid search, **we don't have an option of number of iterations. Now to estimate the number of iterations we have to take the multiply count of each feature in parameter grid** [11].

(i) Facebook Prophet

Facebook's Prophet is an open source library that is very powerful and easy to use. It is used on the dataset to automate forecasting by picking a best set of features for the model (hyper-parameter tuning). The library is rich in methods and allows tuning of efficient parameters. It handles the trend and non-stationarity in the best way possible [5].

(j) LightGBM

LightGBM is best used for handling large-scale data, which is a gradient booting framework that deploys tree-based algorithms. Here the tree grows horizontally and not vertically.

(k) Long Short-Term Memory (LSTM)

Classical linear methods are a bit complicated and difficult to deploy multivariate dataset. LSTM is a type of Recurrent Neural Network (RNN) and can be seamlessly used to operate on large dataset with vivid nature [7]. It uses back propagation and has memory blocks that connect layers together. It is easy to use and reduces the size of the network to a larger extent.

(l) Bayesian Optimization: Automate Hyperparameter Tuning

Domain space, Optimization algorithm and objective function are the three prime factors for this methodology. It aims to give minimum of the output function by evaluating the function with lowest possible probability value.

(m) Deep Belief Network

A deep belief network (DBN) is made of a sequence of confined Bolzmann machines which are related in a particular order. The end result of the 'output' layer of the Boltzmann device is then furnished as an entrance to the following Boltzmann device within the sequence, that's then trained till convergence, and so on, till the entire network has been trained [8].

3. Design/Prototype

A solid data-driven approach is proposed to make temperature predictions using neural networks. The proposed system considers the possible downsides of the existing works and seeks to bridge the gap between current and the goal condition. Fig. 1 depicts the detailed procedure of experiment and suggests apprehensions required to improve the current models.

The block diagram of the model creation and prediction is divided into 4 primary stages which subsume the arbitrary processes. Stage I revolves around sources and preparedness of dataset to be process. The data could be directly transferred from a stored repository or from the processed physical layout (usually structured data from hardware reading). Stage II includes the pre-processing of the dataset using different mechanisms suited for the nature of the data. Feature engineering using Lag features and Rolling window is done [5]. Linear interpolation, KNN, and Cubic Spline are exclusively used to preprocess the data. In the latter part of the stage we divide the preprocessed data into training, validation, and testing. Feature tuning is specifically done using Bayesian optimization and Grid search. In Stage III, the respective algorithms such as SARIMAX, Prophet, LightGBM, LSTM, and DBN are deployed and the model is built to test the performance of the data. L1, L2, and Batch regularization is done as and when required to add extra information to the performance of the algorithms. Stage IV covers the overall visualization using vivid charts and alongside predictions are possible as the model is trained and tested based on the dataset fed. The functionality of the entire system is shown in Fig. 2.

The picture above represents the architecture outlook of the entire system with abstract inputs to the explicit outputs of the given problem. The physical source here represents the layout of actual hardware that is used to capture data directly.

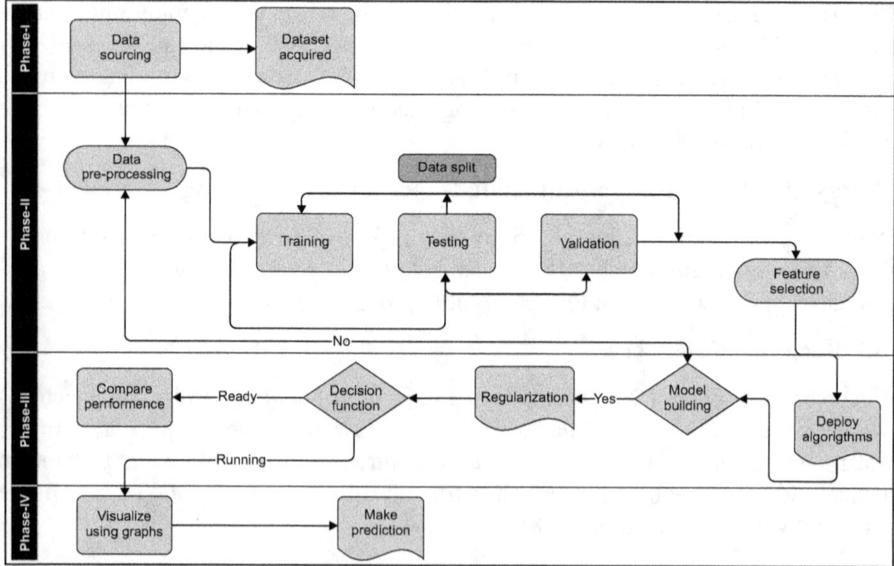

Fig. 1: Block diagram of model creation and preparation.

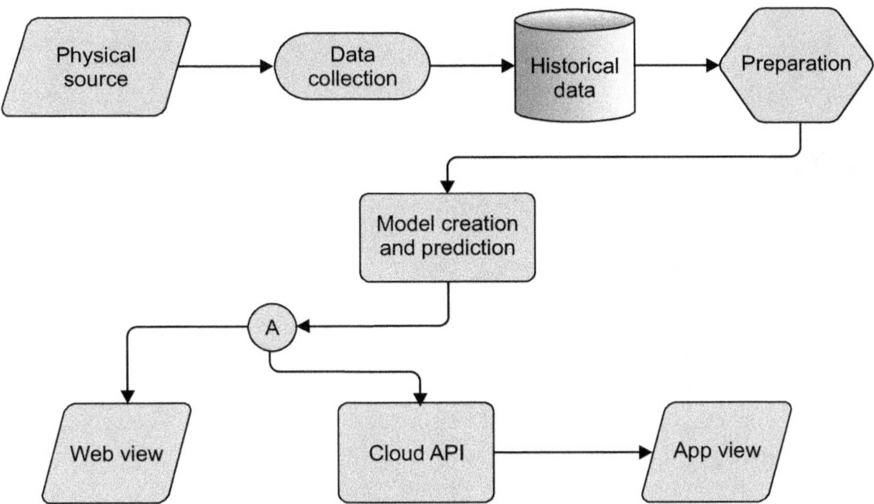

Fig. 2: Proposed architecture of the system.

The data is then regressively filtered and pushed further to the following steps. The model creation and prediction is where the entire 4-stage process takes place (Fig. 1). The results are then made available for end users through web page or on mobile application. The output data is stored in cloud storage are, maybe a simple storage service (S3) using a bucket.

4. Results and Discussion

The results of the experiment were quite reinforcing, as it allowed us to understand the benefits of using appropriate methods to preprocess the data. It also showed excellent performance when algorithms were implemented on it. All these implementations are done in Google Colab, and Kaggle notebook using available libraries in TensorFlow and Keras.

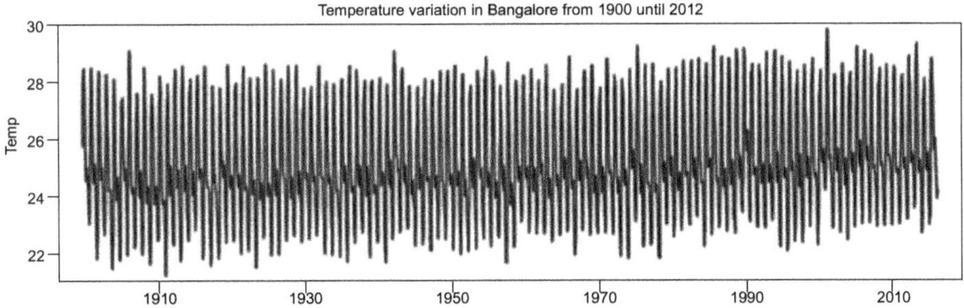

The graph above depicts the temperature variation in Bangalore from 1900. It is quite evident that there exists a sequential pattern in the variation and we can easily have a line of separation. This is a part of Exploratory Data Analysis (EDA) of the preprocessed dataset.

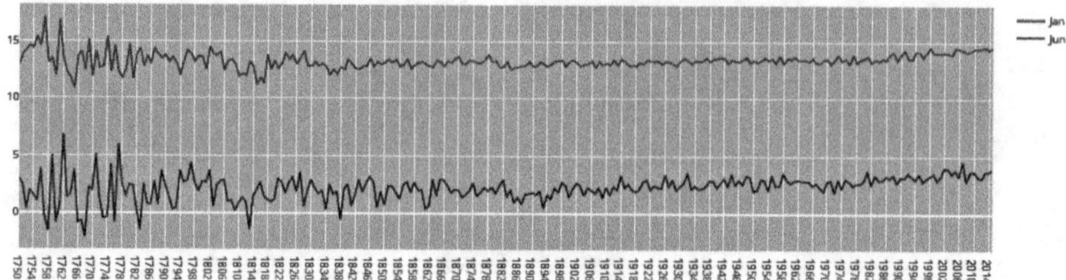

Comparison of any two months showcase the increase in temperature over the years. The trend of increasing temperature is a key aspect to understand the underlying variation in the pattern. To understand the pattern in the long run, we need to consider the seasonality and stationarity of the dataset. This helped us to distinguish between medium range and the long range forecast.

When did Global warming begin?

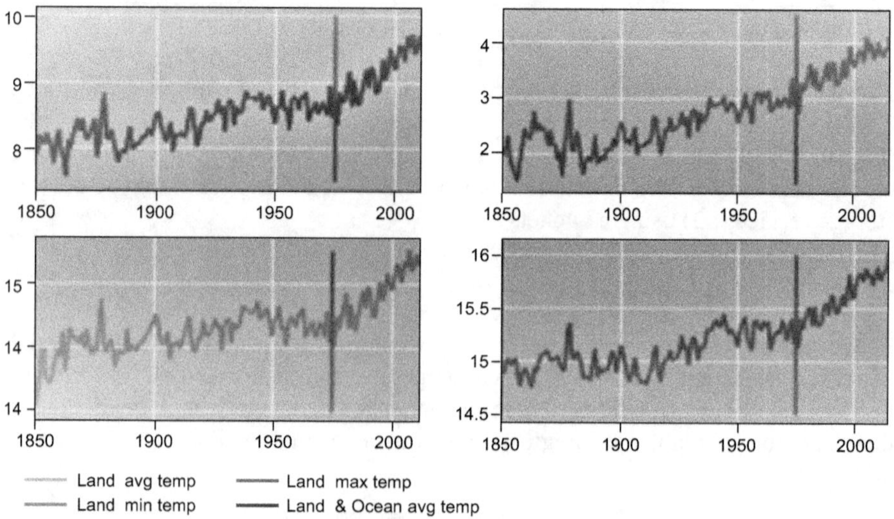

The pictorial presentation of land segment of the dataset is a clean indication of increase in the temperature, the causes of which could differ based on the adventitious events.

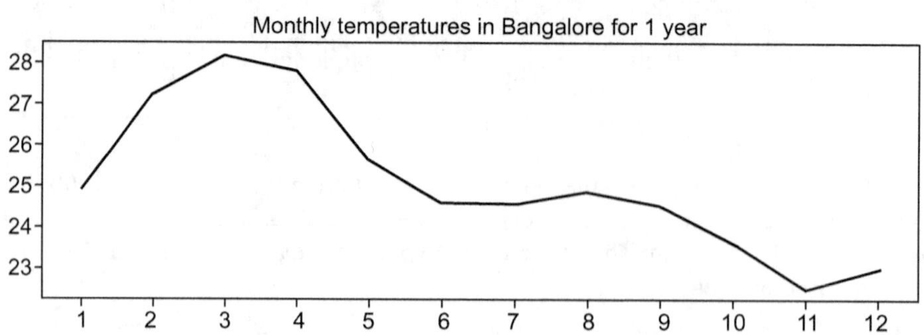

SARIMAX helped us to predict the monthly temperature in Bangalore for next 12 months. However, it was not that good of a choice for long-term forecasting. We had to choose neural network methods for it. The prediction with SARIMAX was precise due to the preprocessing and later tuning using the grid search method. Our model was successful in forecasting the previous 12 and upcoming 12 months' temperature/weather.

Baseline model and RMSE

```
test_baseline = test['Temp'].shift()

test_baseline[0] = test['Temp'][0]

rmse_test_base = measure_rmse(test['Temp'],test_baseline)
rmse_test_extrap = measure_rmse(test['Temp'], test['Pred'])

print(f'The baseline RMSE for the test baseline was {round(rmse_test_base,2)} celsius degrees')
print(f'The baseline RMSE for the test extrapolation was {round(rmse_test_extrap,2)} celsius degrees')
print(f'That is an improvement of {-round((rmse_test_extrap/rmse_test_base-1)*100,2)}%')

The baseline RMSE for the test baseline was 1.01 celsius degrees
The baseline RMSE for the test extrapolation was 0.68 celsius degrees
That is an improvement of 33.4%
```

Our model had a strong baseline model and the RMSE rate was significantly reduced because of the data handling before deployment. The results show an improvement of 33.45% from the existing model implementation.

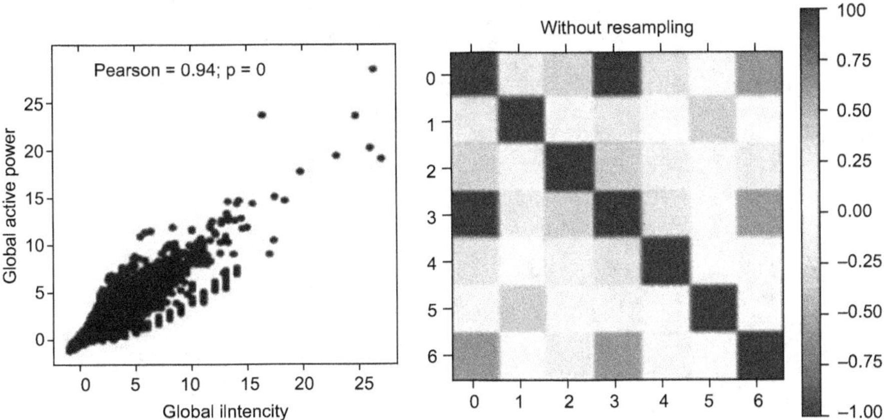

From above two plots it is seen that "Global Temperature By City" and "Global Temperature By Country" are correlated. But "Ocean Temperature" and "Global Temperature By City' are less correlated. This is an important observation for Ml purpose. Then the resampled features were gathered with the help of certain feature engineering methods like rolling window. The LSTM network was deployed on a pretrained set and the lag feature boosted the performance of the same. The ACF and PACF let us estimate the working of the model by considering the Auto encoding entropy. The lag feature correlation (cross validation) can be seen in the following picture.

```
   var1(t-1)  var2(t-1)  var3(t-1)  var4(t-1)  var5(t-1)  var6(t-1)  \
1   0.636816   0.295738   0.337945   0.631157        0.0   0.011366
2   0.545045   0.103358   0.335501   0.541487        0.0   0.144652
3   0.509006   0.110073   0.283802   0.502152        0.0   0.030869
4   0.488550   0.096987   0.315987   0.481110        0.0   0.000000
5   0.455597   0.099010   0.434417   0.449904        0.0   0.008973

   var7(t-1)   var1(t)
1   0.782418  0.545045
2   0.782676  0.509006
3   0.774169  0.488550
4   0.778809  0.455597
5   0.798917  0.322555
```

```
Epoch 2/10
151/151 [==============================] - 24s 160ms/step - loss: 0.4386 - acc: 0.8604 - val
_loss: 0.1476 - val_acc: 0.9590
Epoch 3/10
151/151 [==============================] - 24s 162ms/step - loss: 0.3276 - acc: 0.8960 - val
_loss: 0.1146 - val_acc: 0.9674
Epoch 4/10
151/151 [==============================] - 24s 160ms/step - loss: 0.2716 - acc: 0.9144 - val
_loss: 0.1002 - val_acc: 0.9705
Epoch 5/10
151/151 [==============================] - 24s 162ms/step - loss: 0.2458 - acc: 0.9228 - val
_loss: 0.0858 - val_acc: 0.9750
Epoch 6/10
151/151 [==============================] - 25s 163ms/step - loss: 0.2242 - acc: 0.9315 - val
_loss: 0.0804 - val_acc: 0.9762
Epoch 7/10
151/151 [==============================] - 24s 158ms/step - loss: 0.2076 - acc: 0.9361 - val
_loss: 0.0733 - val_acc: 0.9781
Epoch 8/10
151/151 [==============================] - 25s 164ms/step - loss: 0.1981 - acc: 0.9387 - val
_loss: 0.0705 - val_acc: 0.9781
Epoch 9/10
151/151 [==============================] - 24s 161ms/step - loss: 0.1886 - acc: 0.9423 - val
_loss: 0.0647 - val_acc: 0.9817
Epoch 10/10
151/151 [==============================] - 25s 163ms/step - loss: 0.1713 - acc: 0.9462 - val
_loss: 0.0594 - val_acc: 0.9836
```

The LSTM network produced an accuracy of **0.9836,** which was later analyzed to check for over fitting of the model. The regularization methods aided the process. L1, L2, and Batch regularization is done as and when required to add extra information to the performance of the algorithms. The results are operated on the historical data to forecast the temperature from 1960 to 2010 (50 years).

As stated earlier, FB's Prophet is really a powerful open source that unveiled the hidden insights and patterns. As a result, the model loss was exponentially reduced and the RMSE showed up small. The very reason for this behavior of the model is

FB prophet model - multivariate
Original
Predicted

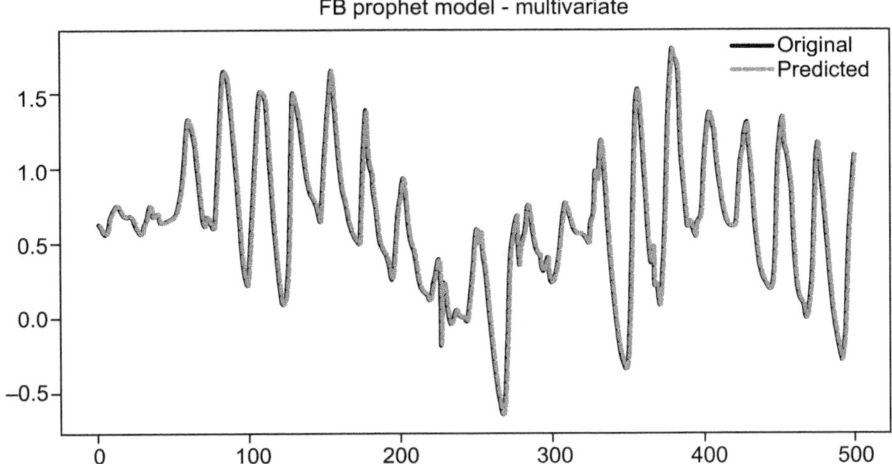

the Bayesian optimization and Linear interpolation. Prophet's RMSE is **0.003** and MAE is **0.001.**

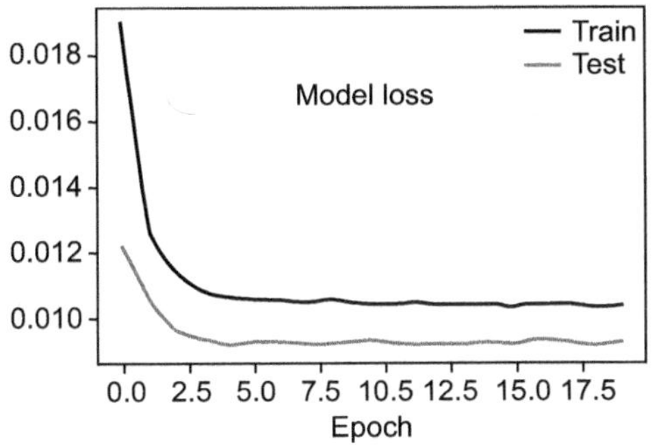

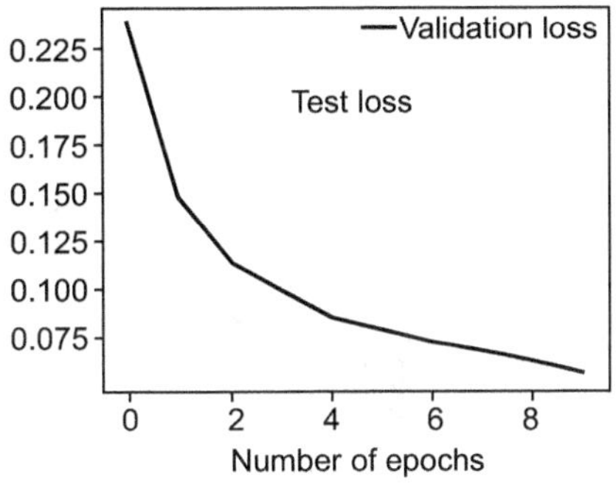

Train
Test
Model loss
Epoch

Validation loss
Test loss
Number of epochs

ROC AUC: 0.94

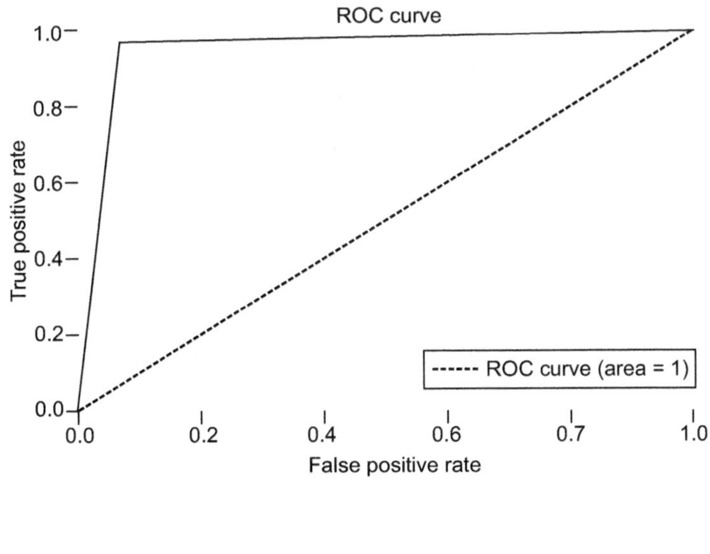

ROC curve
True positive rate
False positive rate
ROC curve (area = 1)

DBN can be viewed as a composition of simple, unsupervised networks such as restricted Boltzmann machines (RBMs) or autoencoders. Due to its vanishing gradient boosting feature, this ANN accounted to an accuracy of **0.94.** Though LSTM produced the highest accuracy, we were able to improve the performance of both Prophet and DBN by using effective methods.

5. Conclusion

The current work implements the AI and ML mechanisms to improve the conventional performance. Hyper tuning of parameters in neural networks can certainly enrich the current performance. The results suggest that our proposed framework outperformed other existing works. Prediction of weather and climate would help governments and authoritative bodies to take necessary precautions accordingly. We are able to assess the earth's climate sensitivity using ML and also find the rate at which it is warming. Using LSTM and Lag correlation, we are able to understand the trend in the temperature change. The current implementation is based on historical data, however, if present data is fed into the system it would add more value to the work and help in creating a dynamic repository. The future improvement would be to accurately forecast atmospheric conditions by considering various events like wind direction, speed, cloud distribution, and radiations etc.

References

[1] Rasp, S., Dueben, P.D., Scher, S., Weyn, J.A., Mouatadid, S., Thuerey, N. WeatherBench: A benchmark data set for data-driven weather forecasting. *Journal of Advances in Modeling Earth Systems*, 12(11), 2020. https://doi.org/10.1029/2020ms002203.

[2] Chantry, M., Christensen, H., Dueben, P., Palmer, T. Opportunities and challenges for machine learning in weather and climate modeling: Hard, medium, and soft AI. *Philosophical Transactions of the Royal Society A: Mathematical, Physical and Engineering Sciences*, 379(2194), 2021. https://doi.org/10.1098/rsta.2020.0083.

[3] Ott, J., Pritchard, M., Best, N., Linstead, E., Curcic, M., Baldi, P. A fortran keras deep learning bridge for scientific computing. *Scientific Programming*, 2020, pp. 1–13, 2020. https://doi.org/10.1155/2020/8888811.

[4] Kuremoto, T., Kimura, S., Kobayashi, K., Obayashi, M. Time series forecasting using a deep belief network with restricted Boltzmann machines. *Neurocomputing*, 137: 47–56, 2014. https://doi.org/10.1016/j.neucom.2013.03.047.

[5] Haupt, S.E., Chapman, W., Adams, S.V., Kirkwood, C., Hosking, J.S., Robinson, N.H. Towards implementing artificial intelligence post processing in weather and climate: proposed actions from the Oxford 2019 workshop. *Philosophical Transactions of the Royal Society A: Mathematical, Physical and Engineering Sciences*, 379(2194), 2021. https://doi.org/10.1098/rsta.2020.0091.

[6] Grönquist, P., Yao, C., Ben-Nun, T., Dryden, N., Dueben, P., Li, S. Deep learning for post-processing ensemble weather forecasts. *Philosophical Transactions of the Royal Society A: Mathematical, Physical and Engineering Sciences*, 379(2194), 2021. https://doi.org/10.1098/rsta.2020.0092.

[7] James, S.C., Zhang, Y., O'Donncha, F. A machine learning framework to forecast wave conditions. *Coastal Engineering*, 137: 1–10, 2018. https://doi.org/10.1016/j.coastaleng.2018.03.004.

[8] Singh, S., Gill, J. Temporal weather prediction using back propagation based genetic algorithm technique. *International Journal of Intelligent Systems and Applications*, 6(12): 55–61, 2014. https://doi.org/10.5815/ijisa.2014.12.08.

[9] Weather Forecasts Based on Rainfall Prediction Using Machine Learning Methodologies. *Adalya Journal*, 9(6), 2020. https://doi.org/10.37896/aj9.6/009.

[10] Fu, L., Cao, Y., Kuang, S.Y., Guo, H. Index for climate change adaptation in China and its application. *Advances in Climate Change Research*, 12(5): 723–733, 2021. https://doi.org/10.1016/j.accre.2021.06.006.

[11] Dengler, K., Keil, C., Frech, M., Gerz, T., Kober, K. Comparison and validation of two high-resolution weather forecast models at Frankfurt Airport. *Meteorologische Zeitschrift*, 18(5): 531–542, 2009. https://doi.org/10.1127/0941-2948/2009/0399.

[12] Thinsungnoen, T., Kerdprasop, K., Kerdprasop, N. A deep learning of time series for efficient analysis. *International Journal of Future Computer and Communication*, 6(3): 123–127, 2017. https://doi.org/10.18178/ijfcc.2017.6.3.503.

[13] https://towardsdatascience.com/end-to-end-time-series-analysis-and-forecasting-a-trio-of- sarimax-lstm-and-prophet-part-1-306367e57db8.

[14] shorturl.at/ahmDH.

[15] https://public.wmo.int/en/resources/united_in_science.

CHAPTER 20

Demand Forecasting Methods

Using Machine Learning to Predict Future Sales

*Sunanda Vincent Jaiwant** and *Joseph Varghese Kureethara*

1. Introduction: Demand Forecasting

To thrive in the market today, businesses must increase the effectiveness, dependability, and accessibility of their services. Sales estimation and operative demand scheduling definitely impact the end result of the organizations, influencing their procurement process, production, delivery, supply chain, marketing communications, etc.

Demand forecasting refers to the assessment of a possible forthcoming demand for a product or service. It is the course of approximating the forthcoming demand for a company's offerings. It is a critical activity for businesses as it allows them to plan and make decisions about production, inventory, and sales. The phrase is often used as a substitute word for demand planning and demand sensing. Demand planning aims to create a forecasting model that supports decision-makers while making decisions regarding the procurement of necessary raw-materials, production, distribution, and sales process. Forecasting helps in providing fundamental action plans to different departments of the organization at varied planning levels.

There are several methods of demand forecasting, including quantitative and qualitative approaches. Quantitative approaches include the usage of statistical models to analyze historical data and recognize patterns or trends. These methods

CHRIST (Deemed to be University), Eamil: frjoseph@christuniversity.in

* Corresponding author: sunanda.vincent@christuniversity.in

include time-series analysis, regression analysis, and exponential smoothing. Qualitative methods involve gathering information from experts, customers, and other stakeholders to make informed judgments about future demand. These methods include market research, surveys, and expert opinion.

Both methods have their strengths and weaknesses, and a combination of both approaches is often used to achieve more accurate demand forecasts. Additionally, demand forecasting is an ongoing process, and forecasts must be updated regularly to reflect changing market conditions and other factors that may impact demand.

Traditionally, marketers used to estimate the demand for the product and arrive at the sales targets for a particular time period depending upon their individual discernments, instincts, understanding, and past experiences. These kind of forecasts are not very reliable, take lot of time to arrive at the end results, and are not very efficient ways of forecasting. This problem was solved by the use of computer networks that go a long way in assisting the decisions regarding future sales. Advanced technology like ML is now solving issues encountered during the traditional forecasting methods and can be easily used to generate operative sales prediction models exploiting the massive volume of data and relevant information.

Demand forecasting has been revolutionized due to the use of the smart devices and advanced technologies. Demand forecasting techniques heavily rely on the data that is made accessible due the upsurge of smart devices, machines, and internet [7]. Sales forecasting is a field that is very significant for all the marketers and is immensely in focus. Forecasting the future demand is extremely essential for the marketers because an effective and optimum sales forecast can empower them to fight the competition battles and remain alive and powerful in the market. Sales forecasts helps the marketers and vendors to estimate, produce, and stock the right quantities of goods to fulfill the expected demand. Meeting the demand is fundamental to any business. For this they need to employ and exercise an efficient system of predicting the demand in real-time. Real-time forecasting encompasses understanding the consumer behavior trends and analysis of data generated or acquired from varied sources. ML algorithms have the potential to empower the organizations to utilize their funds and other resources in an optimum manner. The forecasting by ML can be employed for several uses, namely, forecasting the future demand of the goods or services, foreseeing the sales of goods in a given time-period. ML is the domain where machines have the ability to outperform humans in specific tasks. ML provides more accurate results and quicker predictions.

1.1 Demand Planning: Understanding Market Needs

Demand planning is the process of forecasting future customer demand for a product or service, and using this information to make informed business decisions. It involves analyzing historical sales data, market trends, and other factors to create a forecast of future demand. The goal of demand planning is to ensure that the right amount of product is available at the right time and in the right place, to meet customer demand while minimizing costs and reducing waste. This involves balancing the costs of holding inventory against the risks of stockouts or overstocking. Demand planning is an important process for businesses of all sizes and types, as it can help improve supply chain efficiency, reduce costs, and increase customer satisfaction. It is typically performed by a team of analysts or planners who use statistical models and other tools to create accurate forecasts of future demand.

Demand planning is the process of predicting the future demand for a product or service. It involves analyzing historical data, market trends, and other factors to predict the quantity of goods or services that customers will want to buy in the future. The goal of demand planning is to guarantee that the accurate quantity of stock is made accessible at the exact time to fulfill the consumers' requirement and simultaneously, the surplus stock units is minimized along with the accompanying outlays.

Demand planning typically involves a range of activities, including:

- Data collection: Gathering data on historical sales, market trends, and other relevant factors.
- Forecasting: Using statistical models and other techniques to predict future demand based on the available data.
- Collaboration: Working with sales, marketing, and other stakeholders to incorporate their insights and knowledge into the forecast.
- Review and adjust: Continuously reviewing and adjusting the forecast based on new data and feedback from stakeholders.

Demand planning is the foundation for a number of other processes, including transportation, logistics, pricing projections, and delivery schedules, which tries to meet demand and needs information on anticipated consumer demands. Effective demand planning can help businesses improve their customer service levels, reduce inventory carrying costs, and increase operational efficiency. It is a critical process for any organization that wants to manage its supply chain effectively and competitively keep up with the rapidly changing corporate world of today.

1.2 Demand Sensing: Generating Interim Forecasts

Demand sensing is a comparatively novel terminology in the planning process, which refers to an estimating technique that makes use of unconventional analytical procedures and methods to apprehend real-time variations in purchase behavior. This technology carries boundless benefit for organizations, functioning in a volatile market. Demand sensing is a process of using real-time data and analytics to detect and respond to changes in demand for a product or service. It involves analyzing various sources of data, such as point-of-sale (POS) data, social media trends, weather forecasts, and other external factors to anticipate changes in consumer demand.

The goal of demand sensing is to improve forecasting accuracy and reduce inventory levels, resulting in better customer service, reduced costs, and improved overall performance. By using advanced analytics and ML algorithms, demand sensing can help companies predict demand patterns and respond quickly to fluctuations in demand.

To detect an increase or reduction in sales by comparing current patterns to historical ones, demand sensing technologies take everyday statistics from POS systems, facilities, and alternative entities. The structure mechanically analyzes the importance of every divergence examines the influences, and provides suggestions for short-term plan modifications.

According to many studies, demand sensing relatively decreases prediction inaccuracies by 30–40%. It enables businesses to quickly respond to unexpected changes in client needs and makes it easier to create a supply chain that is data-based. However, this approach does not work for mid- or long-term scheduling, so one cannot rely all the decisions solely on it. Yet, it might be a useful addition to conventional prediction models.

Demand sensing can also be used to improve supply chain management by enabling companies to adjust production schedules, optimize inventory levels, and improve delivery times. This can help companies respond speedily and efficiently to changes in demand, reduce the risk of stockouts, and minimize the cost of carrying excess inventory.

Overall, demand sensing is an important tool for companies looking to maintain competitiveness in current expeditious and continuously fluctuating commercial setting. By leveraging real-time data and advanced analytics, companies can better understand their customers, anticipate their needs, and respond quickly to changes in demand.

2. Literature Review

Barnett, 1988 [1] there have been many erroneous guesses and suppositions that the earlier marketers made not because they did not know about or possess any predicting tools or techniques like regression analysis, past trend analysis, and likewise. In fact, they all showed a misguided important supposition that associations fueling demands in the earlier times will remain unchanged. The businesses neither foresaw variations in behaviors of their customers, nor understood the saturation level of the markets of their products or services. Marketers failed to realize that past sales, demands, or trends cannot be a reliable guide because of globalization and that local economies and markets were becoming more global, new technologies were emerging, and industries were evolving. This made the marketers doubt the conventional methods of forecasting the demands of their products and services.

Marketers geared to create valued understandings into forthcoming market environments and demand-levels grounded on a profound reasoning and identification of the drivers behind the overall market demand. Such developed innovative visions and institutions possess the potentials of making some strategies win while others fail and flounder.

Predicting the demand of the overall market may not always promise a positive and effective strategic decision. However, lack of the prediction may adversely affect the decisions on investments, marketing-support, and allocating resources. Marketers are able to gauge total market-demand unmistakably, and thereby adding more likelihood to managing and maneuvring the outcome and performance of the organization. Overall, market predictions are just the primary step in developing a plan to success.

Smirnov and Sudakov, 2021 [6] believe that the issue with prediction of demand for any new product founded on its features and descriptions is considered vital for many manufacturing organizations because this kind of issue when solved can lead to optimization of the manufacturing process, management operations, supply chain procedures, and my more exercises and activities with the aim to increase revenues and lessen expenses. Standard demand estimating approaches undertake the accessibility of sales statistics for a particular past time duration, which clearly does not happen when introducing a new product.

The researchers proposed using ML systems to predict demand. Several applications of the gradient boosting algorithm were employed in the research study, algorithms like XGBoost, Light GBM, Cat Boost [11]. Managing the massive volume of data has always been very important and equally complex and business organizations require an effective demand forecast structure. Data analyzing techniques should be accurate and give results speedily to facilitate more informed and strategic organizational choices. The study discusses about the ML approaches that focus on the data tuning and decision-making by offering an efficient mechanism. Businesses must be equipped with contemporary methods to accommodate various customer behavior patterns by projecting favorable sales turnover to be successful [11]. E-commerce platform development is accelerating due to the growing impact of the Internet on people's lives, with both users and revenue on these platforms increasing. The e-commerce industry has recently developed in a favorable environment thanks to the robust support of national policies. The quantity and competitiveness of e-commerce systems and e-commerce companies are growing in varied circumstances. A platform should be able to more effectively to encounter user needs, undertake well in all areas of coordination, and sustain its competitive advantage. The fundamental principle of XGBoost's algorithm is to grow a tree by continuously adding new trees and performing feature splitting. To be a part of the remaining section of earlier predictions, one actually learns a new function f(x) each time one adds a tree. After training, one has to foresee the score of a sample when one has k trees. In fact, each tree will denote to a relating leaf node, and each leaf node resembles a score, based on the characteristics of this sample. Finally, to obtain the predicted value, one simply needs to add the scores assigned to each tree [8]. In particular, an organization's (restaurant's) future earnings estimates, can be easily derived with accuracy using the offer details, and weather gauge setting. The needs of the customers are a major focus of the turnover. In any case, over the last few years, the performance has changed as a result of the presentation of massive volumes of data and calculations when it comes to gaining the upper hand. Learning and comprehending the significance of the data that will be used in any commercial operations is essential. Besides, the organizations can use climate forecasting in conjunction with commercial opportunities [10]. Market basket analysis (MBA) creates a frequent item set, or association rules, that can easily reveal customer purchasing patterns. The retailer can set up his retail store with the assistance of these constructs and grow his business in the future. The Apriori algorithm is the primary one used in market basket analysis. This could be a very effective technique for evaluating consumer purchasing trends. Support and confidence are the multiple useful parameters used in MBA. In contrast to confidence, which assesses the algorithm's accuracy or predictive power, support assesses how frequently an item appears in an assigned transactional dataset. Predictive modeling that uses transactional data produce specific issues that required to be judiciously tackled to create effective models. Using MBA, prominent venders can initiate better lucrative advertisements and promotional campaigns, fascinate more buyers, thereby increasing the value of market basket.

Business organizations and retailers are in the struggle to extract maximum information about their customers' needs and demands beforehand. This gives them the advantage of avoiding any kind of scarcity during sale and help them to maintain optimum inventory. Everyday predictions of the consumers' demand empower the enterprises and venders to be more precise in their judgments of sale, and generate increased revenue. Accurate forecasts result in minimum wastage and minimum losses. Organizations use different ML algorithms like Random-forest regressor, XGBoost Regression, Artificial Neural Network (ANN), etc., to forecast sales in the future [13].

3. Machine Learning

ML emphasizes on getting the machines take decisions after doing data analysis and then generating results. This process does not require much of programming. ML algorithms make predictions about customers with the help of data about past trends and consumer behaviors, sales-transactions, and demographic data. Organizations have been benefiting by using ML algorithms in forecasting for their sales, including some well-known companies like Amazon, Airbnb, Facebook, Netflix, and Uber [9].

ML is used in creating the chatbots and predictive text in generating language conversion assistance in language translation applications. The ML technologies are also utilized by Netflix where it is used to recommend entertainment that the user might like to watch. Social media feeds are also generated and offered using ML. ML is increasingly used in the medical field to control automated vehicles and machines that diagnose the medical state of the patients. ML is a subfield of artificial intelligence (AI) that empowers the computers to acquire information in the absence of any human involvement or any pre-created programs [2].

ML is a type of AI that enables computers to learn and improve from experience without being explicitly programmed. It involves the use of algorithms and statistical models to analyze and draw predictions or insights from data, which can be used to automate tasks or make informed decisions.

ML can be classified into three categories: supervised learning, unsupervised learning, and reinforcement learning. A labeled dataset is used in supervised learning, where algorithms learn to anticipate outcomes based on the data. Unsupervised learning requires the computer to analyze an unlabeled dataset in order to find patterns or structures via interaction with the environment and learning via trial and error; reinforcement learning involves giving or taking away incentives for specific behaviors.

Being a subcategory of AI, ML enables computers to absorb from experience and become improved versions without explicit programming. There are various varieties of ML, such as:

- Supervised Learning: In supervised learning, the algorithm picks up new information from a labeled dataset in which the input data is annotated with the desired result. This labeled data is then used by the algorithm to make predictions about fresh, unforeseen data.

- Unsupervised Learning: In unsupervised learning, the algorithm picks up knowledge from a dataset that has not been labeled with the intended results. The system then finds correlations and patterns within the data, assembling related data and identifying outliers.
- Semi-Supervised Learning: Unsupervised and supervised learning are combined in semi-supervised learning, a kind of ML. It entails using a smaller amount of labeled data and a greater amount of unlabeled data to train an algorithm. The program then makes predictions on fresh data using the labeled data, while simultaneously making patterns and relationships using the unlabeled data.
- Reinforcement learning: Reinforcement learning is a sort of ML in which an agent learns to make decisions in a given environment by getting feedback in the form of rewards or penalties. The algorithm gains knowledge through trial and error, modifying its actions to increase rewards and decrease penalties.
- Deep Learning: A branch of ML called deep learning use neural networks to learn from data. It is especially helpful for projects like audio and image recognition, recommendation systems, and natural language processing.
- Transfer Learning: Transfer learning is the process of training a model to solve one problem, then using that model to complete another task that is similar, but different. Because the previously trained model may be used as a jumping-off point for the current assignment, this enables quicker and more effective model training.
- Online learning: Online learning is building a model using data that is continuously sent throughout time, as opposed to data that is only available at one time. The model is updated as fresh data is received, enabling it to adjust to shifting data patterns.

ML is used in a variety of applications, including natural language processing, image and speech recognition, predictive maintenance, autonomous vehicles, and fraud detection. It has the potential to revolutionize many industries and fields, as it enables computers to make complex decisions and automate tasks that would otherwise require human intervention (Fig. 1).

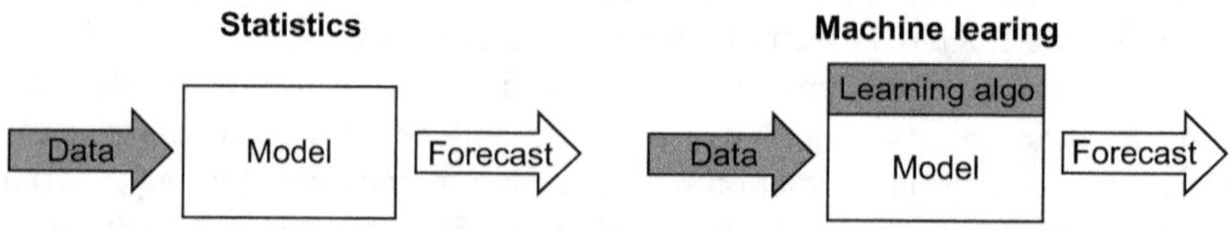

Fig. 1: Statistical models vs. Machine Learning models. *Source:* (Vandeput, 2021) [12].

3.1 Machine Learning for Demand Planning: Improved Precision at the Cost of Greater Intricacy

Prerequisites for broader use of ML to develop forecasts were generated by increases in both computer power and demand volatility. To produce short-term predictions in reaction to various market developments, let's use the demand sensing example that we stated previously. Beyond only predicting demand, ML also powers predictive analytics. It mixes past and present data to produce insights into patterns and specific behavior under certain circumstances.

Predicting sales is a popular and necessary application of ML. Forecasted demand can be employed to set standards and quantify the progressive benefits of innovative efforts, allocate assets responding to predicted sales, and make an estimation of upcoming appropriations. Because it has revolutionized the selling process, ML aids in increasing sales conversions. ML systems will be able to identify, act on, and close high-potential sales far faster and more scalable than human salespeople. High-performing sales teams see the value of ML; they are 4.1 times more likely to use it than their lower-performing counterparts [1].

Machine Learning Forecasting is unquestionably the best strategy to achieve higher prediction accurateness and moving forward. Without the need for human interaction, ML forecasting allows processors to learn from enormous volumes of huge data, producing unparalleled client demand information. Extrapolative analytics and algorithms such as Neural Networks (NNs), Recurrent Neural Networks (RNNs), and Support Vector Machines (SVMs) have emerged as the most recent topic of discussion among senior management sides. When compared to traditional forecasting techniques such as average, moving average, trend, and multiple linear regression, Machine Learning Forecasting is auto-regulating and robust.

ML approaches can forecast the number of products/services that will be purchased in the future (Fig. 2). Unlike traditional demand forecasting methodologies, ML allows to:

- Increase the speed of data processing.
- Give a more precise forecast.
- Update forecasts automatically based on the most recent data.
- More data should be analyzed.
- Recognize hidden patterns in data.
- Build a strong system.
- Improve your ability to adjust to changes.

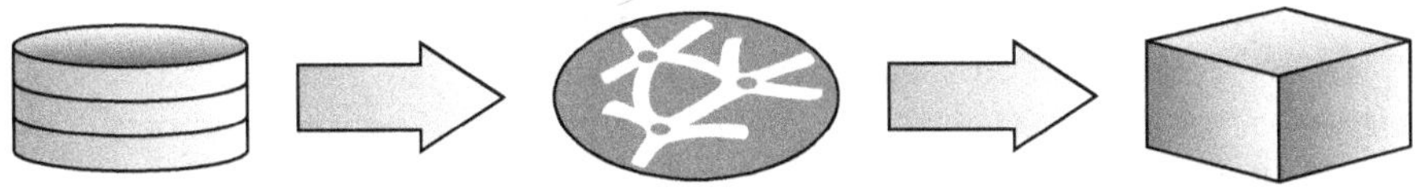

Fig. 2: Sales forecasting predictive model. *Source:* https://www.neuraldesigner.com/solutions/sales-forecasting [4].

3.2 Machine Learning Models Used in Sales Forecast

ML is a speedily expanding area, which means that innovative models are getting produced and tested on a regular basis on various datasets. The selection of ML frameworks is influenced by a number of criteria, and also includes the business purpose, data category, data volume and quality, prediction time, and so on. Ml is a trending area which develops number of models that analyze different datasets to predict everyday operation of organizations. Some of the popular models used in the area of sales predictions are regression algorithm modeling and multivariate modeling.

Many ML methods can be used to sales forecasting. Here are a few of the most typical:

3.2.1 Regression algorithms

In ML, regression techniques are widely used to forecast sales volume. They accomplish this by analyzing enormous amounts of data from the past to identify patterns that can be used to forecast future events. To achieve the best results, these algorithms are generally utilized with incomplete data.

For example, a clothing company can utilize regression algorithms to forecast sales volume based on historical data. ML may generate models that analyze data and generate forecasts that can be used internally to estimate cash flow or externally to launch a new advertising campaign.

Regression algorithms are quite common tools employed by ML experts to predict sales for an organization. ML examines and processes the past data bulk which reflect some kind of patterns and on the basis of these, future actions or demands are determined.

3.2.2 Multivariate Models

Multivariate modeling is an extension form of regression algorithm modeling (Fig. 3). Multivariate approach makes use of several variable quantities to arrive at further precise estimates. It is widely used to generate sales prediction for big volumes of sales and access information about the profitability percentages of specific goods or services.

Regression procedures are expanded in multivariate models. They provide more accurate forecasting by employing a huge data set by using numerous factors. Multivariate models are most commonly used to forecast total sales volume, but they can also be used for estimating the how much a business offering is profitable.

A clothes company, for example, can utilize these algorithms to determine the profit margin of specific products. They can anticipate future profitability and sales volume for each product by analyzing historical data from previous transactions. This will assist them in determining which products to advertise based on the best return on investment.

Multivariate models can be useful for predicting sales in an ML context. By analyzing multiple variables simultaneously, multivariate models can capture complex relationships between different factors that may influence sales. One common type used for sales prediction is regression analysis. This involves analyzing the relationships between a few independent variables (for example, advertisement budget, price, promotions) and the dependent variables of sales. By examining such relationships, a regression model can be trained to predict future demand based on changes in these independent variables. Other multivariate models which are

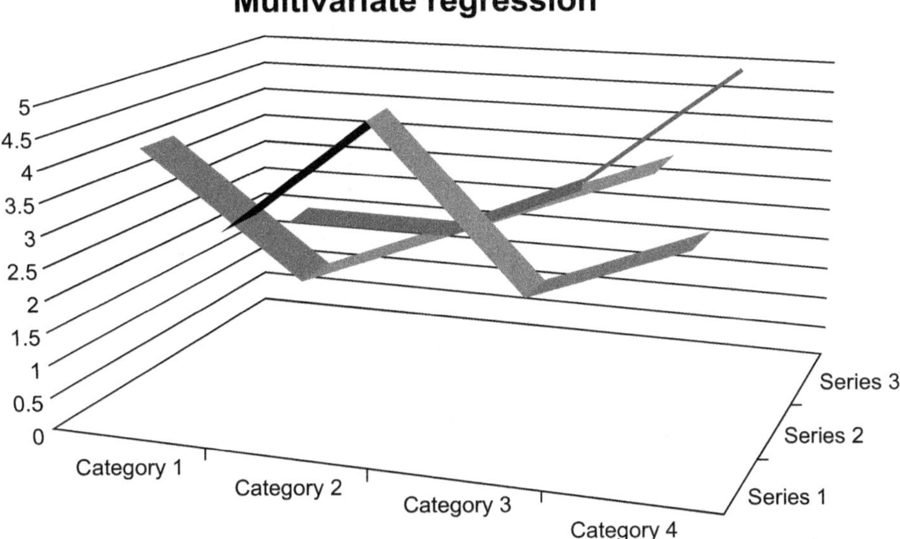

Fig. 3: A multivariate regression model. *Source*: https://www.educba.com/multivariate-regression/ [5].

employed for sales prediction comprise of NNs, decision trees, and SVMs. These models can capture nonlinear relationships between variables and may be useful in cases where the relationship between sales and independent variables is complex and difficult to model using regression analysis. Regardless of the specific model used, it is important to select appropriate independent variables that are likely to have a strong influence on sales. Additionally, it is crucial to make sure the model is trained using accurate and representative data in order to produce reliable predictions. Finally, it is important to continuously monitor and update the model as new data becomes available to ensure that it remains accurate over time.

3.2.3 Time-series Approach

Prospective revenue predictions can be made using Time-Series approach by analyzing patterns and trends in past sales data. A time series is a collection of data points gathered at equivalent intermissions over a period. Trends, seasonality, irregularity, and cyclicality are the most important factors to consider. The following are the most suitable time-series models in the retail industry:

- ARIMA (auto-regressive integrated moving average) models are designed to define autocorrelations in time-series data. ARIMA can create reliable predictions when preparing short-term forecasts.
- SARIMA (Seasonal Autoregressive Integrated Moving Average) models are an extension of the ARIMA model that accommodates univariate time-series data with seasonal period back-shifts.
- Exponential Smoothing models forecast new values by employing weighted averages of previous observations. The key of these frameworks is the seamless calculation of error, trend, and seasonal components.

3.2.4 Linear Regression

Linear regression is an uncomplicated and effective ML algorithm that may be utilized to forecast sales based on historical data. It works by creating a linear relationship between the independent variables (such as time or advertising spend) and the dependent variable (sales), and then using this relationship to make predictions. This model is used to find the linear relationship between two variables, in this case, sales and time. Linear regression may be applied to forecast future sales built on past information.

The association amongst a dependent variable and a number of independent variables can be modeled statistically using linear regression. It is a method that is frequently used in statistics, ML, and data analysis. There exists only one independent variable in basic linear regression, and it is assumed that the relationship between the independent and dependent variables is linear. Finding the most suitable line to represent this relationship is the aim of linear regression. The line that reduces the total of the squared variances among the dependent variable is observed values and anticipated values are referred to as the best-fit line. In multiple linear regression, the connection between the dependent variable and two or more independent variables is examined. In multiple linear regression, the association between the dependent variable and the independent variables continues to be presumed to be linear even though there are two or more independent variables. Finding the most suitable plane or hyperplane that represents the relationship in question is the aim of multiple linear regression. Many possibilities can be predicted using linear regression, including the price of a house based on its size, the performance of a stock according on its historical pricing, and the likelihood that a consumer will buy an item depending on their age, income, and other demographic factors.

3.2.5 Decision Trees

Decision trees are one more widespread ML algorithm for sales forecasting. They function by breaking down the sales figures into reduced and smaller subgroups based on different criteria, such as time, geography, or customer segment. Each subgroup is then analyzed to determine the most important factors that drive sales, and these factors are used to make predictions. A decision tree is a type of ML algorithm applied for categorization or regression analysis. It is a tree-like model where a specific internal node signifies a test on a characteristic, every branch denotes the result of the examination, and each leaf node signifies a class label or a numerical value. The decision tree algorithm begins with the entire dataset at the root node, and then separates the data based on the most significant attribute which delivers the best separation of the classes. This course stays recursively till an ending condition is reached, for example, all instances in a leaf node belonging to the same class or a maximum tree depth has been reached.

Decision trees have several advantages, such as being easy to understand and interpret, handling both numerical and categorical data, and being able to handle missing data. They are also able to handle irrelevant features and can be used for feature selection. However, decision trees can also be inclined to overfit where the framework develops more intricacy and matches the training data too precisely, subsequently becoming weak in labeling of the new data. This can be mitigated through systems like pruning, ensemble methods like random forests, and using other decision tree algorithms such as gradient boosting.

3.2.6 Random forest

Random forest is an ensemble ML algorithm that combines multiple decision trees to create a more accurate and robust model. It works by creating a large number of decision trees on randomly selected subsets of the data, and then combining the results of these trees to make predictions. This model is used for both classification and regression tasks. It is particularly useful when there are many features or variables that could be affecting sales. Random forest works by creating many decision trees and combining their results to make a prediction.

Random forest is a popular ML algorithm that belongs to the class of ensemble methods. It is used for both classification and regression tasks. The basic idea of a random forest is to create multiple decision trees using a subset of the features and a random subset of the training data. Each decision tree is trained independently, and then their predictions are combined to make the final prediction. During the training process, the algorithm randomly selects a subset of features at each node of the decision tree, rather than considering all the features. This helps to reduce overfitting and increase the generalization of the model.

It is a powerful algorithm that can handle a large number of features and data points, and it is less prone to overfitting compared to other ML algorithms. It is widely used in various applications such as image recognition, natural language processing, and recommendation systems.

3.2.7 Neural Networks

NNs are a kind of deep learning algorithms which may be applied in sales forecasting. They function by creating a network of interconnected nodes that are trained on historical sales data. The network learns to categorize forms and trends in the data and might often utilize such knowledge in generating estimates. They are a set of algorithms designed to recognize patterns in data. They can be used to predict future sales by analyzing past sales data and identifying patterns and trends. They are a popular ML technique used for predicting demand in various industries, such as retail, logistics, and transportation. They represent a kind of deep learning algorithm which have the capacity to understand and acquire multifaceted patterns in data and make accurate predictions based on historical data.

To use NNs for predicting demand, one will need to gather historical data about the demand for the product or service one is interested in. This data can include information such as sales data, customer behavior data, and market trends. Once the marketers have collected the data, they will need to preprocess it to make it suitable for input to the NN. This may include techniques such as normalization, scaling, and encoding categorical variables. Next, one can build their NN model. A typical NN model for predicting demand will consist of several layers, including an input layer, several concealed layers, and an output layer. The input layer will take in the preprocessed data, and the hidden layers will exercise nonlinear transformations to the input data to extract useful features. The output layer will produce the predicted demand value.

The marketers will then need to train their NN using the historical data. During training, the NN will learn to optimize its internal parameters to lessen the difference amongst its projected result and the real demand values in the training data.

Once the NN is trained, the marketers may utilize it in making estimates on new, unseen figures. This may help the marketers forecast demand for their product or service and make informed decisions about inventory management, pricing, and marketing strategies.

3.2.8 Support Vector Machines (SVMs)

SVM is an ML algorithm that may be employed in making estimates of the sales based on historical data. It functions by creating a hyperplane which divides the data into two groups—one for positive sales and one for negative sales. The algorithm then finds the hyperplane that increases the margins amongst these two groups, and uses this hyperplane to make predictions. This model is employed for both classification and regression exercises. It functions by discovering the greatest line or hyperplane that parts the data into two or more modules. In sales forecasting, SVMs may be utilized for estimating the upcoming sales grounded on past figures.

SVM is an administered ML algorithm employed for sorting and regression analysis. It is constructed on the impression of identifying the optimal hyperplane that separates data points belonging to different classes. In SVM, the goal is to find the hyperplane that maximizes the margin, which is the distance between the hyperplane and the closest data points from each class. The data points that are closest to the hyperplane are called support vectors, and they state the location and alignment of the hyperplane.

An SVM can be utilized for both linear and nonlinear classification problems. For linear classification, a linear hyperplane is used, whereas for nonlinear classification, a nonlinear kernel function is used for transforming the data into a higher-dimensional space where a linear hyperplane can be used for separating the modules. SVM possesses numerous benefits over other classification algorithms. It is highly operative in high-dimensional places, can handle nonlinear decision boundaries, and is less prone to overfitting. Additionally, SVMs have been shown to function well in a diverse set of utilizations, for example, text classification, image classification, and bioinformatics.

Nevertheless, SVM possesses a few disadvantages also. An SVM can be computationally expensive, especially for large datasets. Also, it is sensitive to the choice of hyperparameters, such as the kernel function and the regularization parameter, and selecting these hyperparameters can require some expertise. Overall, SVM is an influential and extensively applied ML algorithm for classification and regression analysis.

3.2.9 XGBoost

XGBoost is a prevalent ML algorithm that is widely employed for classification and regression problems. In the context of sales forecasting, XGBoost may be applied to create a framework which foresees future sales based on past sales figures and other relevant aspects.

One can create an XGBoost model for sales forecasting using following steps:

- Gather and clean data: Collect historical sales data along with other factors that can influence sales such as pricing, promotions, marketing campaigns, weather, and seasonality. Clean the data by removing any outliers and missing values.
- Split the data: Split the data into training and testing datasets. The training dataset will be used to train the XGBoost model, while the testing dataset will be used to evaluate the model's performance.
- Define the model: Define the XGBoost model and its hyperparameters that can be tuned to achieve the best performance on the training data.
- Train the model: Train the XGBoost model using the training dataset. The framework will learn the patterns and relationships in the data that are associated with sales.
- Evaluate the model: Estimate the XGBoost model's functioning on the testing dataset. Calculate the model's accuracy, precision, recall, and F1 score to assess its performance.
- Predict sales: Use the XGBoost model to predict future sales based on new data that includes the relevant factors.
- Refine the model: Refine the XGBoost model by tweaking the hyperparameters and adding new features to improve its performance.

4. Machine Learning Techniques and Demand Forecasting in Different Domains

There are many ML techniques which are employed for estimating the demand in the forthcoming times periods and some of the domains are mentioned below where ML technology may be employed effectively:

- Supply-chain demand prediction
- Airline demand prediction
- Water demand prediction
- Retail demand prediction
- Marketing demand prediction
- E-commerce demand prediction
- Travel/riding demand prediction
- Electricity demand prediction
- Tourism demand prediction

In the retail industry, demand forecasting helps retailers to predict customer demand for products and services, optimize inventory levels, and improve supply chain management. ML systems such as time-series analysis, regression analysis, and clustering can be utilized to analyze past sales data and recognize trends and patterns which may assist in predicting future demands. In the e-commerce industry, demand forecasting helps online retailers to optimize pricing strategies, manage inventory levels, and improve customer experience. ML techniques such as collaborative filtering, matrix factorization, and deep learning can be used to analyze customer behavior, purchase history, and product attributes to predict future demand.

In the transportation industry, demand forecasting helps transportation companies to optimize their route planning, improve fleet management, and reduce operational costs. ML techniques such as regression analysis, decision trees, and NNs can be used to analyze historical passenger and cargo data and predict future demand. In the healthcare industry, demand forecasting helps healthcare providers to optimize staffing levels, manage patient flow, and allocate resources efficiently. ML techniques such as time-series analysis, logistic regression, and clustering can be used to analyze historical patient data and predict future demand for healthcare services. In the energy industry, demand forecasting helps energy providers to optimize energy generation and distribution, improve energy efficiency, and reduce costs. Techniques such as regression analysis, decision trees, and NNs can be used to analyze historical energy consumption data and predict future energy demand. In summary, ML techniques can be applied to demand forecasting in different domains to optimize business operations, improve customer experience, and reduce costs.

5. Conclusion

Demand forecasting using ML has been established as a valuable instrument for businesses in a wide range of industries. By leveraging historical data, businesses can build predictive models that help them better anticipate customer demand, optimize their inventory levels, and ultimately improve their bottom line. One of the significant benefits of using ML for demand forecasting is its ability to handle large and complex data sets. ML algorithms can analyze data from a variety of sources, including POS systems, social media, and web analytics, to generate more accurate forecasts. Additionally, ML models can be updated in real-time, allowing businesses to quickly adapt to changes in demand patterns. This is particularly important in industries where demand can fluctuate rapidly, such as retail or hospitality.

Overall, demand forecasting using ML has the potential to provide businesses with a competitive edge by enabling them to make more informed decisions about inventory management, production planning, and sales strategies. However, it's important to note that ML models are not a silver bullet, and they must be developed and implemented carefully to ensure their accuracy and effectiveness. ML can offer several advantages for demand forecasting compared to traditional statistical methods. Some of these advantages include: (1) Improved accuracy: Machine learning algorithms can learn from historical data and patterns, which can result in more accurate demand forecasts. By using sophisticated techniques like NNs and deep learning, ML models can identify nonlinear relationships in the data and make more precise predictions. (2) Scalability: ML algorithms can handle large amounts of data, making it easier to forecast demand across multiple products or regions. This scalability is particularly useful for businesses with complex supply chains, where demand forecasting for one product can affect the demand for other products. (3) Speed: ML algorithms can process data and generate forecasts much faster than humans. This speed can be particularly useful for businesses that need to make decisions quickly in response to changes in demand. (4) Flexibility: ML algorithms can adapt to changing conditions and trends, making them more flexible than traditional forecasting methods. For example, if a new product is introduced, an ML algorithm can quickly incorporate data from the new product and adjust the forecast accordingly. (5) Reduced errors: ML algorithms can help reduce errors in demand forecasting by identifying outliers and anomalies in the data. This can help businesses avoid overstocking or understocking products, which can result in lost sales or excess inventory. Overall, ML can help businesses improve their demand forecasting accuracy, scalability, speed, flexibility, and reduce errors. Marketers have begun using ML in demand forecast processes because it can help them to make more accurate predictions about future customer demand. By analyzing large amounts of data, ML algorithms can identify patterns and trends that may be difficult or impossible for humans to spot. This allows marketers to better understand their customers' behavior, preferences, and buying habits, and to use that information to anticipate future demand.

ML can also help marketers to optimize their marketing campaigns and promotions by providing insights into which products or services are likely to be most popular with their target audience, and when and how to promote them. Additionally, it can help marketers to identify potential new markets and customer segments, allowing them to expand their reach and grow their business. Overall, using ML in demand forecast processes can give marketers a competitive edge by allowing them to make data-driven decisions, improve customer satisfaction, and increase revenue.

References

[1] Altexsoft.com. (2022). *Demand Forecasting Methods: Using Machine Learning to See the Future of Sales*. 2022. Retrieved from https://www.altexsoft.com/blog/demand-fo recasting-methods-u.

[2] Barnett, W. Four steps to forecast total market demand. *Harward Business Review*. 1988, July. Retrieved from https://hbr.org/1988/07/four-steps-to-forecast-total-market-demand.

[3] Brown, S. *Machine Learning, Explained*. MIT Management Sloan School. 2021. Retrieved from Sara Brown.

[4] Demand Forecasting Methods: Using Machine Learning to See the Future of Sales. 2022. Retrieved from https://www.altexsoft.com/blog/demand-forecasting-methods-using-machine-learning/.

[5] https://www.neuraldesigner.com/solutions/sales-forecasting.

[6] https://www.educba.com/multivariate-regression/.

[7] Smirnov, P.S., and Sudakov, V.A. Forecasting new product demand using machine learning. 2021. *Journal of Physics: Conference Series*. doi: 10.1088/1742-6596/1925/1/012033.

[8] Purvika Bajaj, et al. Sales predicion using machine learning algorithms. *International Research Journal of Engineering and Technology (IRJET) e-ISSN: 2395-0056*, 7(6), 2020.

[9] Sakib, S.M. Restaurant Sales Prediction Using Machine Learning. 2021. doi:https://doi.org/10.31224/osf.io/wa927.

[10] Smolic, H. Sales Forecasting: How to Apply Machine Learning. 2022, September 1. Retrieved from https://graphite-note.com/machine-learning-sales-forecasting.

[11] Soham Patangia, et al. Sales prediction of market using machine learning. *Internationnal Journal of Engineering Research & Technology (IJERT)* 2020, September 1. doi: 10.17577/IJERTV9IS090345.

[12] Sunitha Cheriyan, et al. Intelligent sales prediction using machine learning. *International Conference on Computing, Electronics & Communications Engineering (iCCECE).* 2018. doi:10.1109/iCCECOME.2018.8659115.

[13] Vandeput, N. How To: Machine Learning-Driven Demand Forecasting. 2021, October 11. Retrieved from, https://towardsdatascience.com/how-to-machine-learning-driven-demand-forecasting-5d2fba237c19.

[14] Varshini, Preethi. An Analysis of Machine Learning Algorithms to Predict Sales. 2022, June. *International Journal of Science and Research (IJSR),* 11(6): 462–466. doi: 10.21275/SR22601144946.

[15] Zixuan Huo, et al. Sales Prediction based on Machine Learning. *2nd International Conference on E-Commerce and Internet Technology (ECIT).* 2021. doi:10.1109/ECIT52743.2021.00093.

Chapter 21

Demand and Supply Forecasts for Supply Chain and Retail

Astha Gupta,[1] *Preethi Nanjundan*[2,*] and *Jossy P George*[2]

1. Introduction

Demand and supply forecasts serve as the backbone of strategic decision-making in today's rapidly changing business environment, assisting organizations in optimizing inventory levels, production planning, and pricing strategies. The ability to forecast demand and supply accurately is critical for effective supply chain and retail management. This chapter provides a comprehensive overview of supply chain and retail demand and supply forecasts. It discusses various forecasting methods and techniques, as well as related concepts. In addition, the chapter emphasizes the significance of accurate forecasting in optimizing supply chain and retail operations, as well as emerging trends and future directions in demand and supply forecasting.

1.1 Market Forces

A market force is a factor that significantly affects how the market changes. Market forces control the price and accessibility of a good or service in a market. In a mixed economy, market forces are managed by government interference, whereas in a free-market economy, they operate organically. In a socialist economy, only government control is evident [1].

[1] School of Commerce, Finance and Accountancy, CHRIST (Deemed to be University), Lavasa, Pune, Maharashtra, India, Email: astha.gupta@bcomfan.christuniversity.in

[2] Department of Data Science, CHRIST (Deemed to be University), Pune, Lavasa. Email: frjossy@christuniversity.in

* Corresponding author: preethi.n@christuniversity.in

During times of low supply and high demand, prices are driven up by market forces; during times of high supply or low demand, prices are driven down by market forces. When supply and demand for an item or service are equal, it is said that the market has reached equilibrium.

To supply means to provide something that is wanted, i.e., to make it available [2]. Whereas, to demand something refers to the willingness to buy backed by the capacity to buy it.

The law of supply and demand is a crucial and prominent aspect of the global economy, playing a vital role in making it more profitable and beneficial. It is useful in various ways and profitable for consumers from different countries and fields and those involved in different businesses. Management of the movement of resources, including information and money, used in the creation of a good or service, from the acquisition of raw materials to delivery to the ultimate location, is known as supply chain management.

So, understanding demand and supply patterns by analyzing historical data can greatly aid us in identifying or predicting trends that are likely to be repeated in the future. Supply refers to a variety of goods and services that companies and warehouses provide. Meanwhile, demand focuses on the consumers and their needs, and it is necessary to understand the demand for products in various areas of business. Overall, understanding the concepts of supply and demand is crucial for the proper working, future planning and development of any business [3].

1.2 Forecasting

The purpose of forecasting is to give management information to aid in decision-making. Almost every company, whether it be public or private, works in a dynamic environment that is uncertain and where the future is only partially known. Making the proper decisions and being able to predict the future are key components of effective leadership. Even though there may still be some degree of uncertainty in these business decisions, forecasting can be utilized as a tool to help [4].

1.3 Supply Chain

A supply chain can be understood as a connection between the user and supplier of materials that consists of extraction of raw materials, conversions into finished goods and final supply and replenishment and it also includes all the procedures and information resources used in the process.

The goal of supply chain management is to effectively integrate suppliers, manufacturers, warehouses, and retail outlets so that goods are produced and distributed in the appropriate quantities, at the appropriate times, and in the correct locations in order to reduce system costs and meet service level requirements [6].

1.4 Retail

The term 'retail' covers all business activities, including the sale of products and services to customers for their own, individual, or family use. This includes sales of anything, including autos, clothing, restaurant meals, and movie tickets. The final phase of the distribution route is retail. The task of retailing is to strike a balance between the wants of the final consumer and the supplies of various marketers. The ability to provide customers with more convenient access to a wide range of goods, freedom of choice, and a wide range of services is what makes the retail sector such a fiercely competitive one.

Retail refers to sales that are made in the final stages to customers or other end users.

However, retail supply chains are a source of purchases for businesses as well. Retail transactions (B2B) come in two varieties: business-to-business and business-to-consumer (B2C). Given the length of consumer supply chains, there are both B2B and B2C ties, such as those between first- and second-tier suppliers and between merchants and customers.

Three things are mastered by effective retailers. The store must first decide which products to sell to each target market category, ideally in those where demand is increasing.

Second, retailers need to create and refine a successful store concept. The retailer needs to create a long-lasting competitive advantage. Seldom is price enough to compete; something more must be provided [5].

In cumulation with the two theories presented above, it can be understood that a retail supply chain is a process that is used to get your products to final consumers. The retail supply chain, which goes beyond stores, officially starts with the initial acquisition of raw materials for product manufacturing. The supply chain will alter based on the kind of product and the preferences of the consumer and end user [5].

2. Need of Forecasting

Forecasting can be defined as predicting future actions or patterns on the basis of studies carried out on past actions or patterns. There is a likelihood of repetition of certain happenings under the conditions that influenced them previously. Since market forces are driven by an aggregate of decisions taken by the public on a daily basis, forecasting can be done by studying the pattern of past behavior and response to them in terms of demand and supply to form a basic idea of the future. The retail supply chain has some integral parts such as inventory management and stock storage which can be managed effectively using the forecasting study. The aim of a business prediction is to employ statistical analysis and domain knowledge to provide trustworthy estimates that will ultimately guide subsequent corporate planning initiatives under this subject. The last step of the distribution chain is retail.

Forecasting can provide significant advantages at various business levels such as planning, coordinating, controlling, and decision-making. Having a predetermined response to actions can make it easy for retailers to take action as per their capabilities.

Demand and supply forecasting will help users to integrate the demand of consumers with the supply capabilities and provide insights to formulate strategic and tactical decisions about what to do in the future.

2.1 Need of Demand Forecasting

By embracing demand forecasting, businesses can reduce uncertainty, optimize their supply chain infrastructure, increase revenue, reduce the need for safety and surplus stock, and improve their order fulfillment process. All of these benefits can help retail businesses realize their growth ambitions and build a loyal customer base. Retailers require a number of forecasts with various levels of precision and time horizons to execute restocking, capacity planning, and other business decisions effectively. Because of this, a company's ability to use the same demand estimate across all of its supply chain planning depends on its ability to flexible aggregate across products or over several planning horizons.

By employing time-series modeling and historical data to predict future demand, merchants have historically produced their baseline forecasts. These forecasts were frequently modified using user input and causal modelling. Yet, modern businesses now use machine learning (ML) to replace these antiquated methods of demand forecasting.

Demand forecasting can be generalized under 3 categories:

1. **Time-series forecasting:** It arises when scientific hypotheses are founded on time-stamped historical data. It comprises developing models through historical research and applying them to make conclusions and guide future strategic decision-making. This type of model allows for the prediction of demand swings based on a common pattern, such as peak vs low times, by tracking specified independent variables over a specific period.
2. **Qualitative forecasting:** Market analysis and professional projections are the foundation of qualitative forecasts. For instance, surveys of public opinion offer a sense of consumer confidence. Consumer buying intentions can be revealed by more targeted polls. On the basis of their experience working inside the industry, consultants, analysts, and other specialists offer insights. Qualitative projections are influenced by the leadership of a retailer, who offers insights based on their years of expertise.
3. **Casual modeling:** Simulations are used in causal modeling to produce more precise demand estimates. Regression techniques are used by analysts to develop equations that link different inputs to sales outcomes. Future appliance sales will be impacted by leading signs like new housing starts, for instance. Sales are impacted by store-specific elements like competitors' close proximity. Internal variables with quantifiable effects include headcount trends and overall ad spend.

2.2 Demand Forecasting Using ML

A system must be able to examine vast volumes of data on a range of aspects that may have an impact on demand in order to estimate it accurately. Modern demand planning systems are capable of completing millions of estimation computations per minute, taking into consideration more variables than ever before, thanks to the development of large-scale data processing and in-memory computing.

ML is a branch of artificial intelligence (AI), that uses algorithms and statistical models to help computers learn from data without explicit programming. ML can be used to estimate demand and improve supply chain operations in the context of retail.

Retailers should employ ML for demand and supply forecasting because it can improve inventory levels, uncover trends and patterns in sales data, and assist predict customer demand more accurately. This may result in greater operational effectiveness, higher sales, and lower expenses.

By analyzing vast volumes of data from numerous sources, such as sales history, consumer behavior, weather trends, and social media activity, ML addresses the issues of demand forecasting in retail. As a result, retailers are better able to recognize and forecast demand trends, seasonality, and other elements that could influence consumer behaviour.

The following are some specific ways that ML can assist with the difficulties of demand forecasting:

- A higher degree of accuracy: It is possible because ML algorithms can evaluate enormous amounts of data and spot patterns and trends that might be hard for people to notice. This can result in better decision-making and more accurate demand estimates.
- Faster forecasting: Because ML algorithms can swiftly evaluate data and produce forecasts in real time, businesses can react to changes in consumer demand more quickly.
- Personalization: Retailers can offer more individualized recommendations and promotions by using ML to better understand individual customer behavior and preferences.
- Demand shaping: Using targeted promotions, pricing plans, and other marketing initiatives, ML can assist merchants in influencing consumer behavior and shaping demand.

Overall, by accurately estimating demand and making data-driven decisions, ML may assist retailers in streamlining their supply chain processes, lowering costs, and increasing customer happiness.

Think about the three main sources of variability that affect demand continuously: periodic changes in baseline demand patterns, your own internal business choices, and external variables like the weather or regional events.

1. **Periodic changes in baseline demand patterns:** These are the daily, weekly, or seasonal fluctuations in demand that occur over time. To make accurate predictions about future demand, ML algorithms can be trained to detect and analyze these patterns in historical sales data. A retailer, for example, may use ML to determine that demand for certain products increases during the summer months and adjust their inventory levels accordingly.
2. **Internal business decisions:** These are decisions made by a retailer that have the potential to affect demand, such as changes in pricing strategies or promotional campaigns. ML can assist retailers in analyzing the impact of these decisions on customer behavior and adjusting their forecasting models as needed. For example, a retailer may use ML to determine whether a specific promotion

increases sales of a specific product and then adjust future demand forecasts accordingly.

3. **External variables such as weather or regional events:** This refers to external factors that can influence demand, such as weather conditions, economic trends, or regional events. ML algorithms can be trained to analyze data from various sources, such as weather forecasts or social media activity, to determine how these external variables are likely to impact demand. For example, a retailer may use ML to predict that a heatwave will increase demand for cold beverages and adjust its inventory levels accordingly.

ML can provide retailers with more accurate and robust demand forecasting models by accounting for these three sources of variability. This allows retailers to optimize inventory levels, reduce stockouts, and improve overall supply chain operations. Furthermore, by continuously monitoring and analyzing demand patterns with ML, retailers can quickly adapt to changing market conditions and maintain a competitive edge.

3. Success in Retail

Forecasting can help us to identify the factors required to achieve success in retail.

The main objectives that a retailer dwells on to achieve can be as follows-

- Delivering the right products in the right quantities is essential to keep customers satisfied and loyal. By understanding their needs and preferences, you can tailor your offerings and avoid overstocking or understocking.
- Timely delivery of products is critical in today's fast-paced business environment. Customers expect to receive their orders quickly, and delays can lead to frustration, negative reviews, and lost sales. To meet their expectations, you need to have a well-organized supply chain and efficient logistics.
- Inventory turns are a key performance indicator (KPI) for supply chain management. They measure how often a company sells and replaces its inventory during a given period, and they reflect the effectiveness of its procurement, production, and sales processes. High inventory turns indicate that a company is agile, responsive, and able to adapt to changing demand patterns. They also allow companies to generate higher margins by reducing their carrying costs and minimizing waste.
- Waste reduction is an important aspect of sustainable business practices. By minimizing the amount of materials, energy, and other resources used in the supply chain, companies can reduce their environmental impact and improve their profitability. This can be achieved through better inventory management, process optimization, and collaboration with suppliers and customers.
- Prompt updates are critical to staying competitive in today's digital marketplace. Customers expect to be informed about product availability, pricing, promotions, and other relevant information in real-time. By using technology tools such as ERP systems, CRM software, and e-commerce platforms, companies can streamline their communication channels and provide timely updates to their

stakeholders. This can improve customer satisfaction, reduce errors and delays, and enable faster decision-making.

To achieve these objectives forecasting in two aspects plays a crucial role: financial statement analysis and product replenishment and budgeting. These are discussed further in the chapter.

3.1 Financial Statement Analysis

The study of financial statements is a useful forecasting technique for the retail and supply chain sectors. Here are a few explanations:

1. **Identifying trends:** Analysis of financial statements can be used to spot trends in sales, costs, and profits over time. This can help in making strategic decisions based on historical trends and forecasting future performance.
2. **Evaluating financial health:** Financial statement analysis can provide light on a company's liquidity, solvency, and profitability as well as its overall financial health. Potential dangers and possibilities can be predicted with the aid of this knowledge.
3. **Assessing efficiency:** Analysis of financial statements can be used to assess a company's operational efficiency, including inventory turnover, asset usage, and cash conversion cycle. Forecasting possible areas for improvement and optimisation can be done using this knowledge.
4. **Performance benchmarking:** Financial statement analysis can serve as a foundation for comparing a company's performance to industry norms or rivals. This data can be used to anticipate possible weaknesses and growth possibilities.
5. **Decision-supporting:** Financial statement analysis can give decision-makers the knowledge they require to decide on investments, growth, and other strategic activities. By doing so, risks can be reduced and future outcomes can be predicted.
6. Due to the complexity of the retail and supply chain industries, financial statement analysis can be particularly beneficial in these sectors. Stakeholders may improve forecasting and management of inventory levels, streamline supply chain operations, and make wise choices on pricing, promotions, and other elements that may affect revenue and profitability by analysing financial data.

3.2 Product Replenishment and Budgeting

In the retail and supply chain industries, product replenishment and budgeting are critical aspects of forecasting. Here are some of the reasons:

1. **Managing inventory levels:** Product replenishment is critical for maintaining optimal inventory levels in the retail and supply chain industries. Companies can avoid stockouts, minimize overstocking, and optimize supply chain operations by forecasting product demand and ensuring that stock levels are replenished on a regular basis.

2. **Increasing customer satisfaction:** Product replenishment can help ensure that customers have access to the products they require when they require them. This can boost customer satisfaction and loyalty while also helping the company build a positive reputation.
3. **Cost reduction:** In the retail and supply chain industries, proper budgeting can help companies manage their expenses and reduce costs. Companies can ensure that they stay within their financial constraints and optimize their resource allocation by forecasting expenses and setting budgets for various activities such as marketing, promotions, and logistics.
4. **Increasing profitability:** In the retail and supply chain industries, effective budgeting and product replenishment can help increase profitability. Companies can improve their margins and achieve long-term profitability by optimizing inventory levels, lowering costs, and increasing revenue.
5. **Risk mitigation:** Forecasting product demand and budgeting can help companies in the retail and supply chain industries mitigate risks. Companies can adjust their strategies and take proactive measures to minimize potential risks by anticipating changes in demand, market trends, and other external factors.

To summarize, in the retail and supply chain industries, product replenishment and budgeting are critical components of forecasting. Companies can optimize their operations and achieve long-term success by managing inventory levels, improving customer satisfaction, lowering costs, increasing profitability, and mitigating risks.

References

[1] Nordqvist, C. What are market forces? Definition and... Market Business News. *Market Business News*. 2015. August31.https://marketbusinessnews.com/financial-glossary/market-forces/.
[2] Mankiw, N.G. *Principles of Economics* (9th Edn.). 2021a. Cengage Learning.
[3] Can Akdeniz. *Law of Supply and Demand.* (2019). (n.p.), 40pp.
[4] Reza Hoshmand, A. *Business Forecasting,* (2nd Edn.). Routledge, 2009.
[5] Ayers, J.B., and Mary Ann Odegaard. *Retail Supply Chain Management.* CRC Press. 2021.

CHAPTER 22

Role of Artificial Intelligence in Weather Forecasting and Climate Behavioral Analysis

Mahesh Kumar S V [1,*] and *Sreya U Parvathi* [2]

1. Introduction

1.1 Overview of Weather Forecasting and Climate Behavioral Analysis

Analysis and prediction of weather with the help of environmental conditions is known as weather forecasting. Weather forecasting models predict the weather conditions by the statistical analysis of weather data collected from different sources. These models collect quantitative weather data, that represents the present state of the atmosphere, land, and ocean. The collected weather data has been used to predict future atmospheric changes in the specified place with the help of meteorology concepts. Consistent weather forecasting is vital for farming, and food security; it is also an essential need of human society. Moreover, weather forecasting is strongly related to human life, including transportation safety, preventing wildfires, sea navigation, fishing, and aviation schedules. Several important events such as social events, sports, and military operations are also organized and rescheduled based on weather forecasting reports.

Meteorologists developed several methods for weather forecasting. The traditional synoptic and Numerical Weather Prediction (NWP) approaches are the

[1] Department of Electronics and Communication Engineering, Amrita College of Engineering and Technology, Nagercoil, Tamil Nadu, India.

[2] Department of Soil Science and Agricultural Chemistry, College of Agriculture, Vellayani, Thiruvananthapuram, Kerala, India.

* Corresponding author: maheshyesvee@gmail.com

most vital methods. The meaning of the word synoptic is viewed together. Hence, the synoptic method is mainly focused on weather viewing at common time and points. This method forecasts the weather using the analysis of a synoptic chart series [1]. The NWP is another broadly used weather forecasting approach, which uses a lot of mathematical equations to describe fluid flow. These mathematical equations were interpreted into the computer code and used for governing type equations, numerical approaches, parameterizations of physical processes, and combined with the boundary conditions before running over a specific domain or geographic area. The NWP approaches are mainly developed for forecasting the weather in different time horizons ranging between hours to days [2].

1.2 Issues in Conventional Weather Forecasting and Climate Behavioral Analysis

Reliable weather forecasting is essential for socioeconomic development. It is also a vital need for food security. In conventional weather forecasting, meteorologists used weather charts and forecasted the weather based on their experience attained through years of observations, and weather theories [3]. Before the invention of computers, the conventional method was the only offered method for weather forecasting. Meteorologists can achieve significant forecasts using this conventional method, up to 24 hours in advance. However, beyond this time duration, conventional forecasting is not suitable. Moreover, precise weather and climate condition forecasting is a challenging task due to the dynamic atmospheric nature. Hence, analytic solutions are not only sufficient for forecasting the weather. Hence, numerical techniques are also preferred in weather forecasting. In 1922, Lewis Fry Richardson attempted the first manual NWP, and later early-stage computer-aided weather forecasting, climate analysis models were developed in 1950 [4]. Recently, computer-aided models are broadly used for weather forecasting and climate analysis.

1.3 Need for Artificial Intelligence in Weather Forecasting

Recently, weather forecasting is done effectively with the help of data mining and artificial intelligence (AI) techniques. Machine learning (ML) algorithms, Deep learning (DL) algorithms are broadly used in AI-based weather forecasting models. These AI-based models can rapidly analyze highly complicated weather data, within a short time duration [5]. Hence, meteorologists can attain accurate predictions and save human lives. Moreover, ML-based models can also be used to predict other climatic parameters such as temperature, wind speed, and humidity. The improved computational power, wide availability of larger weather databases, and enormous development of ML algorithms contributed to the current achievement in AI-based weather forecasting. In recent years, convolutional neural networks (CNNs) are largely used in weather and climate forecasting. The CNN models are trained to identify the spatial features from the satellite images, and produce accurate forecasting model output [6,7]. The AI-based models largely reduced human errors in weather forecasting and rapidly produced the prediction outcome even for a large amount of weather data.

1.4 Organization of the Chapter

The overview of AI-based systems used for weather forecasting and climate behavioral analysis is shown in Fig. 1. The details about the meteorological satellites, satellite remote sensing, and the impact of satellite data on weather prediction modeling are briefed in Section 2. Various ML- and DL-based weather forecasting models, and AI-Powered IoT (internet of things) models used in weather forecasting and climate analysis are discussed in Section 3. Section 4 provides detailed information about the various database sources used for the development of weather forecasting, and climate analysis predictive models, and also compares the performance of various AI-based forecasting systems. Finally, Section 5 concludes the Chapter.

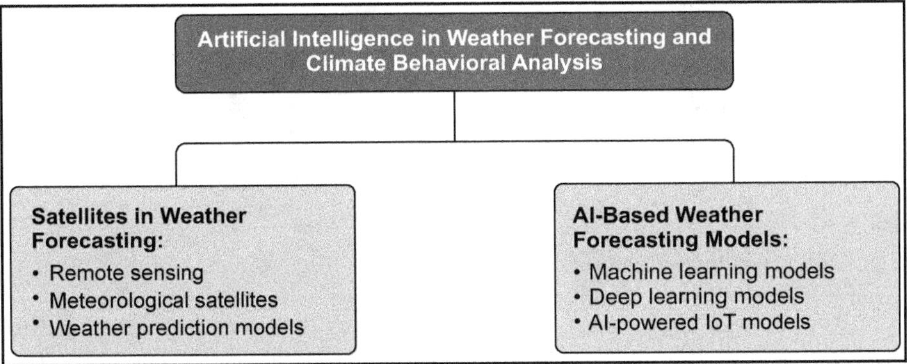

Fig. 1: Overview of AI Systems Used in Weather Forecasting and Climate Behavioral Analysis.

2. Satellites in Weather Forecasting

2.1 Satellite Remote Sensing

Remote sensing is a method of information acquisition from an object using any sensor, that is not directly connected to that object. In meteorological applications, information of interest specifies the location, and the weather forecasting systems development for clouds, cyclones, and rainstorms. The remote-sensing information is acquired using different high-frequency satellites. In satellite remote-sensing systems, image-processing techniques are broadly used to analyze the high-resolution digital images developed from the analysis of various regions of interest. These high-resolution images clearly show the changing weather patterns. Hence, satellite imaging data is largely used for the real-time analysis of various regions of interest of the earth. A satellite remote sensing-based weather forecasting system is shown in Fig. 2. In this system, high-resolution satellite images created from the region of interest were transferred to the remote server for further analysis of weather data. Usually, a software-based image analysis tool can be used to interpret the weather pattern from high-resolution images. The data acquired using satellite imaging systems can be combined with other data types also for effective visualization and further analysis, with the help of a Geographic Information System (GIS).

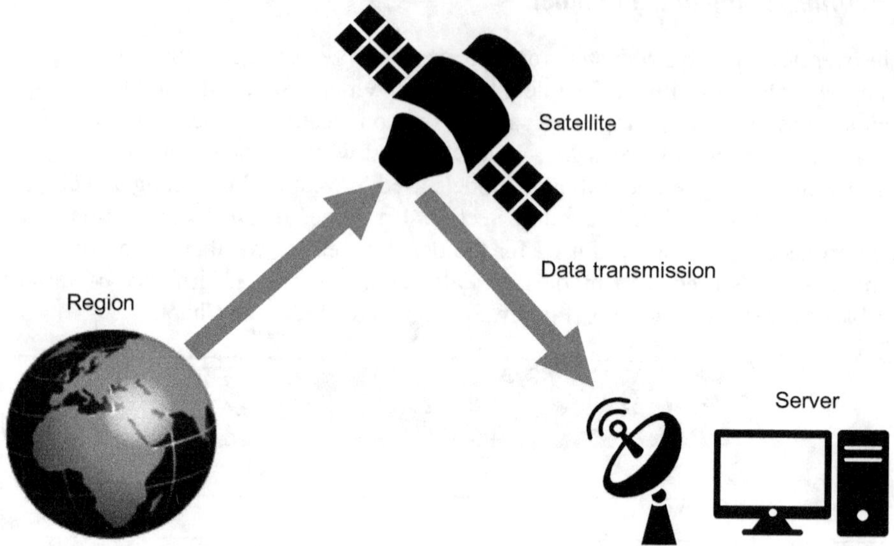

Fig. 2: Satellite Remote Sensing-based Weather Forecasting System.

2.2 Meteorological Satellites

Numerous forecasting systems were developed over the decades to infer weather patterns from satellite images. In synoptic and conventional weather forecasting methods, satellite data is broadly used along with traditional meteorological observations, for attaining information related to agriculture. Various meteorological satellites were used for global, and regional-level observations. The currently used six types of satellite imaging systems are listed as (1) Visible/Infrared water vapor imaging systems, (2) Infrared sound systems, (3) Microwave imaging systems, (4) Microwave sound systems, (5) Scatterometers, and (6) Radar altimeters [8]. Water vapor imaging is commonly used in geostationary satellites. The visible/infrared imaging systems are used in geostationary and polar-orbiting satellites. The remaining four imaging systems are used in polar orbiting satellites only. Commonly, a global space observation system is evolved with these geostationary, polar-orbiting satellites.

World's first meteorological satellite TIROS-1 was launched in April 1960 [9]. The Indian National Satellite (INSAT) program [10] is the major satellite series used for weather surveillance in India. In 1980s, this program was initiated by launching the INSAT-1 satellite series. After that, the INSAT-2 series was developed based on user feedback. The INSAT-2A and 2B series were also launched in 1992 and 1993, with an enhanced resolution of up to 02 km in the visible spectrum and up to 08 km in the thermal band. The imaging competence of these satellites was included in three modes such as the full frame, normal mode, and the sector mode with 05-minutes time duration for fast coverage of critical weather conditions. The INSAT-2E was launched in the year 1999 and can be operated in three channels as a visible spectrum (02 km), a thermal band, and a water vapor channel (08 km). Moreover, it has a CCD camera equipped with three channels as visible wavelength, near-infrared, and short-wave infrared of resolution 01 km, which can be used to cover the vegetation map. The INSAT images were broadly used in weather forecasting in India. The geostationary meteorological satellite (METSAT) is another satellite (also known as Kalpana-1), which was launched in the year 2002, and used by the India Meteorological Department (IMD). The INSAT-3D is the currently used INSAT series satellite for weather forecasting.

2.3 Impact of Satellite Data on Weather Prediction Models

In the past 60 years, meteorological satellites developed largely. Hence, the observations of meteorological satellites play a major role in weather forecasting. Moreover, these observations are essential for atmospheric measurements and fed into the NWP models for precise weather prediction, also crucial for tracking weather condition changes such as tropical storm movement and other severe weather conditions. During extreme weather condition changes like hurricanes, satellites are the only reliable tool available to track the location. They are also used to identify storm intensity, and communicate to the weather prediction models on a real-time basis to give more accurate weather forecasting reports for the upcoming days. Satellite data along with the NWP models, and data integration approaches emerged as essential elements of the Earth System (ES) framework [11]. Recently, ML techniques are also used for targeted satellite data assimilation [12]. The ML-based satellite data assimilation process can optimize huge data usage in weather prediction models.

3. Review of AI-Based Weather Forecasting Systems

3.1 ML-based Weather Forecasting Models

ML technique gives systems with the ability to learn from the experience and produce better outcomes without any human programming. The ML-based models are considered as an alternative approach for dynamic and nonlinear conditions because these models have effective computing structures, and also contain parameter estimation features [13,14]. The ML-based models use past data to forecast future data. Hence, it is not essential for understanding all physical processes which control the atmosphere. Therefore, ML-based models are effectively used for weather prediction [15]. Vasantha et al. [16] developed a simulated framework that employs data analysis, and ML techniques to predict various climate conditions over the Indian subcontinent. In this framework, climate parameters like temperature, and humidity were utilized for model fitting, and ML techniques were used for inferring the weather pattern information.

A study conducted by Hanoon et al. [17] addressed different ML-based models such as Linear regression (LR), Random forest (RF), Gradient Boosting Tree (GBT), and Artificial Neural Network (ANN) for predicting the air temperature (T) and the relative humidity (Rh) parameters. Kaps et al. [18] developed a new ML framework based on satellite observations of the cloud, and the related progressions in climate. This method can assign the cloud distribution types to the rough data. Moreover, this method performed an objective valuation of the clouds in the ES model and also improved the consistency of the cloud analysis process. Recently, Quantum ML-based approaches were also used in weather forecasting. Surendiran et al. [19] conducted a study on quantum ML-based approaches and discussed the usage of these techniques in climate transformation, and weather forecasting. Moreover, this study addressed the latest trends and developments in Quantum ML and also discussed how it improved the performance of conventional ML approaches.

3.2 DL-based Weather Forecasting Models

The development of satellite imaging supports meteorologists to study large-scale weather patterns. However, predicting weather extremes using massive weather databases is still a challenging task for scientists researching climate behavioral analysis. To account for the nonlinearity characteristics of the time-series data in weather forecasting, various DL architectures were developed. These DL-based models were used as an alternative system for recognizing extreme weather changes. Goel et al. [20] developed a DL-based model, that produced high-level representations of numerous patterns using the labeled data. In this model, a deep CNN was used for weather forecasting, and this model detected extreme weather changes with an accuracy of 98.5%. The DL-based model proposed by Dabhade et al. [21] significantly improved cyclone prediction performance. A remote-sensing data-based DL model developed by Soni et al. [22] attained 94.2% average forecasting accuracy.

Rajabi et al. [23] developed a weather prediction model using fine-tuning and transfer learning methods of CNN. This model was evaluated using a combined database created using five different databases with five different weather conditions such as cloud, rain, fog, snow, and fine dust. This combined database significantly reduced the mislabeling issue. Moreover, this model attained an accuracy of 99%, which is significantly better than the previous works. Another CNN-based model developed by Zhao and Wu [24] attained a 5.06% improved accuracy compared with other traditional weather recognition methods. A study conducted by Abdalla et al. [25] compared the numerous state-of-the-art DL-based weather forecasting models according to their Neural Network architectures, temporal and spatial scales, databases and benchmarks used for model development. Future-generation radars will improve the performance of severe weather threat detection systems, such as tornadoes, heavy rain, and flood detection systems.

3.3 Intelligent IoT Models for Weather Forecasting

Smart weather prediction systems preferred IoT techniques which are easy to implement and cost-effective [26]. Sadhukhan et al. [27] developed an intelligent weather predictive model using a cost-effective GPS-aided IoT device, that connected with sensors. In this system, the sensed weather information was stored in a server for the prediction of different weather parameters like air pressure, temperature, and humidity of a specific region of interest. Another simple weather station application developed by Bin Shahadat et al. [28] used a microcontroller and IoT technology to measure various meteorological data, such as pressure, humidity, temperature, and rainfall. A simplified and cost-effective weather monitoring application proposed by Murugan et al. [29] used IoT, and AI-based image classification techniques for weather forecasting. This system used an ESP32 camera unit and the acquired images were transferred to a virtual display.

The automated IoT-based weather monitoring embedded system proposed by Mabrouki et al. [30] sensed various climate parameters, like humidity and temperature using different sensors. The sensed climate parameters were wirelessly transferred to remote databases for further data visualization process. This system is compact in size and cost-effective. The IoT-based weather monitoring and forecasting systems needed cloud storage for streaming massive data. Beeharry et al. [31] developed a cloud-based end-to-end weather prediction system focused on real-time data collection and real-time weather prediction. This system was developed using International Business Machines (IBM) cloud environment, and the weather prediction was implemented using an Android application. Verma et al. [32] implemented a real-time system to predict the weather using the data sensed by temperature and humidity sensors. The sensed information was stored in a ThingSpeak cloud-based server, and a logistic regression model was used for predicting the weather. These IoT-based intelligent systems can be used to acquire real-time weather data and predict the weather with help of the sensed data fed into the ML algorithm.

4. Comparison of Weather Forecasting Models

4.1 Weather Databases

ExtremeWeather [33] is a weather database developed to encourage ML-based research in the weather forecasting domain and is also used for understanding the effects of climate changes. This database contains a simulation outcome of 27 years (from 1979 to 2005), and it includes 78,840, 16-channel images with a resolution of 768x1152. WeatherBench [34] is another benchmark database for medium-range of weather forecasting applications. LowaRain [35] is a database that contains the rainfall data of the Iowa state (from 2016 to 2019) collected from the National Weather Service Next Generation Weather Radar (NEXRAD), and this data was processed using a quantitative precipitation estimation system. This database can be useful for disaster monitoring, recovery by predictive, and prescriptive modeling.

CloudCast [36] is a satellite-based database used for forecasting the clouds. This database contains 70,080 images with a spatial resolution of 928x1530 pixels. These images represent 10 cloud types from multiple atmosphere layers. Weather2K [37] is a benchmark multivariate spatiotemporal database used for real-time observation-based weather forecasting using the data attained from different ground weather stations. This database contains the data received from 2,130 ground weather stations of six million square kilometres area. ERA5 [38] is the largely used database that contains the hourly data of single levels from 1940 to the present. The details of the benchmark databases used for weather forecasting model development are summarized in Table 1.

Table 1: Summary of Benchmark Weather Forecasting Datasets.

Name of the Dataset	Data Type	Database Details and Applications
ExtremeWeather [33]	Simulated climate data.	• This database contains simulated data of 27 years (from 1979 to 2005). • This database contains 78,840 (16-channel) 768 × 1152 images. • It is suitable for studies related to climate change projection.
WeatherBench [34]	Data-driven category.	• The database has a 191GB size at a 5.625° resolution. • It is suitable for a medium range of weather forecasting applications.
LowaRain [35]	Observation data.	• This database contains the rainfall data of the Iowa state (from 2016 to 2019) collected from the NEXRAD. • It is suitable for forecasting the rainfall.
CloudCast [36]	Satellite-based database.	• This database contains 70,080 images with a spatial resolution of 928 × 1530 pixels. • It is the first public database containing high-resolution of cloud data. • It is suitable for forecasting cloud types.
Weather2K [37]	Real-time data collected from ground weather stations.	• This database contains the data received from 2,130 ground weather stations. • It is suitable to develop accurate weather forecasting algorithms.
ERA5 [38]	Reanalysis database.	• This is a fifth generation reanalysis database for the global weather and climate data for the past eight decades (from 1940 to the present). • This database is generally used to calculate daily weather statistics.

4.2 Performance Measures

Commonly, the ML and DL-based weather forecasting systems were trained using training weather data. After that, the performance of these forecasting models was tested using testing weather data. Different performance measures such as precision, recall, average precision, F1 score, mean squared error (MSE), and root means squared error (RMSE) were used for evaluating the performance of the AI-based weather forecasting systems. The performance measures were described as follows:

- **Precision:** This performance measure indicates the accuracy of weather prediction (the percentage of correct predictions).
- **Recall:** This measure indicates the accuracy to identify the positive classes in K number of predictions.
- **Average precision:** This performance measure calculates the average precision value for a recall value ranges between 0 to 1.
- **F1 score:** This performance measures combined the precision and recall values.
- **MSE:** It is the mean of the squared difference between the actual and forecasting outcomes.
- **RMSE:** This measure indicates the average of the difference between the value predicted by a forecasting system and the actual value.

4.3 Performance Comparison and Discussions

The comparison of recent AI-based weather forecasting systems is presented in Table 2. Weather forecasting methods proposed in [16,17] focused on specific region-based data for forecasting model development and validation. These methods attained significant prediction accuracy. However, global-level data analysis is required to identify different categories of weather patterns. A satellite observation-based ML model [18] attained 0.886 ± 0.003 accuracy, and 0.472 ± 0.005 F1-score in cross-validation. This method identified the cloud-related progressions in climate. A Quantum ML-based approach [19] investigated how it improved the performance of conventional ML-based weather forecasting approaches. DL-based approaches [20–24] forecasted the weather using the automated features extracted from the weather data with significant prediction accuracy.

CNN is a mainly used approach in DL-based weather forecasting. A fine-tuning and transfer learning CNN-based method [23] attained 99% of accuracy in

Table 2: Comparison of Recent AI-based Weather Forecasting Systems.

Authors	Technique Used	Issues Addressed	Observations
Vasantha et al. [16]	Data analysis and ML techniques.	Various climate conditions prediction.	• This method addressed the prediction of various climate conditions over the Indian subcontinent. • The precision outcome was proposed to be expand by 70%.
Hanoon et al. [17]	LR, RF, GBT, and ANN.	Prediction of the air temperature (T), and the relative humidity (Rh) parameters.	• This study addressed the temperature and the humidity forecasting at the region of Terengganu state, Malaysia. • This method has good potential in prediction of daily as well as monthly T and Rh with a significant range of accuracy.
Kaps et al. [18]	ML framework based on satellite observations.	Cloud-related progressions in climate.	• This method obtained an accuracy value of 0.886 ± 0.003, and F1-score of 0.472 ± 0.005 (in cross validation).
Surendiran et al. [19]	Quantum ML-based approach.	Climate transformation, and weather forecasting.	• This study addressed how the Quantum ML improved the performance of conventional ML approaches.
Goel et al. [20]	Deep CNN	Prediction of weather changes.	• This model detected the extreme weather changes. • This model attained 98.5% accuracy.
Dabhade et al. [21]	DL-based model.	Cyclone prediction.	• This model significantly improved the cyclone prediction performance.
Soni et al. [22]	Remote sensing data-based DL model.	Weather forecasting.	• This method attained 94.2% average forecasting accuracy.
Rajabi et al. [23]	Fine-tuning, and transfer learning methods of CNN.	Weather prediction.	• This model attained an accuracy of 99%.
Zhao and Wu [24]	CNN	Weather forecasting.	• This method attained a 5.06% improved accuracy.

weather prediction. Moreover, CNN-based weather forecasting systems are also suitable to identifying extreme weather changes. A CNN-based model proposed in [20] detected extreme weather changes with 98.5% accuracy. Another CNN-based model [24] attained a 5.06% improved weather forecasting accuracy. The DL-based model along with satellite data can attain significant prediction accuracy. A satellite data-based model developed in [22] attained 94.2% average forecasting accuracy. Moreover, DL-based models are used to predict cyclones also. A DL-based model [21] significantly improved cyclone prediction performance. The above comparative analysis shows the efficiency of DL-based models in weather forecasting.

5. Conclusions and Future Research Scope

Weather forecasting is essential to plan people's daily activities, protect human life and property. Accurate prediction of weather and climate conditions is challenging due to the dynamic atmospheric nature. Therefore, weather forecasting and climate behavioral analysis become an important research domain, and many researchers were involved in the development of precise weather and climate prediction models. Satellite imaging and AI techniques were largely used for the development of weather forecasting systems. Various AI-based systems used for weather forecasting, and climate analysis were discussed. Advanced intelligent techniques such as ML, DL, and AI-powered IoT models used for weather forecasting were also discussed in detail. Moreover, this chapter also focused on the satellite remote sensing systems used for weather and addressed various databases used for weather forecasting model development.

The comparative study of this chapter illustrates that the satellite imaging and AI-based systems performed significantly in weather forecasting and climate analysis. However, the result of AI-based weather forecasting systems is significantly dependent on the satellite imaging models, weather database used for training, and technique used for predictive analysis. Future research works will focus on weather forecasting using multidimensional data. Furthermore, satellite remote sensing-based forecasting models can improve the weather predictive accuracy. Future research works may be focused on reinforcement learning-based classification techniques for effective weather forecasting.

References

[1] Byers, H.R., Landsberg, H.E., Wexler, H., Haurwitz, B., Spilhaus, A.F., Willett, H.C., and Houghton, H.G. *Compendium of Meteorology*. American Meteorological Society, 1951.

[2] Ramírez, L., and Vindel, J.M. Forecasting and nowcasting of DNI for concentrating solar thermal systems. *In*: *Advances in Concentrating Solar Thermal Research and Technology*, Elsevier Inc., 2016, pp. 295–310.

[3] Goswami, B.N. Old and modern ways of weather forecasting. *In*: *The Challenge of Weather Prediction. Resonance*, Springer, 1997, pp. 8–15.

[4] Lynch, P. The origins of computer weather prediction and climate modeling. *J. Comput. Phys.*, 227(7): 3431–3444, Mar. 2008. doi: 10.1016/J.JCP.2007.02.034.

[5] Schultz, M.G., Betancourt, C., Gong, B., Kleinert, F., Langguth, M., Leufen, L.H., Mozaffari, A., and Stadtler, S. Can deep learning beat numerical weather prediction? *Philos. Trans. R. Soc. A*, 379(2194), Apr. 2021. doi: 10.1098/RSTA.2020.0097.

[6] Zhu, X.X., Tuia, D., Mou, L., Xia, L., Zhang, G.S., Xu, F., and Fraundorfer, F. Deep learning in remote sensing: A comprehensive review and list of resources. *IEEE Geosci. Remote Sens. Mag.*, 5(4): 8–36, Dec. 2017. doi: 10.1109/MGRS.2017.2762307.

[7] Gagne, D.J., Haupt, S.E., Nychka, D.W., and Thompson, G. Interpretable deep learning for spatial analysis of severe hailstorms. *Mon. Weather Rev.*, 147(8): 2827–2845, Aug. 2019. doi: 10.1175/MWR-D-18-0316.1.

[8] Kalsi, S.R. Satellite based weather forecasting. *In*: *Satellite Remote Sensing and GIS Applications in Agricultural Meteorology*, 2005, pp. 331–346.

[9] Garner, R. TIROS 1, World's 1st Weather Sat.—This Week in Goddard History. 2019. http://www.nasa.gov/feature/goddard/2019/launch-of-tiros-1-worlds-1st-weather-satellite-this-week-in-goddard-history-march-31-april-6 (accessed Mar. 19, 2023).

[10] Satellite Images. http://www.satellite.imd.gov.in/insat.htm (accessed Mar. 19, 2023).

[11] Collins, N., Gerhard Theurich, Cecelia DeLuca, Max Suarez, Atanas Trayanov, Balaji, V., Peggy Li, Weiyu Yang, Chris Hill, and Arlindo da Silva. Design and Implementation of Components in the Earth System Modeling Framework. http://dx.doi.org/10.1177/1094342005056120, 19(3): 341–350, Aug. 2005, doi: 10.1177/1094342005056120.

[12] Lee, Y.J., Hall, D., Stewart, J., and Govett, M. Machine learning for targeted assimilation of satellite data. *Lect. Notes Comput. Sci. (including Subser. Lect. Notes Artif. Intell. Lect. Notes Bioinformatics). 11053 LNAI*, pp. 53–68, 2019. doi: 10.1007/978-3-030-10997-4_4/COVER.

[13] Abobakr Yahya, Abobakr Saeed, Ali Najah Ahmed, Faridah Binti Othman, Rusul Khaleel Ibrahim, Haitham Abdulmohsin Afan, Amr El-Shafie, Chow Ming Fai, Md Shabbir Hossain, Mohammad Ehteram, and Ahmed Elshafie. Water quality prediction model based support vector machine model for ungauged river catchment under dual scenarios. *Water*, 11(6): 1231, Jun. 2019. doi: 10.3390/W11061231.

[14] Najah Ahmed, A., Faridah Binti Othman, Haitham Abdulmohsin Afan, Rusul Khaleel Ibrahim, Chow Ming Fai, Md Shabbir Hossain, Mohammad Ehteram, and Ahmed Elshafie. Machine learning methods for better water quality prediction. *J. Hydrol.*, 578: 124084, Nov. 2019. doi: 10.1016/J.JHYDROL.2019.124084.

[15] Salman, A.G., Kanigoro, B., and Heryadi, Y. Weather forecasting using deep learning techniques. *In*: *ICACSIS 2015–2015 International Conference on Advanced Computer Science and Information Systems, Proceedings*, Feb. 2016, pp. 281–285, doi: 10.1109/ICACSIS.2015.7415154.

[16] Vasantha, B., Tamilkodi, R., and Kiran, L.V. Rainfall Pattern Prediction Using Real Time Global Climate Parameters Through Machine Learning. Mar. 2019, doi: 10.1109/VITECON.2019.8899463.

[17] Marwah Sattar Hanoon, Ali Najah Ahmed, Nuratiah Zaini, Arif Razzaq, Pavitra Kumar, Mohsen Sherif, Ahmed Sefelnasr, and Ahmed El-Shafe. Developing machine learning algorithms for meteorological temperature and humidity forecasting at Terengganu state in Malaysia. *Sci. Reports*, 2021, 11(1): 1–19, Sep. 2021. doi: 10.1038/s41598-021-96872-w.

[18] Kaps, A., Lauer, A., Camps-Valls, G., Gentine, P., Gomez-Chova, L., and Eyring, V. Machine-learned cloud classes from satellite data for process-oriented climate model evaluation. *IEEE Trans. Geosci. Remote Sens.*, 61, 2023. doi: 10.1109/TGRS.2023.3237008.

[19] Surendiran, B., Dhanasekaran, K., and Tamizhselvi, A. A study on quantum machine learning for accurate and efficient weather prediction. *In*: *Proceedings of the 6th International Conference on I-SMAC (IoT in Social, Mobile, Analytics and Cloud), I-SMAC*, 2022, pp. 534–537. doi: 10.1109/I-SMAC55078.2022.9987293.

[20] Goel, S., Markanday, S., and Mohanty, S. Analysis of multi-class weather classification using deep learning models and machine learning classifiers. *In*: *2022 OITS International Conference on Information Technology (OCIT)*, Mar. 2023, pp. 223–227. doi: 10.1109/OCIT56763.2022.00050.

[21] Dabhade, A., Roy, S., Moustafa, M.S., Mohamed, S.A., El Gendy, R., and Barma, S. Extreme Weather Event (Cyclone) Detection in India Using Advanced Deep Learning Techniques. 2021. doi: 10.1109/ICOT54518.2021.9680663.

[22] Soni, S., Vashishtha, K., and Bhandubey, C. Deep learning based weather forecast: A prediction. *In*: *Proceedings of the 5th International Conference on Trends in Electronics and Informatics, ICOEI 2021*. Jun. 2021, pp. 1736–1742. doi: 10.1109/ICOEI51242.2021.9452855.

[23] Rajabi, F., Faraji, N., and Hashemi, M. An intelligent method for detecting weather conditions using deep learning on a combined dataset. 2022. doi: 10.1109/ICSPIS56952.2022.10043995.

[24] Zhao, X., and Wu, C. Weather classification based on convolutional neural networks."*In: Proceedings of the 2021 International Conference on Wireless Communications and Smart Grid, ICWCSG 2021.* 2021, pp. 293–296. doi: 10.1109/ICWCSG53609.2021.00064.

[25] Abdalla, A.M., Ghaith, I.H., and Tamimi, A.A. Deep learning weather forecasting techniques: literature survey. *In: - Proceedings of the 2021 International Conference on Information Technology, ICIT 2021.* Jul. 2021, pp. 622–626. doi: 10.1109/ICIT52682.2021.9491774.

[26] Shivanshu, Nagwanshi, P., and Chauhan, A. Smart real time weather forecasting system. *In: Proceedings: 2021 3rd International Conference on Advances in Computing, Communication Control and Networking, ICAC3N 2021.* 2021, pp. 558–562. doi: 10.1109/ICAC3N53548.2021.9725697.

[27] Sadhukhan, M., Dasgupta, S., and Bhattacharya, I. An intelligent weather prediction system based on IOT." *In: Proceedings of 4th International Conference on 2021 Devices for Integrated Circuit, DevIC 2021.* May 2021, pp. 528–532. doi: 10.1109/DEVIC50843.2021.9455883.

[28] A.S. Bin Shahadat, A.S., Ayon, S.I., and Khatun, M.R. Efficient IoT based Weather Station. *In: Proceedings of 2020 IEEE International Women in Engineering (WIE) Conference on Electrical and Computer Engineering, WIECON-ECE 2020.* Dec. 2020, pp. 227–230. doi: 10.1109/WIECON-ECE52138.2020.9398041.

[29] Murugan, K., Tiruveedhi, R.K., Ramireddygari, D.R., Thota, D., and Neeli, C. AI based Weather Monitoring System. *In: 2nd IEEE International Conference on Advanced Technologies in Intelligent Control, Environment, Computing and Communication Engineering, ICATIECE 2022.* 2022, pp. 1–5. doi: 10.1109/ICATIECE56365.2022.10047380.

[30] Mabrouki, J., Azrour, M., Dhiba, D., Farhaoui, Y., and El Hajjaji, S. IoT-based data logger for weather monitoring using arduino-based wireless sensor networks with remote graphical application and alerts. *Big Data Min. Anal.*, 4(1): 25–32, Mar. 2021. doi: 10.26599/BDMA.2020.9020018.

[31] Beeharry, Y., Fowdur, T.P., and Sunglee, J.A. A cloud-based real-time weather forecasting application. *In: 2019 14th International Conference on Advanced Technologies, Systems and Services in Telecommunications, TELSIKS 2019:- Proceedings*, Oct. 2019, pp. 294–297. doi: 10.1109/TELSIKS46999.2019.9002327.

[32] Verma, G., Mittal, P., and Farheen, S. Real time weather prediction system using IOT and machine learning. *In: 2020 6th International Conference on Signal Processing and Communication, ICSC 2020.* Mar. 2020, pp. 322–324. doi: 10.1109/ICSC48311.2020.9182766.

[33] Racah, E., Beckham, C., Maharaj, T., Kahou, S.E., and Pal, C. ExtremeWeather: A large-scale climate dataset for semi-supervised detection, localization, and understanding of extreme weather events. *In: 31st Conference on Neural Information Processing Systems (NIPS 2017).* 2017, pp. 1–12. Accessed: Apr. 08, 2023. [Online]. Available: https://github.com/eracah/hur-detect.

[34] Rasp, S., Dueben, P.D., Scher, S., Weyn, J.A., Mouatadid, S., and Thuerey, N. WeatherBench: A benchmark data set for data-driven weather forecasting. *J. Adv. Model. Earth Syst.*, 12: 1–17, 2020. doi: 10.1029/2020MS002203.

[35] Sit, M., Seo, B.C., and Demir, I. IowaRain: A statewide rain event dataset based on weather radars and quantitative precipitation estimation. *Tackling Climate Change with Machine Learning Workshop at ICML 2021*, pp. 1–5, 2021.

[36] Andreas Holm Nielsen, Alexandros Iosifidis, and Henrik Karstoft. CloudCast: A Satellite-Based Dataset and Baseline for Forecasting Clouds. *IEEE J. Sel. Top. Appl. EARTH Obs. Remote Sens.*, 14: 3485–3494, 2020. Accessed: Apr. 09, 2023. [Online]. Available: https://ieeexplore.ieee.org/ielx7/4609443/9314330/09366908.pdf.

[37] Zhu, X., Xiong, Y., Wu, M., Nie, G., Zhang, B., and Yang, Z. Weather2K: A multivariate spatio-temporal benchmark dataset for meteorological forecasting based on real-time observation data from ground weather stations. *In: 26th International Conference on Artificial Intelligence and Statistics (AISTATS) 2023.* Feb. 2023, pp. 1–19. Accessed: Apr. 09, 2023. [Online]. Available: https://arxiv.org/abs/2302.10493v1.

[38] Hersbach, H., Bill Bell, Paul Berrisford, Shoji Hirahara, András Horányi, and Joaquín Muñoz-Sabater. The ERA5 global reanalysis. *Q.J.R. Meteorol. Soc.*, 146(730): 1999–2049, Jul. 2020. doi: 10.1002/QJ.3803.

Index

About the Editors

Dr. Sachi Nandan Mohanty, He received his PostDoc from IIT Kanpur in the year 2019 and Ph.D. from IIT Kharagpur, India in the year 2015, with MHRD scholarship from the Govt. of India. He has authored/edited 28 books, published by IEEE-Wiley, Springer, Wiley, CRC Press, NOVA, and De Gruyter. His research areas include Data mining, Big Data Analysis, Cognitive Science, Fuzzy Decision Making, Brain-Computer Interface, Cognition, and Computational Intelligence. Prof. S.N. Mohanty has received 4 Best Paper Awards during his Ph.D at IIT Kharagpur from International Conference at Beijing, China, and the other at International Conference on Soft Computing Applications organized by IIT Roorkee in the year 2013. He has been awarded the Best Thesis award first prize by the Computer Society of India in the year 2015. He has guided 9 Ph.D Scholars. He has published 120 International Journals of International repute and has been elected as FELLOW of the Institute of Engineers, European Alliance Innovation (EAI), and Senior member of the IEEE Computer Society Hyderabad chapter. He is also the reviewer of the *Journal of Robotics and Autonomous Systems* (Elsevier), *Computational and Structural Biotechnology Journal* (Elsevier), *Artificial Intelligence Review* (Springer), *Spatial Information Research* (Springer).

Dr. Preethi Nanjundan is an Associate Professor (SRG) in the Department of Data Science at Christ University, Pune, Lavasa campus, Maharashtra, India. She received her Doctorate degree from Bharathiyar University, Coimbatore in 2014. She received her Master of Philosophy in computer science from Bharathiar University in 2007 and earned her Master's degree in Computer Applications from Bharathidasan University in 2004. Her research and teaching experience spans 18 years. Besides publishing over 20 papers in international refereed journals, she has contributed chapters to various books and published 5 books. Four of her patents have also been granted. In 2020, she received the Best Professor award from Lead India and Vision Digital India. Her contributions to a book titled *Covid 19 and its Impact*' have been inducted into the Indian and Asian books of records. Her research area includes machine learning, natural language processing, and Neural network, etc. She is a lifetime member of professional societies including Computer Society of India (CSI), International Association of Computer Science and Information

Technology (IACSIT), Computer Science Teachers Association, and Indian Society for Technical Education (ISTE).

Tejaswini Kar received her B. Tech degree in Electronics and Telecommunication engineering from B.P.U.T. in 2003, received her M. Tech degree in communication system engineering from KIIT deemed to be university in 2008 and Ph.D degree in Electronics and Telecommunication engineering in 2018 from KIIT deemed to be university, Bhubaneswar, India. She has a total of 18 years of teaching experience. She is currently an Assistant Professor with the School of Electronics Engineering, KIIT deemed to be University. She has published many research papers in refereed international conferences and journals. She served as a reviewer of many peer reviewed journals and conferences. She received the Best paper award in ICDMAI 2019 held in Malaysia. She has been awarded with a Certificate of Excellence award as a mentor by Samsung for Samsung Prism project in 2022. Her current research interests include image processing, video processing, Machine Learning, and Deep Learning.

For Product Safety Concerns and Information please contact our EU representative GPSR@taylorandfrancis.com
Taylor & Francis Verlag GmbH, Kaufingerstraße 24, 80331 München, Germany

www.ingramcontent.com/pod-product-compliance
Lightning Source LLC
LaVergne TN
LVHW020609110826
845149LV00002B/419

* 9 7 8 1 0 3 2 5 0 6 1 8 0 *